D0214556

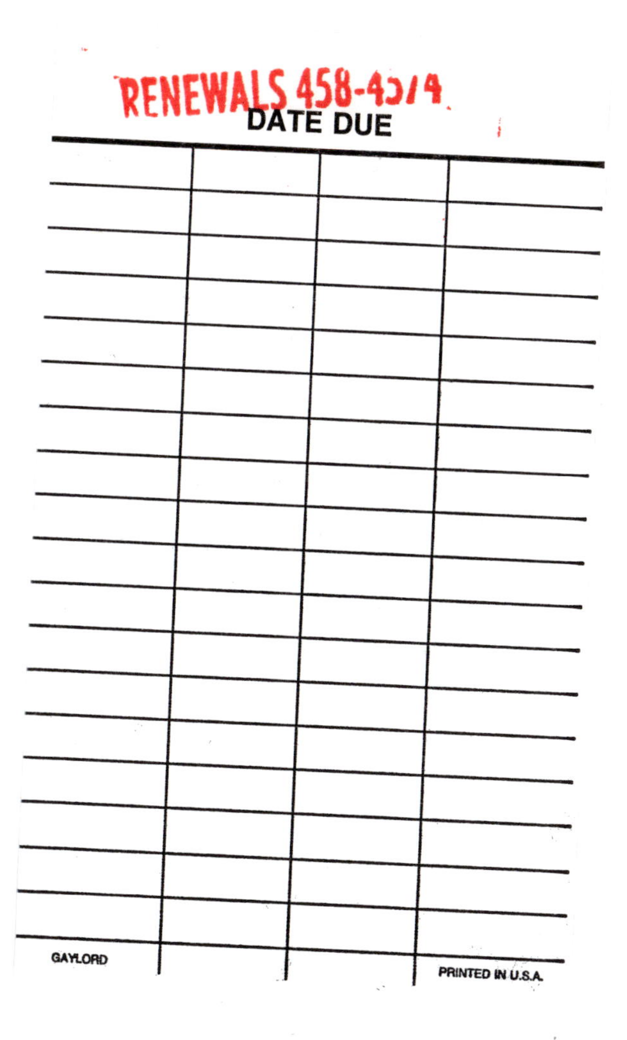
RENEWALS 458-4574
DATE DUE
GAYLORD
PRINTED IN U.S.A.

Occupational Health and Safety in Construction Project Management

Occupational Health and Safety in Construction Project Management

Helen Lingard and Steve Rowlinson

LONDON AND NEW YORK

First published 2005
by Spon Press
2 Park Square, Milton Park, Abingdon, Oxon OX14 4RN

Simultaneously published in the USA and Canada
by Spon Press
270 Madison Avenue, New York, NY 10016

Spon Press is an imprint of the Taylor & Francis Group

Typeset in Sabon by
Integra Software Services Pvt. Ltd, Pondicherry, India
Printed and bound in Great Britain by
MPG Books, Bodmin, Cornwall

British Library Cataloguing in Publication Data
A catalogue record for this book is available from the British Library

Library of Congress Cataloging in Publication Data
Lingard, Helen.
Occupational health and safety in construction project management / Helen Lingard & Steve Rowlinson.
p. cm.
Includes bibliographical references and index.
ISBN 0–419–26210–5 (hb: alk. paper)
1. Building—Safety measures. I. Rowlinson, Stephen M.
II. Title.
TH443.L56 2004
690′.22—dc22

2004001226

ISBN 0–419–26210–5

Contents

Preface

The construction industry performs poorly in occupational health and safety (OHS). Despite many OHS campaigns and initiatives, the statistics reveal that construction workers continue to be killed or injured at work each year. More insidious is the large number of construction workers who suffer impaired health or long-term illness caused or made worse by their work. Many of these illnesses only manifest themselves years after exposure and many are ultimately fatal. We believe that these injuries, deaths and illnesses can and should be prevented. Their persistence indicates a serious management failure in an industry that prides itself on having considerable management expertise and utilising state-of-the-art project management tools and techniques.

Construction contractors have traditionally borne the responsibility for OHS on site. Their site-based project managers and site staff are responsible for the day-to-day management of OHS. However, site-level managers and professionals usually have a limited understanding of their legal obligations relating to OHS or OHS principles and practices – which we suggest is, to a large extent, the result of a serious gap in most construction management and engineering degree courses. This book aims to provide construction professionals and students of construction project management with an understanding of theory pertaining to OHS as well as to introduce a range of tools and techniques representing best practice in the management of OHS.

The activities of site-based project managers do not occur in isolation. The book also argues that OHS is a strategic issue for management in all construction organisations, irrespective of size. As such, corporate OHS issues are addressed. As legal entities, corporations have responsibilities for OHS and may be subject to serious criminal charges in the event of serious injury or death. It is noteworthy that the first (and to our knowledge, the only) conviction of a corporation for manslaughter in Australia, occurred as the result of a construction plant operator's death in Melbourne. The present-day OHS legislation requires that companies implement OHS management systems and proactively manage the OHS risks posed by their operations, even those performed by self-employed persons or subcontractors.

It is unacceptable to assume that worksites are safe because they have been accident-free.

There is also a growing understanding that OHS in construction should not be solely a matter for contractors. Construction projects are complex socio-technical systems, and can take many organisational forms. The features of the contractual relationships, implicit in different forms of construction procurement, have an impact upon OHS. Furthermore, all parties to a construction project, including clients, designers, specialist consultants, specialist subcontractors and suppliers have a role to play in ensuring OHS risks are controlled by the best possible available methods. This book adopts a total project management approach, in that it does not prescribe actions to be taken by the contractor but describes a management process, in which construction project parties work collaboratively to ensure that OHS is managed in a holistic manner. The book emphasises the need for a safety management infrastructure within an organisation which deals with both the 'hard' and 'soft' issues inherent in keeping the workplace safe.

We hope that this book will prove useful to students and practitioners in the area of construction project management. For the former, it will provide an understanding of the complex interaction of legal, organisational, technical and psychological influences on OHS and the ways in which sources of occupational injury and illness can be identified and controlled – before workers are affected. For the latter, the book explains how modern OHS theory and practice can be applied in the dynamic and uncertain environment of a construction project.

Thus far, the industry's appalling OHS record has proved to be resilient to change.

It is our goal, by highlighting the nexus between OHS and project management concepts, that practitioners, both present and future, will be equipped to reduce the number of injuries, illnesses and deaths in the construction industry.

Helen Lingard and Steve Rowlinson

Acknowledgements

The authors would like to thank the following people and organisations for their assistance and expert advice provided to us while writing this book. Their generosity and willingness to provide comments and information is greatly appreciated.

Mr Chris Carstein – Baulderstone Hornibrook Pty Ltd
Mr Michael Kalinowski – Baulderstone Hornibrook Pty Ltd
Mr Graeme Tessier – Department of Main Roads, Queensland
Mr Dino Ramondetta – Master Builders' Association (Victoria)
Mr Earl Eddings – ARK Consulting Group Pty Ltd
Mr Mike West

Chapter 1

Introduction

Occupational health and safety in the construction industry

In most industrialised and developing countries, the construction industry is one of the most significant in terms of contribution to GDP and also in terms of impact on the health and safety of the working population. The construction industry is both economically and socially important. In Australia, the value of construction work performed in residential, non-residential and engineering construction sectors in 1996–1997 was AU$40.5 billion (ABS 1997a,b). On average, the industry amounts to between 6.5 and 7 per cent of Australia's GDP, and between 1995 and 1996, 682,000 people were employed in the construction industry in Australia (AEGIS 1999).

The industry provides the homes we live in, the buildings we work in and the transport infrastructure we rely upon. The construction industry contributes a great deal to improving our quality of life. However, for many workers and their families and friends, involvement in the construction industry leads to the unimaginable pain and suffering associated with an accidental death or serious injury. The construction industry continues to kill and maim more of its workers each year than almost any other industry. Typical figures for accidents and fatalities worldwide are shown in Figures 1.1 and 1.2. It should be noted that there is a huge range of incidence rates, varying from country to country. Part of this variation can be accounted for in terms of differing rates of economic and infrastructure development, but not all of it. In a project-based industry, accident rates will vary from project to project. Each project is unique, and each project type (for example, a road or a bridge or a house) has its own characteristics, methods of working, materials employed and techniques for construction. These characteristics, materials and techniques also vary from country to country. Take for example the Hong Kong Special Administrative Region of China and China itself. Hong Kong makes extensive use of bamboo scaffolding for building construction whereas China, the source of Hong Kong's bamboo, does not.

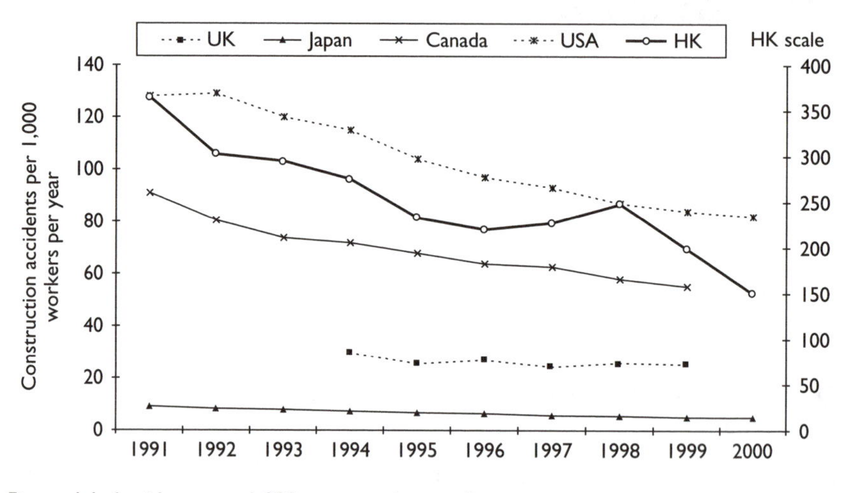

Figure 1.1 Accidents per 1,000 construction workers per year.

The way in which construction work is organised makes the management of occupational health and safety (OHS) more challenging than in other industries (Ringen *et al.* 1995). The industry has diffused control mechanisms, temporary worksites and a complex mix of different trades and activities. Much work is subcontracted out and workers are employed on short-term,

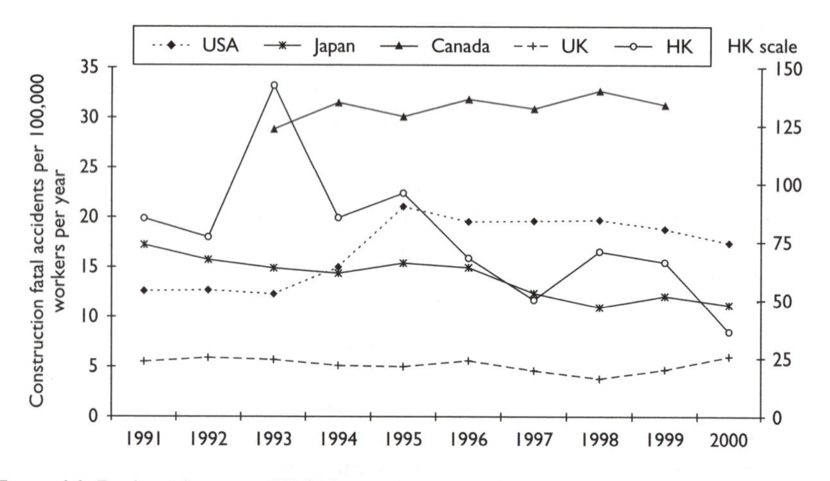

Figure 1.2 Fatal accidents per 100,000 construction workers per year (Source: as Figure 1.1). (Source: *Japan Statistical Yearbook*, Ministry of Public Management, Home Affairs, Posts and Telecommunications; Japan International Center for Occupational Safety and Health (JICOSH); *Fatal Injuries to Civilian Workers in the United States, 1980–1995*; *United States Bureau of Labor Statistics Data*, U.S. Department of Labor; *Health and Safety Statistics*, HSE publication; *ILO Yearbook of Labour Statistics 2001*, International Labour Office, Geneva; Occupational Safety and Health Branch of the Labour Department, Hong Kong Special Administrative Region.)

fixed contracts and then released at the end of a project. Such employment arrangements are associated with higher incidences of industrial accidents than where permanent employment exists (Guadalupe 2002).

The fact that construction is a project-based industry is an important contextual issue. When attempting to manage a dynamic, changing environment, such as a construction site and, indeed, a construction firm, it should be borne in mind that there needs to be an appropriate organisation structure to deal with the changing nature of the project. As it moves from design to construction to in-use phases, and as problems arise (such as late delivery of materials or labour shortages) on a day-to-day basis, there is a need for rapid, decentralised decision-making, contingency planning and an appropriate, organic form of organisation. This engenders a free, independent spirit in construction site personnel and has, traditionally, led to a disregard for authority and regulations. In many instances this disregard has been taken too far and descended into unacceptable corruption and malpractice. The Housing Authority short piling case in Hong Kong (HKLEGCO 2003) and the practices revealed during the Royal Commission inquiry into the Building and Construction Industry in Australia in 2002/2003 are examples of how malpractices can drastically effect the well-being of the industry and those working in it: the cherished characteristic of independence and initiative have also given the industry a bad name.

In turn, these characteristics also make it difficult to implement programmes across the whole of the industry, and safety and health on construction sites are badly affected. Indeed, the Royal Commissioner in the Australian inquiry named OHS as the most important issue facing the building and construction industry and reported that a change in attitude was needed to reduce the number of deaths and injuries on construction sites (Feehely and Huntington 2002). Commissioner Cole argued that a new paradigm must be established by which projects are completed safely, on time and within budget, rather than just on time and within budget. The report recommended 19 changes for workplace safety and the establishment of a special industry commissioner to improve health and safety standards (*Sydney Morning Herald*, 28 March 2003, p. 10).

The construction industry is recognised internationally as one of the most dangerous industries in which to work. For example, McWilliams *et al.* (2001) report that between 1989 and 1992, 256 people were fatally injured in the Australian construction industry. The fatality rate is 10.4 per 100,000, which is similar to the fatality rate for road accidents. The construction industry's rate of occupational injury and disease is 44.7 per 1,000 persons, which is nearly twice the all-industry rate. The Royal Commission inquiry into the Building and Construction industry reported that there are on average 50 deaths a year on Australian construction sites and that, in 1998–1999, accidents and deaths cost $109 million a year and almost 50,000 weeks of lost working time (Lindsay 2003). Similar figures are reported for the USA. Gillen *et al.* (2002) report that in 1998, the construction industry reported the largest number of workplace fatalities compared to any other industry. Construction accounted for 20 per cent of total work-related deaths. In 1997, the non-fatal injury rate for construction workers was 9.3 per 100 full-time workers, which is considerably higher than the all-industry figure of 6.6 per 100 full-time workers. Zwerling *et al.* (1996) report that in Iowa state, construction workers' incidence of occupational injuries is 2.5 times greater than that of other occupational groups. In the UK, a similar situation exists.

Apart from the alarming number of deaths, many construction injuries lead to long-term disability. Guberan and Usel (1998) followed a cohort of 5137 men in Geneva over twenty years and report that only 57 per cent of construction workers reached 65 without suffering a permanent impairment. Falls and manual handling are prominent risk factors associated with serious injuries and long-term disability in construction (Gillen *et al.* 1997; Nurminen 1997). In a study of workers' compensation data in Victoria, Australia, Larsson and Field (2002) report eleven prominent construction trades to have average injury durations longer than the all-industry average. In addition to falls and manual handling, which affected almost all construction trades, power tools, falling objects and electricity were significant risk factors in some trades. In terms of injury severity (taken as a combined index of days compensated, hospital costs and nature of injury), steelworkers, painters and rooflayers are reported to be in the most risky construction trades (McWilliams *et al.* 2001).

The industry's unenviable safety statistics are often explained in terms of the construction industry's inherently hazardous nature. The use of the term *accident*, which is defined in the *Oxford English Dictionary* as 'an event without an apparent cause', perpetuates the belief that occupational injuries and fatalities are chance events that cannot be prevented. This conventional wisdom is simply not true. What is of greater concern than the construction industry's appalling OHS record is that the same *types* of accidents occur in construction year after year. A review of the international literature reveals that the same type of work-related deaths, injuries and illnesses occur in construction industries all over the world. Many construction hazards are well-known and, in some cases, have been studied in great depth. What is apparent is that the construction industry fails to learn from its mistakes. We understand where deaths, injuries, and, to a lesser extent, illnesses occur in the construction industry but we still fail to prevent them. The same methods of working that have been used for generations are still being used, giving rise to the same hazards and ultimately resulting in the same incidence of death, injury and illness. Furthermore, the construction industry's organisation, structure and management methods militate against the identification and implementation of innovative solutions to the industry's OHS problems. Improvements in

OHS will not occur unless new methods of working that reduce known OHS risks are developed. However, this is likely to require that the industry's structural and cultural barriers to the adoption of new methods of working be overcome. These structural and cultural barriers are discussed in greater detail in Chapters 3 and 10.

Some of the more common occupational health and safety hazards in construction are described below, followed by an examination of the structural and organisational factors that currently impede the improvement of OHS within the construction industry. The remaining chapters of this book suggest ways in which OHS could be better managed at project, corporate and industry levels.

Construction as a process

By way of introduction, the following paragraphs highlight some of the 'idiosyncrasies' of the construction industry that lead to the typical safety and health issues that the industry faces. Currently, the whole process of production in the construction industry is undergoing a radical rethink with the introduction of 'lean and agile production', 'supply chain' and 'business process re-engineering' paradigms. These are briefly reviewed by Rowlinson *et al.* (1999), who state:

> It is argued that the construction industry needed to shift its focus to the underlying philosophy of lean production by recognising construction as a flow process in which construction should be seen as a hierarchical collection of value generating flows and achieve the goals of lean construction, hence the move to the term value chain rather than simply supply chain which emphasises a holistic view based around the value concept... It has been proposed that the lean revolution is essentially a conceptual revolution, at the heart of which are the flow and value models. The characteristics that the construction industry possesses are 'one-of-a-kind' nature of projects, site production and temporary multi-organisation. Because of this the construction industry is often seen as being different from manufacturing. Indeed, these characteristics may prevent the attainment of flows as efficient as those in manufacturing. However, the general principles of flow design and improvement apply for construction flows and in spite of these characteristics, construction flows can be improved to reduce waste and increase value in construction. As it is not possible to change the circumstances of the construction industry to fit a theory that is useful in a more stable environment such as manufacturing, other approaches are necessary. Initiatives undertaken in several countries have been trying to alleviate related problems associated with construction's peculiarities. The one-of-a-kind feature of the construction industry is

> reduced through standardisation, modular coordination and widened role of contractors and suppliers. Difficulties of site production are alleviated through increased prefabrication, temporal decoupling and through specialised or multi-functional teams. Lastly the number of liaisons between organisations is reduced through encouragement of longer term strategic alliances and partnering.

All of these issues have implications for health and safety on construction sites.

Autonomous working and the subcontracting system

Wilson (1989) discusses the problems inherent in the subcontracting and labour employment practices of the industry. The problem being addressed is an international one and is, in fact, a structural problem of the economics of the construction tendering system and the organisation of small and medium-sized companies. It is obvious now that many of the problems concerning site safety on construction sites are attributable to subcontractor performance. No safety improvement scheme can be successful unless it includes subcontractors. Hence, a system must be devised to assess the performance of subcontractors, and it has been recommended that subcontractors be actually registered to work on construction projects. The Singapore List of Trade Subcontractors (SLOTS) system in Singapore provides one example of how this can be undertaken. As of 23 April 1999, all subcontractors engaged in public sector projects, with the exception of ceiling glazing and metal work trades, were required to be registered under SLOTS. SLOTS membership also determines eligibility to obtain work permits for foreign workers. The SLOTS system therefore presents an opportunity to vet the OHS management activities of subcontractors and provides opportunities to intervene where necessary. However, despite this, incident rates have not fallen substantially in Singapore since 1999.

It is essential that the level of performance of subcontractors be appraised and the level of performance of individual contractors and subcontractors together be analysed. If it is found that certain subcontractors are regularly performing badly, then it is essential that a system be in place that allows for them to be suspended from operation. This is possible only if a clearly defined registration system for subcontractors exists. It is not possible for individual contractors to substantially improve the performance of subcontractors without some form of sanctions such as a registration system and suspension from it. A more detailed discussion of subcontracting issues follows later in this chapter.

The role of client and designer in safety

Traditionally, it has been the role of the main contractor to deal with all safety aspects of construction. However, in recent years, and with the advent of self-regulation, it has become apparent that the contractor alone cannot deal with safety issues. In fact, many safety issues arise due to design considerations. Take, for example, the decision to construct a post-tensioned, pre-cast concrete bridge over a road. This requires the provision of falsework and formwork that has to be kept in place for a substantial period of time before construction is complete. This brings with it a whole series of access-to-height problems. However, if a pre-cast, pre-tensioned concrete bridge is to be constructed, then the bridge deck components can be constructed at ground level, even off site. They then can be lifted into position using a crane. Obviously, the site safety issues can be greatly reduced, although different risks occur, by using the alternative method of construction. Hence, it is important that the designer is aware of the role he or she can play in enhancing construction site safety. By paying attention to details, such as provision for changing light bulbs in high ceilings, safety can be built into the project during the design, construction, maintenance and operation phases. This, of course, has a price. By designing safety into the product, more careful consideration has to be made of decisions during the design phase. This may be more expensive initially, but in the long term, if the site is kept accident-free, and the facility is also kept accident-free in use, then the savings can be enormous. Hence, responsibility moves not solely to designer, but to the client, to insist that safety be designed into the project. Thus, the philosophy of the UK's Construction Design and Management (CDM) Regulations (described in Chapter 2) incorporates such ideas. However, this requires a change of mindset on behalf of clients and designers. A new philosophy of design is required, and this is discussed briefly below.

Designing safety and health into the process

The logical extension of the argument above is the designing of health and safety into the production process. Later in this chapter, this very issue is addressed from a historical/traditional perspective. However, in order to break the current mould and promote open thinking on this issue, an example from another industry is introduced. General Electric developed guidelines for product design in 1960, but a significant benefit was not realised until systematic design for assembly protocol, DFA, was introduced in the 1970s. Other 'design for somethings', such as manufacturability, inspectability and quality, were developed in the 1980s. Now, *design for x (DFX)* has been used as an umbrella for these terms. DFX aims to design a product from many viewpoints or characteristics (Gutwald) for such benefits as:

- achieving a product exhibiting better qualities of *x* (for example, Design for Manufacturability (DFM), Design for Assembly (DFA), Design for Disassembly, Design for Quality and Design for Environment); and
- early failure detection.

The potential of DFX is enhanced by the availability of a powerful representation tool such as virtual reality, where design can be represented in three-dimensional graphical detail, and walkthrough functions enable the user to discuss any aspect of the object in a virtually real location. The virtual reality of products and processes of construction projects is being used to develop a Design-For-Safety-Process (DFSP) methodology, which aims at pointing out the safety hazards inherited from the construction components and activities. This development is discussed in Chapter 8.

Codes of practice and standards

Codes of practice and standards have a particularly important role to play in the construction industry, especially when being used in conjunction with performance-based safety legislation. Because there are no prescriptive procedures to be followed with such legislation, it is important that an example of best practice be given. This is the role of the code of practice, and any employer providing a system of working that is as safe as the code of practice can generally be assumed to be complying with the legislation. The same applies to standards that cover levels of attainment to which we have to aspire. The key concept, really, with performance-based legislation is setting goals and monitoring the achievement of those goals. If achievement is not taking place then changes have to be made. The codes of practice give us a starting point; one would expect to see a process of continuous improvement so that organisations would enhance the standards presented by the code and, eventually, the codes should be revised to reflect this enhanced performance.

Falls

Construction workers are a high-risk population for falls. Studies suggest roofing workers may be particularly at risk (Janicak 1998; Derr *et al.* 2001). Incidents involving roofing workers falling through fragile materials, such as asbestos cement sheeting, are widely reported (Suruda *et al.* 1995). Gillen *et al.* (2002) report that 383 of the construction industry deaths, occurring in the USA in 1998, occurred as the result of falls from roofs, scaffolds, ladders, girders or other steel structures, but Kines (2001) demonstrates that falls are also involved in many serious and minor lost-time injuries.

Bobick *et al.* (1994) differentiate between *primary* fall protection, designed to prevent workers from falling to a lower level, such as crawling boards and planks, and *secondary* fall protection, designed to reduce the impact of a fall after it has occurred, such as lifelines and safety net.

Kines (2002) examined the causes of fatal and non-fatal falls in the Danish construction industry and reports that a disproportionate number of fatal falls occurred in the afternoon, in the absence of secondary fall protection devices, such as safety nets. Non-fatal falls were more likely to occur in the morning, when safety nets were more likely to be present. Kines (2002) suggests that the absence of safety nets enhances workers' perception of risk and increases safe behaviour but that, as the working day wears on, the concentration required gives way to worker fatigue and an increased likelihood of fatal slips or lapses. Also, if a fall occurs, workers' reactions and reflexes are negatively affected by fatigue and the injury outcome is likely to be worse. Fatally injured workers were found to have head injuries, such as skull fractures or brain injuries, whereas workers who sustained non-fatal serious injuries were more likely to suffer injuries to the spine, shoulder, hip, lumbar, arm, hand or leg. This finding led Kines (2002) to recommend the use of strapped-on safety helmets designed to provide protection against non-repetitive impact when working at height.

Mechanical equipment

Power tools

Due to the nature of the industry, a wide range of specialist plant, equipment and tools are used in the construction industry. This can be confusing for the worker and supervisor, as when they move from job to job they may encounter new and different kinds of plant and tools. Take, for example, foundation construction of piles. Piles can be drilled, bored, vibrated or hammered into place and can be made of timber, steel or concrete (or a combination of the latter two). Steel piles can be tubes, H-shaped or 'sheet' piles. Special procedures should be in place for dealing with mechanical equipment, plant and materials. Plant and equipment are generally beyond the control of the individual worker. Hence, planned maintenance and set operating procedures are essential if each worker is to safely work with, or work next to, the plant or equipment. The supply of spare parts needs to be properly organised, as does the provision of instructions for new workers and the provision of updating training for existing workers when new models of a particular plant or equipment arrive. It is in such areas that certification and registration of workers is an important aspect.

Hand tools

A wide range of powered and non-powered hand tools is used on construction sites, and these have the potential to cause serious injury if not operated or handled correctly. However, control and maintenance of these tools is difficult. Due to the nature of construction and construction sites and the autonomous nature of the workers involved, tracking and tracing the tools is a difficult problem. This is exacerbated by the 'disappearance' of many items from construction sites. It can be easily imagined that maintaining the integrity of a safety and maintenance system in such circumstances is an arduous task.

Hands tools are well-known as a major source of injury on construction sites. Lingard and Rowlinson discussed this in relation to the accident record of the Hong Kong Housing Authority (HKHA). Most non-power-driven hand tool accidents and portable power-driven hand tool accidents result in abrasion injuries. The HKHA studies noted that an overwhelming proportion of non-power-driven hand tool accidents occur to carpenters/formworkers (70 per cent in Hong Kong). Carpenters and formworkers also suffer a significant proportion of all accidents involving portable power-driven hand tools. However, research indicates that lack of experience was not associated with either type of hand tool injuries as the majority of injuries occurred to experienced (over ten years) workers. It should be noted that years of experience does not, of course, equate to good working practice. Indeed, these figures indicate that bad habits, irrespective of experience, may be a factor associated with these accidents. In the Hong Kong case it was reported that 70 per cent of workers injured using non-power-driven hand tools were untrained in the use of these tools and 80 per cent of those injured using portable power-driven hand tools were untrained. One might well conclude that the use of initiative and 'learning on the job' is inherently dangerous.

Most injuries involved hammers, drills, spanners and nail guns. Almost all of the injuries involving hammers were related to the construction of traditional, actually old-fashioned, wooden formwork, in-situ. These injuries took place when workers were dealing with difficult sections which had gradients or curves and access was very difficult to these locations. Hence, in this instance one might conclude that the hand tool injury was caused at its root by poor access to the location resulting from a failure to consider safety in design.

The drill injuries occurred, in the main, when workers were accessing areas which were difficult to reach or get to. Hence, the injury was not related to the drill itself but, as in the previous case, by the design of the workplace and resulting difficulty in gaining access to the work area.

The injuries involving spanners were almost exclusively related to fixing or mechanical works and the construction of traditional, old-fashioned, wooden formwork. Again, as in the cases above, these hand tool injuries related to the inability to access the areas easily from the platforms provided and, again, this relates back to the original design of structures and temporary works and the adequacy of method statements produced in order to deal with these issues.

Nail gun injuries are a different matter. Most of the nail gun injuries occur when apparently inexperienced users attempt to use the nail gun in either inappropriate circumstances or nail into inappropriate materials. In most instances, the nail propels through the material and into a worker below or adjacent to where the nail gun was working.

A smaller subset of hand tool injuries is related to the maintenance of the tool itself. This includes the splitting or breaking of the shaft of a hammer or shovel and the inadequate provision of insulation to electrical equipment. Such injuries are almost certainly a failure of the safety management system in assessing and removing from use any tools which are in a dangerous condition. This should be a daily inspection routine which can be easily incorporated into toolbox talks.

When dealing with larger machinery, such as bar-bending machines and table-mounted circular saws, the issues relating to safety again follow a quite similar pattern. With bar-bending machines, the machinery often fails to respond to the controls used by the operator. This indicates either a poor ergonomic design or, more likely, poor quality gear and brake mechanisms combined with poor quality materials in the construction of the machine. The most common failures in table-mounted circular saw incidents were the failure to use a guard and, slightly more commonly, the failure to use a pusher in order to ensure that the worker's hands remain away from the saw blade.

Essentially, one can trace all of these accidents back to a lack of management oversight and, second, to the lack of awareness of workers as to the dangers that they experience on a daily basis.

Accidents involving hand tools are not unique to Hong Kong. A Swedish study of construction accidents found that 18 per cent of accidents involved hand-held machines or tools Helander (1991). Myers and Trent (1988) found that construction was the industry with the second highest number of non-power-driven hand tool accidents (the first was agriculture) and construction accounted for more power-driven hand tool accidents than any other industry. Also, Olsen and Gerberich (1986) reported the construction industry as having a higher rate of finger amputations than other industries, many of which are the result of incidents involving hand tools.

Temporary works

Temporary works are used in order to facilitate the construction process, much more so than in any other industry. Temporary works do not form part of a finished product but are essential for the product to be completed. Generally, the main contractor is expected to design what temporary works are necessary but, in the case of particularly complex elements, it is essential that the designer consider temporary works in the design process. Hence, the argument again surfaces for systems whereby design and construction are well and truly integrated into a seamless process. The design-build process achieves this. It would be instructive to attempt a study in this field, and compare safety performance on design-build compared to traditional contracts. This is the essence of the *DFX* production philosophy, and is a direction in which the organisation and management of the construction process are moving. The sooner the concept of design for safety is accepted and used throughout the industry, the better it is in terms of the overall safety performance of the industry.

Manual handling risks on construction sites

Manual handling risks are now legislated for in most countries. The legislation basically requires that a risk analysis be undertaken of all manual handling activities. This is a sensible approach, as some activities are repetitive, and the analysis can be very quickly accomplished, and standard tools, plant or equipment provided in order to ensure safe performance. However, when an unusual or cumbersome load is being dealt with, a more proactive approach is required than simply producing standard solutions. In order to deal with this, a thorough analysis of the load to be lifted and the method of lifting should be undertaken. An example might be the problem of lifting pre-cast facade units on to a tower block or pre-cast concrete bridge beams onto a bridge. If workers are expected to manhandle these elements into the final location, then a proper system of working should be devised for this to be done safely. It may be necessary for workers to be tied back to the main structure by a parachute harness, special tools may be required for the workers to position the units, and a series of ancillary items such as bolts, spanners and wedges, may be required for the workers to safely complete the work. Hence, a proper assessment of the risks involved in any manual handling operation is essential.

Method statements

As can be imagined from the foregoing discussion of construction plant, equipment and materials, it is extremely important to produce realistic method statements in any safety management system in the construction industry. These method statements should indicate type of materials, the type of plant and machinery, the type of tools and the actual process of working to be undertaken. However, method statements are often produced to impress clients in tender submissions before being filed away in the office never to be seen again. It is important that method statements contain working documents, that are clearly communicated to those performing the work. The method statement should not deal purely with methods, but also with procedures and practices (particularly in relation to safety procedures) any checks that are required, and any permit needed for work to continue. By devising a comprehensive method statement for all elements of construction work, it is possible both to enhance safety and to improve productivity. If these method statements can actually be dealt with during the design process, then one has the opportunity to value-manage the project. By this, it is meant that the improvement in the design, through the consideration of how the design will be constructed, should save time, cost and materials. It should also enhance safety. It is important that method statements are produced on the basis that they aim for efficiency in production and effectiveness in safety management. Both philosophies should be adopted in their production, but no compromise should be made on the safety aspect. It is very important that the system is in place so that the safety, production and management elements contained in the method statements are communicated to all of those on site, at managerial, supervisory and worker levels.

Work-related musculoskeletal disorders

Work-related musculoskeletal disorder is a generic term for long-term discomfort caused by work that is the result of acute or instantaneous trauma of soft tissues or their surrounding structures (Dimov *et al.* 2000, p. 685). In the USA, work-related musculoskeletal disorders are the most frequently cited injury to workers, affecting nearly half of the national work force (NIOSH, cited in Dimov *et al.* 2000). Musculoskeletal problems are common in construction work, which involves awkward static body postures and repetitive dynamic motions (Holstrom *et al.* 1995). However, job tasks vary greatly in construction, and the site environment poses difficulties for analysing the ergonomic aspects of the person/machine interface. Surroundings, weather conditions, pace of work, tools used and postures adopted can vary considerably. Despite this variability, research reveals that certain musculoskeletal problems are characteristic hazards associated with the work of construction trades. For example:

- occupational knee problems are commonly experienced by carpet layers (Bhattacharya *et al.* 1985);
- neck problems are frequently reported by crane operators;
- shoulder symptoms are widely reported among scaffolding erectors; and
- hand/wrist symptoms are common complaints among electricians (Holstrom *et al.* 1995).

Dimov *et al.* (2000) studied body discomfort patterns among carpenters at the end of a working day and found that mid-to-lower back pain was most frequently cited, followed by knee pain. They suggest that trade-specific discomfort data should be collected to identify ergonomic risk factors for work-related musculoskeletal disorders in construction.

This information should then be used in the development of ergonomic awareness training for apprentices in relevant construction trades and technical training courses. Work-related musculoskeletal disorder is a growing problem and must be considered in job analysis and design. The ergonomic aspects of construction tasks are considered in greater detail in Chapter 5.

Hazardous chemicals

An area of increasing concern as a health issue is the use and effect of hazardous chemicals in the construction industry. For example, the International Agency for Research on Cancer has classified some bituminous substances widely used in road building as possible human carcinogens, and animal studies reveal that the dermal application of condensate of bitumen used commonly in roofing is potentially carcinogenic. Burtsyn *et al.* (2000) suggest that the determinants of exposure of road construction workers to harmful bituminous substances depends on the type of asphalt used, the application method, workers' job class and equipment used in the application process. For example, some paving machines can reduce exposure levels by up to 60 per cent because they are equipped with features such as cabins with lockable doors and windows, partial enclosure and ventilation. Such engineering solutions to occupational hazards are important, because even when the emission of asphalt fumes is visible, Greenspan *et al.* (1995) report that protective dust masks are not worn and, to many, potential consequences of exposure do not present themselves immediately.

Noise exposure

There is a clear association between noise exposure and hearing loss. Construction workers are particularly at risk of occupational noise-induced hearing loss (NIHL) because of the intense use of heavy equipment and widespread use of portable power tools by many construction trades, and NIHL compensation claims are increasing among construction trades (Daniell *et al.* 2002). In the USA, the National Institute for Occupational Safety and Health (NIOSH) estimates that approximately 16 per cent of construction workers employed by general building contractors or specialist trades are routinely exposed to noise levels of 85 dBA or over. By comparison, 24 per cent of construction workers employed by heavy construction (civil engineering) contractors are estimated to be exposed to equivalent noise levels. However, Neitzel *et al.* (1999) report higher exposure levels. They suggest that up to 40 per cent of a representative sample of construction workers were exposed to noise levels in excess of 85 dBA. Noise that exceeds an average of 85 dBA over an eight-hour working day (commonly referred to as LAeq.8h 85 dBA) or peak noise of 140 dB (linear) is potentially damaging to hearing.

Furthermore, workers in all of four trades examined (carpenters, labourers, ironworkers and plant operators) were exposed to noise levels that exceed the allowable limits. Neitzel *et al.* (1999) report that noise exposure levels vary according the type of construction and activities being undertaken. Among construction classifications, road construction, installation of machines or equipment and clearing, grading or excavating are reported to account for the largest number of compensation claims (Daniell *et al.* 2002). Noise exposure is also reported to be particularly high during the structural stage of construction and at sites using multiple concrete construction techniques (Neitzel *et al.* 1999). Construction workers are also reported to be lacking in awareness of the risks of noise exposure (Behrens and Brackbill 1993) and most hearing conservation programmes are not widely implemented in construction (Reilly *et al.* 1998). Dosimetry studies in construction reveal that even trades that predominantly use non-power-driven hand-held tools, such as electricians, are exposed to levels of noise in excess of standard exposure limits because of the noise in the general construction environment (Seixas *et al.* 2001). Hearing protection should therefore be available to all workers on site, regardless of their trade or work tasks. However, hearing protection may be ineffective because studies reveal usage rates between 18 and 49 per cent (Neitzel *et al.* 1999). Engineering control strategies are preferred. For example, Schneider *et al.* (1995) suggested that a bulldozer operator's noise exposure could be reduced by 11 IdB at a cost of US$3,450 to $4,300. The lack of significant total p.a. difference between noise exposure in different construction trades means that noise reduction measures should take into consideration the construction site environment as a whole.

Barriers to improvement

The construction industry is organised in a way that does not lend itself to the development or implementation of ways to eliminate hazards or reduce risks to the health and safety of workers to an acceptably low level. Structural and cultural characteristics of the construction industry that currently militate against the improvement of the industry's OHS performance are described briefly below. However, none of these characteristics should ever be used as an excuse for the industry's poor performance because all of these issues are potentially manageable. Chapter 3 of this book discusses the impact of some innovative ways to procure and manage occupational health and safety for construction projects. We argue that these barriers could be overcome if certain business practices were modified, and alternative contracting strategies and management methods were adopted throughout the construction industry.

Traditional separation of design and construction

The Commission of the European Communities claims that over 60 per cent of all fatal construction accidents can be attributed to decisions made before construction work commenced on site (Commission of the European Communities 1993). This suggests that decisions made early in a project's life, particularly during the design stages, may impact upon the health and safety of workers who must then construct the facility in accordance with design and specifications provided by the architect or design consultant. Despite this, the occupational health and safety of construction workers has traditionally been understood to be a matter for the works contractor who has been engaged to undertake construction work for a pre-determined tender price (Williams 1998).

It is also widely understood that 'technological' solutions to OHS risk reduction, that is, those that eliminate a hazard at source, or reduce risks to an acceptably low level through engineering or design solutions, are the most effective. However, technological solutions can often be achieved only if OHS risks are identified and controlled prior to the actual construction of a project (HSE 1995). Thus, the traditional separation of design and construct functions in construction can seriously limit the identification of innovative solutions to OHS problems at the design stage of a project. The European Construction Institute (ECI) (1996) recommends that OHS risks be assessed and control decisions be made in the concept design, project planning and specification stages of a construction project. Incorporating works contractors' experience and knowledge at the design stage can improve project 'buildability' and eliminate OHS problems at source (Hinze and Gambatese 1994). The adoption of the design-and-build contracting approach, in which the design and construction processes are undertaken by the same organisation, has enabled closer attention to be paid to project 'buildability' issues and is likely to have a positive impact upon OHS.

Recent legislative changes in the countries of the European Union have attempted to overcome the traditional separation of design and construction by legislating for OHS to be considered in pre-construction planning and decision-making. For example, in the UK, the Construction (Design and Management) Regulations impose responsibilities upon clients and designers. These statutory changes extend the common law 'duty of care' that designers have always held with regard to their professional practice, and force designers of facilities above a certain contract value to implement OHS risk management activities during construction design work. The impact of these regulations is discussed in greater detail in Chapter 2.

Competitive tendering

Most construction projects are awarded to contractors on the basis of competitive tendering, usually to the lowest bidder (Russell *et al.* 1992). However, in the context of intense competition, competitive tendering places a great deal of pressure on tenderers to keep their bids low, to increase their likelihood of winning work. This pressure can discourage contractors to factor into bids the cost of performing the work safely. Egan (1998) in the UK and Tang (2001) in Hong Kong note that the lowest price does not equate to the best value for a construction project. OHS performance should be considered as part of the overall 'value' of the service provided by a contractor, not least because poor OHS performance in the projects they sponsor reflects badly upon construction clients.

However, the way in which contractors price work often fails to account for OHS requirements. For example, it is common for the unit rate estimated for an activity to ignore safety issues (Brook 1993). Also, despite the importance of pricing OHS into a job, estimators have little or no involvement in pre-construction OHS planning (Oluwoye and MacLennan 1994). Furthermore, OHS advisors are usually excluded from the tendering process (Brown 1996). Research suggests that the inclusion of safety costs in a tender can reduce the 'lost time accident frequency' rate from a range of 2.5–6.0 per 100,000 man hours worked to a range of 0.2–1.0 per 100,000 man hours worked (King and Hudson 1985). This may be because successful bidders are not forced to select cheaper, but less effective OHS risk controls as construction progresses in an attempt to maintain an acceptable profit margin in the project. In an attempt to overcome this practice, the Hong Kong Government Works Bureau introduced a 'pay for safety' scheme into its projects with the express intent of taking safety out of the bidding process. The compliance of the contractor with the 'safety specification' is measured through independent site audits before payment is made.

Construction clients could take greater initiative in managing OHS in the projects they sponsor. For example, if construction clients inviting tenders specified the way in which prospective contractors should allocate OHS costs in their bids, all tenderers would be forced to price OHS into their bids. The costs allocated to OHS could then be easily identified and compared, and performance measured and paid for. This would enable clients to eliminate unscrupulous contractors. Alternative contractor selection methods could also help clients wishing to engage a safe contractor. Selective tendering, whereby contractors are subject to pre-qualification and only those that meet pre-determined performance criteria are invited to bid, allows clients to scrutinise contractors' OHS management systems and past performance and establish minimum standards for the project. Regular construction clients may routinely audit contractors, maintain a list of approved contractors and use this list to exclude contractors whose OHS performance is below standard. In some circumstances, for example high-risk work, clients may decide to engage in negotiation with prospective contractors whose OHS performance is known to be excellent, rather than engage in competitive bidding. This issue is dealt with in some detail in Chapter 3.

The plethora of small businesses

The construction industry in most developed economies is made up of a small number of large firms and a multitude of small and medium-sized enterprises (SMEs). For example, the majority of Australian construction firms are small businesses with 97 per cent of general construction businesses employing less than 20 employees and 85 per cent employing less than 5 employees (ABS 1998). Small businesses are unlikely to have professional OHS advisors on staff and may lack the knowledge and resources they need to implement OHS management activities, such as training, undertaking risk assessments and performing routine inspections and audits. Furthermore, the engagement of expensive OHS advice or services from external consultants may seem to be an unnecessary expense. The pressures associated with cost-cutting and business survival in a cutthroat industry mean that OHS is likely to be low on a small construction firm's list of priorities. International research confirms that small businesses are poorer than larger organisations in implementing formal OHS programmes (Eakins 1992). Unsurprisingly, occupational injury rates vary inversely with the size of construction firms, with smaller firms reporting higher injury frequency rates (McVittie *et al.* 1997).

The problem is illustrated in Australia by the experience of the domestic building sector. This sector has a very poor OHS record and has shown little improvement under the legislation which made risk assessments mandatory (Mayhew 1997). While large construction firms have implemented a wide range of OHS management initiatives, it seems that SMEs in construction undertake little in the way of formal OHS management activities.

Similarly, Lam (2003) reports that the majority of small firms in Hong Kong have limited knowledge of existing legislation and are reluctant to take freely offered occupational safety and health advice. They lack resources to keep themselves abreast of the legislation and technological developments, do not belong to any industry associations and are primarily concerned with profits. He states 'they just remain in their narrow world in which the flow of safety and health knowledge and information is rare; it is no wonder that their safety performance is not up to scratch' (p. 73). However, he also notes:

> There is a general presumption that small contractors are inferior in their implementation of safety management systems because they usually suffer from a lack of resources. This assumption is *found to be* only partially true because, among all the weak elements in the safety management system, some *weaknesses* are suffered by all contractors irrespective of the *size and*... there is a constant proportion of complying contractors (p. iv).

Hence, examples can be found of SMEs which do implement OHS initiatives effectively but it is true to say that the majority struggle to do so.

Subcontracting

The prevalence of subcontracting is often cited as a factor contributing to the construction industry's poor OHS performance. Outsourcing work has implications for the OHS of directly employed workers. For example, if maintenance work is outsourced, poorly maintained equipment could pose a hazard even if the management has passed on the responsibility and risk for its safety. Also, subcontractors and their employees may not be familiar with safety procedures and safe systems of work. Communication is essential, and all subcontractors should be given induction training on commencing work, to ensure they are familiar with site safety rules, emergency procedures and so on. Subcontracting can give rise to risks to both employees and subcontracted workers and it is advisable that principal contractor and subcontractors work together to identify risks and decide how best to control them (Standards Australia 2001). OHS risks experienced by one subcontractor may also be caused by the activities of another subcontractor, and co-ordination is essential, with special attention being paid to manage the interfaces between different work crews and activities. Responsibility for the safety of employees and contractors rests with the person in control of the workplace. Thus, unless subcontractors perform work off-site, the principal contractor remains responsible for the OHS of subcontractors and their employees. Thus, it is advisable that subcontractor selection processes involve vetting subcontractors to ensure that they have acceptable OHS records. Such procedures are rare and subcontractors are more commonly selected upon their price and availability, contributing to the argument that subcontracting is 'bad for safety'.

Emphasis on contractual relationships

Good communication and co-operation are widely acknowledged to be an essential feature of effective OHS management. Open and honest communication is not a feature of many construction projects. Instead, communication between parties to construction contracts is often characterised by conflict and confrontation, making co-operation on matters of OHS difficult. Communication patterns within groups, including project teams, have been described as taking different forms. Cheng *et al.* (2001) suggest that construction projects traditionally adopt a linear 'chain' pattern of communication. Within this model, communication between parties to a construction project is restricted to communication between parties in direct contractual relationships, concerning contractual requirements. In this model, designers who are in a contractual relationship with the client will not communicate directly with contractors, with whom they have no contractual relationship. This restricted communication is associated with low member satisfaction, poor performance and co-ordination problems (Glendon and McKenna 1995). It is unlikely to result in inter-organisational co-operation or optimal OHS performance.

We suggest that a more open communication model is appropriate for managing OHS communication in construction projects. For this to occur, the industry's culture of communication based upon contractual relationships must be overcome, and communication channels opened up between project participants with a role or interest in OHS. These may include clients, designers, suppliers, subcontractors, employees and their trade union representatives.

Stimuli for change

Safety as a social and moral responsibility

Discerning shareholders, clients and employees are increasingly putting pressure on corporations to act in a socially responsible way. This includes managing the health, safety and welfare of their employees.

The relationship between employees and their employers is unequal. Employees are usually in a comparatively low bargaining position, meaning that the possibility exists for various forms of exploitation. It is therefore important to recognise that employees have certain rights that should be respected. One such right is the right to health and safety in the workplace. Rowan (2000) suggests that a right may be understood to be a 'moral claim'. There are two parts to this definition. First, the notion that rights are moral suggests they may not be conventionally recognised. Second, rights are claims that correlate with certain duties on the part of the person against whom these rights are held. Thus, claims are relational because one person's right is also another person's duty. An employee's *right* to a safe and healthy workplace is an employer's *duty* to provide such a workplace.

An important implication of viewing rights as moral claims is that we may be morally obliged to do something, even if it is not required of us by law or corporate policy. Rowan argues that the moral foundation of employees' rights is based upon the fact that employees are persons, and persons have moral importance because they have individual goals and interests. Rowan (2000) suggests that all persons should be free to pursue their goals and interests, and therefore certain fundamental rights exist, such as the right to freedom, the right to well-being and the right to equality, which may broadly be stated as the right to be treated with respect. This is similar to the Kantian notion that it is morally wrong to treat others as a means only (Kant 1981). In an organisational context, it is therefore wrong to treat employees as mere objects or things to be used in order to attain corporate objectives. Some theories of the firm suggest that the separation between work activities and personal activities is artificial and unnecessary, and that work organisations can actually contribute positively to employees' personal lives (Solomon 1992). Organisational wellness programmes and the implementation of family-friendly employment practices are examples of organisational initiatives that recognise the importance of employees' overall physical and mental health, well-being and family functioning.

Theories of accident causation

Blaming the worker or the system?

Efforts to prevent occupational injury and illness are likely to be shaped by assumptions made about how injuries and illnesses occur. Hopkins (1995) identifies two broad sets of assumptions, which he terms *blaming the victim* and *blaming the system*. The first of these approaches explains occupational injury and illness in terms of characteristics of workers themselves that make them particularly susceptible.

The notion that certain workers are accident-prone is one such explanation that was espoused by the Industrial Fatigue Research Board in the 1920s (Industrial Fatigue Research Board 1922). The view of the Board was that personal susceptibility was an important factor in accident causation: that accidents did not happen to workers on an equal basis, and that most accidents happen to a few susceptible workers. This approach led to only one logical prevention measure, which was the scientific selection of suitable workers (Nichols 1997). Such a discriminatory recruitment approach has the potential to disadvantage certain groups of workers and Johnstone (1993) suggests that pre-employment screening out of potential employees on the basis of susceptibility to occupational injury or illness is largely illegal. A further problem with the concept of accident-proneness is its implicit assumption that exposure to danger and injury reporting are equally distributed throughout the population (Sheehy and Chapman 1987). Those who report injuries may, in fact, not be more accident-prone but may just be more conscientious in reporting injuries (Hopkins 1995).

A second 'blame-the-victim' explanation of occupational injury was espoused by the Robens Committee of inquiry into health and safety at work (see Chapter 2). The Robens Committee suggested that the most important causes of occupational injuries were apathy and workers' ignorance of safe working methods. Nichols and Armstrong (1973) mount a compelling argument against this view. They suggest that the Robens Committee provided no sound evidence in support of the apathy/ignorance claim. In fact, the Committee ignored the results of contemporary analyses of industrial accidents, which had one common feature – that the injured workers were all under pressure to keep production going. In ignoring the social relations of production, Nichols and Armstrong argue that the Robens philosophy was fundamentally flawed.

Other explanations of occupational injury and illness that seek to 'blame-the-victim' include the view that 'macho' workers do not want to adopt safe work practices because they fear that they will be seen as effeminate, and the notion that many injury claims are falsely contrived so that workers can have extended time on workers' compensation. There is little evidence to suggest that either of these assumptions is valid.

Alternative explanations of occupational injury and illness focus on social, technological and organisational causes. The social relations of production, such as the pressure to maintain production and bonus or piece-rate payment schemes are seen as playing a key role in encouraging workers to ignore safe work practices. The physical/technological environment, which in many industries presents unusual and sometimes extreme hazardous conditions, is also recognised as a source of occupational injury and illness. Common features of many incidents that lead to occupational injury or illness are organisational breaches of occupational health and safety legislation and codes. In their most comprehensive, 'blaming-the-system' approaches regard accidents as system failures in which accidents are explained in terms of a complex interaction of plant and equipment, management systems and procedures, people and other human factor considerations.

Heinrich's 'domino' theory

In the 1930s, Heinrich, an engineer working for an insurance company in the USA, undertook an analysis of 75,000 accident reports, and attempted to model the causes of accidental injury (Heinrich 1959). This is one of the most frequently cited early attempts to develop a chronological sequence of inter-connected causal factors leading to injury. The theory was likened to dominoes falling, in that if one condition occurred, it would cause the next and so on. Heinrich's 'dominoes' were:

- *Ancestry and social environment.* Someone's family and social background led directly to...
- *Personal factors.* For example, personal characteristics such as greed, stupidity, recklessness could be in a person's nature or learned. These personal factors led directly to...
- *Unsafe acts or conditions.* Unsafe acts referred workers' behaviour whereas unsafe conditions were mechanical or physical hazards. These proximate causes lead directly to...
- *Accidents.* These are events such as collisions, falls and contact with moving machinery, and they led directly to...
- *Injuries.* These can be lacerations, sprains, fractures and so on.

Heinrich's analysis also led him to conclude that 88 per cent of accidents were caused by unsafe acts, and only ten per cent were caused by unsafe conditions. Viner (1991) suggests this approach is welcomed by many who are sympathetic to the view that accidental injury is the consequence of the victim behaving in a way that contravenes accepted behavioural codes. Indeed, many occupational accident prevention programmes still focus on attempts to modify workers' behaviour to eliminate the immediate behavioural causes of accidents (see also Chapter 8). Heinrich's model may be criticised for focusing too much attention on the immediate circumstances surrounding accidents, when it is now recognised that unsafe acts and conditions have systemic and organisational causes. Also, Hopkins (1995) suggests it is misguided to attribute accidents to either an unsafe act or an unsafe condition because most accidents are the result of a complex interaction of multiple causes. The linear, single pathway model developed by Heinrich is therefore over-simplified.

An updated domino sequence

Bird and Loftus (1976) revised Heinrich's domino theory of accident causation by incorporating the role of management in the accident process. Thus, all losses, whether manifest in bodily injury or illness, property damage or other types of wastage of an organisation's assets, are traced back to a lack of management control. The dominoes in this revised sequence are as follows:

- *Lack of control by management.* This leads to...
- *Basic causes of accidents.* These can be personal factors, such as fatigue, lack of motivation or insufficient safety knowledge or job factors, such as unrealistic work schedules, inadequate resources and so on. These lead to...

- *Immediate causes of accidents.* These can be substandard practices, conditions or errors, which lead to...
- *The accident.* This results in...
- *The loss.* This can be minor, serious or catastrophic.

Bird and Loftus (1976) suggest that the traditional focus of accident prevention, the immediate cause of accidents, involves treating the symptoms of the problem rather than its root cause. They suggest that long-term solutions must focus on the first domino in the sequence, management control, without which accidental losses will continue to occur.

Multiple causation models

The theory of multi-causality holds that every accident can be preceded by more than one event. Each of these events can also be preceded by multiple events. Contributing causes combine together in a random manner resulting in an accident. Consider the following example. If a hammer is left carelessly on a scaffold platform and is accidentally kicked by a worker using the platform, the hammer may fall to a lower level. If workers are passing underneath the scaffold, the hammer may strike somebody causing serious injury or even death. Many contributory factors might affect the outcome, such as the presence or absence of protective toe-boards on the scaffold, the degree of care taken by workers working at height, the layout of the site, including the positioning of walkways beneath working platforms, the response of the worker on the ground and the usage of protective safety helmets on the site. Figure 1.3 shows the contributory factors mapped in time. When one of the factors is absent, there is no injury but when all are present an injury is the most probable outcome.

The principle of multi-causality was adopted by the Swedish Board of Occupational Safety and Health in their injury information system, which is described by Andersson and Lagerloff (1983). In this system, injury events are defined as *occurrences*, which denote the direct infliction of an injurious effect on the victim. Incidents can have up to three preceding events. Injury events occur at the same time as a contact event, which is the contact through which the injury occurred. Injury and contact events can be preceded by up to three preceding events. Each of these events can have an agency associated with it. An agency can be a tool, machine, substance, material or other person.

The epidemiological approach

The epidemiological approach to industrial accident analysis was proposed by Gordon (1949) and Suchman (1961). This approach is based upon the idea that occupational injuries bear similar characteristics to infectious and

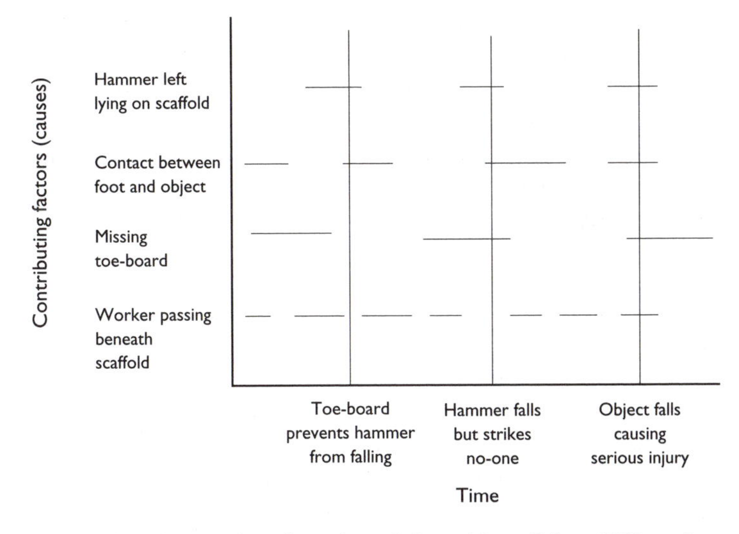

Figure 1.3 Multi-causality of accidents (adapted from Pybus, 1996, p. 4).

non-infectious diseases, and that the same techniques used in the study of diseases and their control, that is, the science of epidemiology, can be usefully applied to occupational accidents and illnesses. Epidemiology has proven useful in delineating patterns in the distribution of diseases, which has led to a better understanding of their causes and possibilities for their control. These methods can, it is argued, provide a similar understanding of the causes of occupational injuries and illnesses and guide prevention efforts.

The public health approach to disease concerns itself with not only treating the consequences of the disease, but also in trying to find out why certain individuals are more susceptible, and what environmental conditions are conducive to the spread and risk of disease. Suchman points out that this is in sharp contrast to the medical approach to health, in which the consequences of disease are considered but rarely is attention given to their aetiology. In safety, this difference is mirrored by short-sighted attempts to address the immediate causes of an accident without searching for the underlying root causes that must be addressed in order to prevent similar occurrences in the long term.

Gordon highlighted the problem that, in accident analysis, too often the agent directly involved in an accident, such as the sheet of glass that led to the laceration, is cited as the only cause of the accident. This is because the sheet of glass is easily identifiable in the immediate aftermath of the accident and, once identified, investigators may look no further. While it is important that the agent directly involved in the accident be identified, there are many other factors that need to be considered as well.

The epidemiological approach is consistent with the concept of multicausality. Gordon describes accident causation as a combination of forces from at least three sources: the host, the agent and the environment.

The *host* refers to the person suffering the illness or injury. Host factors are characteristics that make a person more susceptible to certain types of injury or illness. They could include physiological features, such as strength and size, age, gender; levels of training and competence; and motivation or behavioural issues.

The *agent* is the injury or illness deliverer, and can be physical, chemical or biological in nature. The agent factor could include tools, items of machinery, chemicals or building components.

The *environment* factor is also broken down into the physical, biological and socio-political aspects of the work environment. Physical aspects of the work environment include temperature, ventilation, noise levels, site layout and housekeeping. The biological environment relates to living things. These may pose a health risk when workers are exposed to diseases such as malaria, Weil's disease or anthrax. However, dog bites, snakebites and spider stings are also biological factors that could be related to occupational injuries in some areas. Socio-economic factors are rarely considered in relation to construction site accidents. They include things like the subcontracting and competitive tendering system, group dynamics, the remuneration method and the industrial relations situation.

The epidemiological approach was adopted in an information system developed for the Hong Kong Housing Authority and was used to identify patterns in occupational injury experience on HKHA construction sites (Lingard 1993). This system is described in greater detail in Chapter 9.

Psychological models

Psychological models of accident occurrence focus on the processes of the brain. One such model, developed by Jean Surry, is depicted in Figure 1.4. This model is based upon three stages of the brain's activity. The first is *perception*, in which sensory cues are received. The second is the *cognitive processing* of information received, including deciding how to respond to sensory cues and the application of decision-making rules, often known as *heuristics*. Finally, the model deals with the body's *physiological response* to danger. Surry's model poses a series of questions, which follow the perception, cognition and response sequence.

The model is also in two stages, danger *build-up* and danger *release*. The danger build-up stage refers to a situation in which the possibility of injury is present. For example, when working in a trench, the possibility of collapse is present. Careful workers may watch for warning signs, such

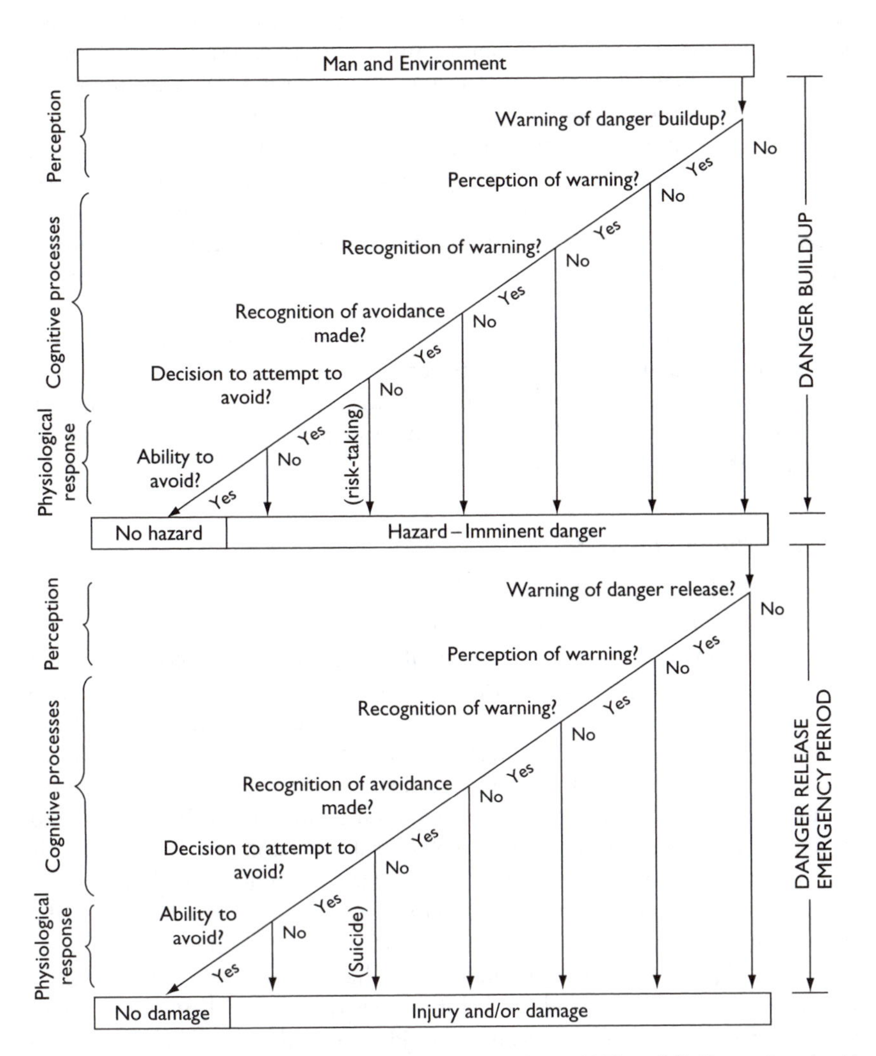

Figure 1.4 Decision model of accident occurrence (Surry 1979, p. 36). Reproduced with kind permission of J. Surry.

as cracks in the ground close to the edge of the trench and respond to these cues by providing a stronger shoring system if necessary. In the danger build-up stage, if danger warnings are perceived and recognised, a decision is made to avoid the danger and the person is physically able to avoid the danger, then the result will be no hazard. If, however, the answer to any of the questions posed is 'no', then exposure to the hazard or imminent danger will result.

In the second stage, the accident process has already begun. For example, the trench may have started to collapse. The potential victim's response to this situation is determined by his or her perception of the imminent danger, decision as to how to respond and ability to respond. For example, the worker in a trench that starts to cave in must respond to warning signs, recognise that to avoid injury he or she must leave the trench and must be physically able to leave the trench before being buried. If the answer to any one of Surry's questions in the danger release stage of an incident is 'no', then injury and/or damage are certain to occur. This model is useful for determining the source of human errors involved in accidents and helps to differentiate between errors arising as a result of lapses in concentration and those resulting from incorrect knowledge or errors of judgement. However, care must be taken when using psychological models not to focus exclusively on individual factors contributing to the occurrence. Psychological models do not adequately account for organisational, social and situational contributing factors, which must be identified if appropriate preventive strategies are to be developed.

Energy-damage models

Haddon (1980) proposed a model of accidental injury that conceptualised hazards as *damaging energies*. According to energy-damage theories, damage or injury arises when a source of energy comes into contact with a recipient, and at the point of contact, the energy exceeds the damage threshold of the recipient. Using this model, hazards are classified as potentially damaging energies, such as electrical, kinetic, gravitational, chemical, acoustical and mechanical vibrations and so on. Energy-damage models do not use the term *hazard* in the way in which it is often used. For example, if an item of equipment is left lying on a scaffold, we commonly refer to this as a hazard. However, the energy-damage model would hold that it is not the equipment's potential kinetic energy that is the hazard; rather, the gravitational potential energy of the person walking on the scaffold, who might dislodge the item of equipment, is the hazard. Energy-damage models are theoretically rigorous and are helpful in classifying incidents for in-depth analysis of patterns of injury occurrence. However, they are likely to be too abstract to provide a helpful basis for routine hazard-spotting exercises.

The socio-technical systems approach

AS/NZS 3931 defines a system as follows:

> [A] composite entity, at any level of complexity, of personnel, procedures, materials, tools, equipment, facilities and software. The elements of this composite entity are used together in the intended operation or support environment to perform a given task or achieve a specific objective.
>
> (Standards Australia 1998, p. 3)

Thus engineering, management and human factors are integrated in systems theories of safety.

James Reason developed a systems-based model of human error (Reason 1997). According to Reason, organisational factors, such as budget allocation, communication, planning, scheduling and unwritten rules about acceptable practices within the company are the starting point for organisational accidents. The consequences of organisational management activities are transmitted throughout the organisation to local workplaces. In the case of construction, these workplaces would be the company's job sites. In local workplaces, these consequences manifest themselves in such factors as unrealistic work schedules, poor maintenance, understaffing, low pay, poor supervisor–worker ratios, ambiguous or unworkable procedures, conflicting goals and so on.

Reason calls these organisational and workplace factors *latent condition pathways*. In a sense, they are 'accidents waiting to happen'. In local workplaces, these latent conditions combine with natural human tendencies and result in human errors or violations. These are unsafe acts committed at the human–system interface. Reason suggests that many unsafe acts occur, but very few of them result in losses because systems have in-built defences. However, in some situations, the system's defences fail as a result of latent conditions – meaning that errors result in organisational accidents. Reason's model, depicted in Figure 1.5, is useful in understanding the complex interaction of organisational, workplace and individual factors in accidents and can be used as a basis for investigating accidents to identify upstream causes. It is particularly useful for understanding the organisational sources of human error.

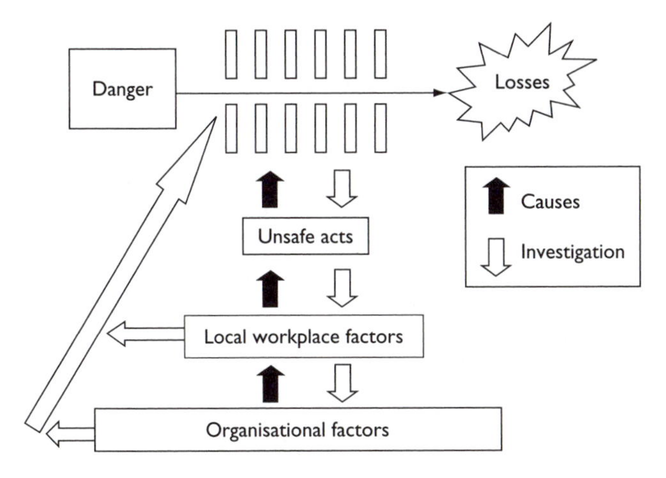

Figure 1.5 The development of an organisational accident (adapted from Reason 1997, p. 17).

Causal factors in construction incidents

A recent report prepared by Loughborough University and University of Manchester Institute of Science and Technology (UMIST) on behalf of the UK's Health and Safety Executive (HSE) sought to test a holistic model of OHS incident causation by carefully investigating the causes of 100 construction incidents. The research team used the information obtained from people involved in selected incidents, including the victims and their supervisors, to describe the processes of incident causation in construction. Figure 1.6 shows the model of incident causation.

The HSE model identifies originating influences affecting incidents in construction as including client requirements, features of the economic climate, the prevailing level of construction education, design of the permanent works, project management issues, construction processes, the prevailing safety culture and risk management approach. In particular, the analysis of the 100 incidents revealed that more than half could have been prevented with alternative design solutions, however, the researchers conclude that many construction designers still fail to acknowledge their responsibility for safety during the construction process. Deficiencies in the risk management system were also apparent in almost all of the 100 incidents studied, which represents a significant management failure. Project management failures were also commonly reported, most of which involved inadequate attention to co-ordinating the work of different trades and managing subcontractors to ensure that workers on site had the requisite skills to perform the work safely.

The next level of contributing causes identified in the HSE model is termed 'Shaping factors' which include issues, such as the level of supervision provided, site constraints, housekeeping and the state of workers' health and fatigue. The study revealed the length of work hours and resultant fatigue to be causes for concern. These issues are discussed in Chapter 10. Poor communication within work teams was also identified as an important shaping factor.

The most immediate circumstances in the HSE incident causation model are the suitability, usability and condition of tools and materials, the behaviour, motivation and capabilities of individual workers and features of the physical site environment, such as layout, lighting and weather conditions. While it is important to identify these immediate circumstances, the model acknowledges that construction incidents occur as a result of a complex process, involving proximal causes as well as distal factors 'upstream' of the construction work.

The HSE research is particularly important because it adopts a similar framework to that presented by Reason (1997), but places it in the context of the construction industry. As such, it is one of the only models of incident causation that adequately addresses some of the unusual organisational features of construction, such as the production of a bespoke product for a

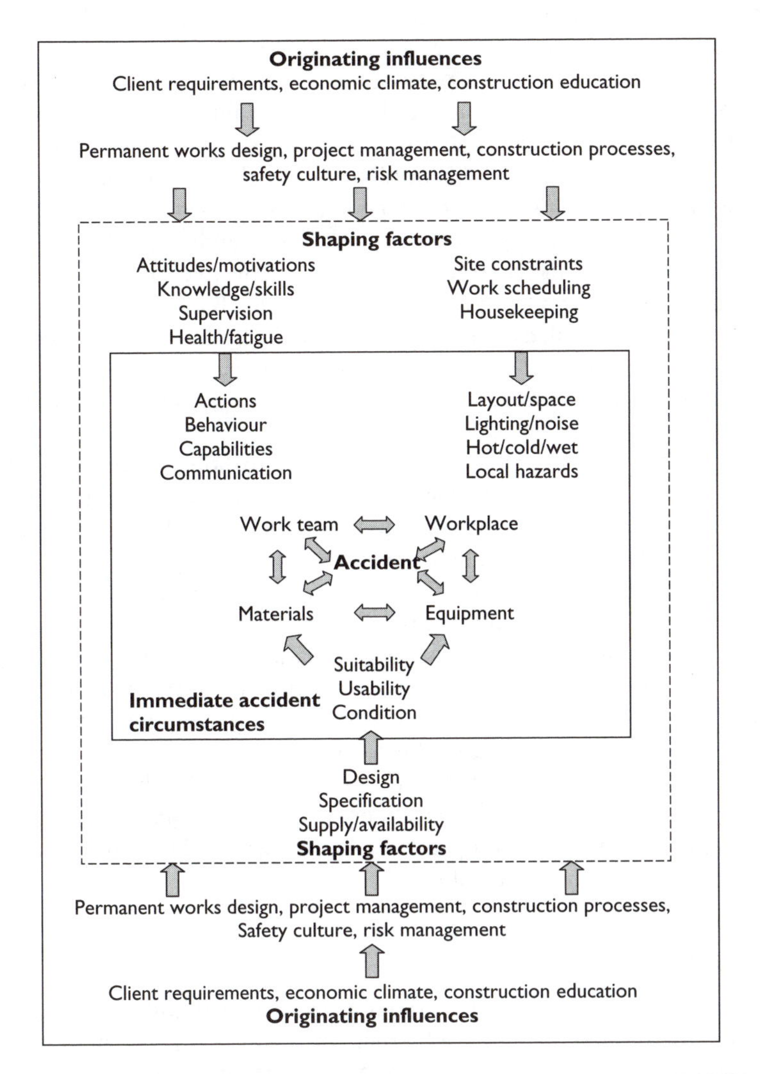

Figure 1.6 Causal factors in construction incidents (adapted from HSE 2003b, p. 59).

particular client, the separation of design and construction and the extensive use of lengthy chains of subcontracting.

OHS as a management responsibility

Assumptions concerning the sources of occupational injury and illness determine our understanding of where responsibility for their prevention should lie. Even though organisational accidents usually involve both individual and system factors, preventive strategies need to reflect the understanding that unsafe acts have systemic or organisational causes and that OHS is a management issue. Addressing the immediate behavioural causes of an accident, for example urging workers to take greater care, will not resolve the latent causes, and is not likely to be the most reliable way of preventing similar accidents from occurring in the future. Conversely, tackling the systemic sources of organisational accidents addresses their latent causes and is far more likely to provide reliable, effective safety solutions. This recognition that accident prevention is more likely to succeed if latent causes are eliminated, places the responsibility firmly on the shoulders of management since it is managers who make decisions about budgets, resource allocation, planning, scheduling, payment and reward structures and so on.

Managers also have control over decisions as to how to control OHS risks. It is recognised that it is better to 'engineer' risks out of a system than to try to protect workers from exposure to these risks, either by implementing safe work procedures or providing personal protective equipment. Managers are the only people who are able to make the decision to eliminate or reduce a risk by physically changing the workplace or process. If managers do not endorse engineering solutions to OHS risks, then the only available options to workers seeking to avoid occupational injury or illness are behavioural – working with care and using protective equipment. While these are important behaviours, we know that humans are prone to error. Far greater system safety and reliability could be achieved if the danger had been engineered out of the work process or work environment. This understanding provides a compelling argument for why OHS must remain primarily a management responsibility.

Mapping the way

Focusing on the management of OHS requires sustained and co-ordinated effort. Construction companies are by their very nature decentralised, with systems and people spanning a wide range of different business activities. Many OHS initiatives are introduced on an apparently *ad hoc* basis, implemented for a finite period, and then abandoned when the next safety 'campaign' comes along. This approach is not likely to lead to long-term or sustained improvements in OHS. Short-term improvements might be observed during the life of a new OHS initiative but soon return to their original levels, especially if the campaign comes to an end. A good example of this effect is in the use of reward schemes such as bonuses or prizes for sites with low lost time injury frequency (LTIF) rates. Such schemes may reduce injuries initially, but improvements are likely to last only as long as the bonus or prize scheme. Such schemes may also have negative effects of encouraging under-reporting of incidents or applying pressure on workers to return to work before they are suitably recovered to reduce the LTIF rate.

The evolution of a culture of safety occurs in three conceptually distinct stages, depicted in Figure 1.7. The occurrence of injuries and ill-health declines in 'steps' corresponding with each of these stages. The traditional approach to OHS is essentially reactive, with hazards being dealt with as they arise and with a strong emphasis on discipline. A transitional approach is more proactive in that hazards are considered before they arise and procedures are established in an attempt to prevent occupational injuries and illnesses. At present, it is postulated that small construction firms are largely in the *traditional* stage in the implementation of their OHS management systems, while the activities of larger construction firms fall mostly in the *transitional* stage. This may explain why construction accident rates have plateaued in recent years.

The construction industry needs to progress to the *innovative* stage of OHS management to achieve further improvement. In this stage, OHS is fully integrated into all business decision-making, and every attempt is made to eliminate hazards or minimise OHS risks through the adoption of technological solutions. Attention is also paid to cultural and motivational issues, and work is organised so as to encourage good OHS performance. This progression will not come about easily. The construction industry is resistant to change. The remaining chapters of this book aim to provide students and practitioners of construction management with the knowledge they need to change the construction industry and move the organisations within it towards an the innovative stage of OHS management.

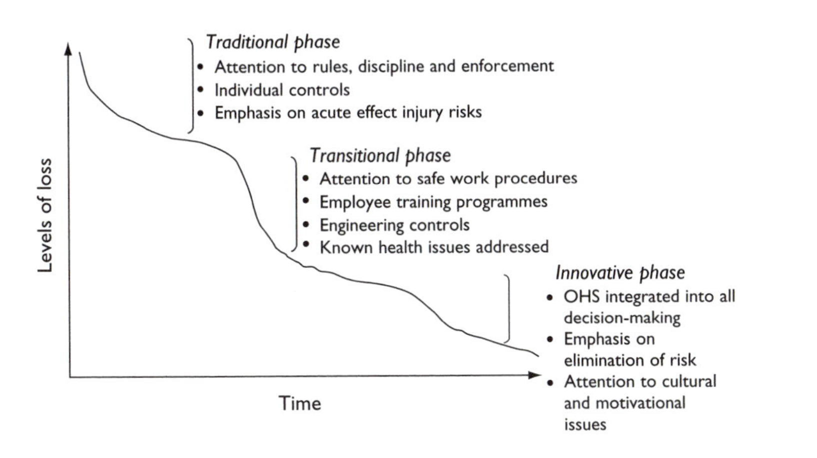

Figure 1.7 Stages in the evolution of a culture of safety (adapted from Pybus 1996, p. 18).

Discussion and review questions

1. Why does the construction industry experience such a high incidence of work-related injuries and illnesses, and why is the industry's OHS problem so intransigent?
2. Critically evaluate the different theories explaining accident occurrence. Is any one approach more useful than the others, and why?
3. Evaluate the role that management can play in improving the construction industry's OHS performance.

Chapter 2

Occupational health and safety law

Introduction

The law has evolved from systems, principles and customs governing conduct in relation to various aspects of people's lives. Rules are established by groups of all sizes to ensure that the rights of members are respected and responsibilities are fulfilled. Small or informal groups tend to adopt simple rules, which are not legally binding, while governments of countries and states develop complex and comprehensive rules, which can be enforced (Morris *et al.* 1996).

The purpose of the law is to ensure that members of a society live and behave according to a set of acceptable rules. Contravention of these rules results in some form of sanction that society deems to be commensurate with the seriousness of the breach.

In relation to OHS, the law exists to identify the responsibility of the parties involved in industrial or commercial activities. It imposes responsibilities on employers to protect the health and safety of their employees when they are at work and protects the right of people to participate in the paid workforce without suffering injury or ill health as a result. As such, the law should be of primary importance in providing a 'level playing field' and ensuring that employers do not profit from failing to provide adequate protection for their employees. OHS law also ensures that, if people do suffer a work-related injury or illness, there are mechanisms by which they may obtain compensation and undergo rehabilitation to enable them to resume participation in the workforce and the community at large. Thus, the aims of OHS law are threefold:

1. to prevent occupational injury and ill-health;
2. to ensure compensation for those who are injured or become ill as a result of their employment; and
3. to rehabilitate workers who suffer injury or ill-health as a result of their work in order that, so far as is possible, they can return to work and resume participation in the community (Quinlan and Bohle 1991).

In this chapter we describe how the law operates to achieve these three aims. We describe how preventive OHS legislation has evolved from a highly prescriptive system to one in which managers have considerable leeway in determining how they will manage OHS risk in their workplaces. We also consider how OHS legislation can meet the needs of the construction industry, in particular, in dealing with contingent workers and complex, multi-organisation environments. We also discuss enforcement approaches and sanctions for OHS offences. The chapter concludes with a description of the way in which the law provides financial assistance to workers who suffer an occupational injury or illness and requires that employers must rehabilitate injured workers. Throughout the chapter, we address the construction industry's response to its legal responsibilities for OHS, and highlight areas in which the industry needs to improve.

Sources of law

Different legal systems have developed in different parts of the world. Many countries' legal systems are based on the English common law. These include Australia, the United States of America, India, Israel, Hong Kong, Canada, New Zealand, Kenya and Malaysia, although in each of these countries the law has developed differently to reflect local values and beliefs about justice. Other countries are civil law countries, meaning that all of their laws are contained in comprehensive legal codes. Most of continental Europe, Thailand, Japan and some South American states are civil law countries.

In common law countries, there are two sources of law: *decisions*, and reasons for those decisions made by judges in the courts (sometimes called *precedent*); and *legal rules* made by parliament (otherwise known as *legislation*). Much of this chapter focuses on legislation because, as Quinlan and Bohle (1991) comment, historically, legislation has been far more important than the common law in the area of OHS. However, the common law does play a role in OHS regulation in that common law definitions of key terms such as 'employee', 'principal' and 'subcontractor' are used to interpret duties owed by the parties under OHS legislation. Also, as is described later in this chapter, in common law there is an implied duty of care in the employment relationship. If an employer is negligent in respect of this duty and death, injury or ill health occurs, employees (and their families) have certain rights to be paid damages by the employer.

Preventive OHS legislation

Gunningham (1996) identified three types of standard, which broadly reflect phases in the implementation of preventive OHS legislation. These are *specification* standards, *principle-based* standards, and *systems* standards.

These standards and their application to OHS legislation are considered below.

Early OHS legislation

The traditional legislative approach to the prevention of work-related injury originated in the Acts passed by the British Parliament in the early 1900s. Following the rapid industrialisation that occurred at the turn of the century, the early 1900s saw the introduction of many Acts of parliament designed to prevent injury associated with hazardous work conditions. This early preventive legislation sought to establish minimum standards, which could be enforced by an independent public inspectorate, with the power to prosecute parties who failed to comply.

Specification standards

The provisions of this early OHS legislation were prescriptive in that they were expressed in the form of specification standards, which clearly informed duty-holders how to achieve an outcome by describing in precise detail what must be done in order to comply. Thus, the early OHS legislation identified the precise type of safeguarding method to be used in a particular situation and specified the design and construction of these safeguards. Under this legislation, there was little scope for employers or other duty holders to design OHS solutions to meet the needs of a particular situation.

Gunningham (1996) suggests that specification standards have certain benefits. First, they are easy to interpret and understand. This makes it easier for employers to know how to comply, which is particularly important in the case of small to medium-sized enterprises which often lack the resources or expertise to deal with less specific types of OHS standards. Specification standards also make it easier for employees' representatives, trade unions and government inspectors to identify breaches in the OHS legislation. In the absence of specification standards, a failure to comply may be evident only after an injury has occurred or an illness becomes evident.

Despite these advantages, the weaknesses of the prescriptive approach to OHS, expressed in specification standards, have been thoroughly documented. Three of the most commonly expressed criticisms are described below.

Unsuitability for certain types of risk

Specification standards contained in the original OHS legislation dealt effectively with immediate effect hazards, for example in specifying the nature and type of machinery guarding, but were not well-suited to protecting against delayed effect occupational illnesses or diseases, such as cancer.

Specification standards are also difficult to apply to hazards that have become more prominent in recent decades. Such hazards include ergonomic hazards, such as occupational overuse syndrome, and psycho-social hazards, such as stress and burnout. It is understood that these hazards are associated with issues of worker-environment 'fit' and prescriptive 'one size fits all' solutions are therefore inappropriate. Manual handling is an example of a hazard that is relevant to the construction industry, which cannot be effectively regulated by specification standards. Indeed, attempts to develop prescriptive standards for manual handling have been criticised because they directly discriminate against women (Quinlan and Bohle 1991). Similarly, in the USA, Bartel and Thomas (1985) have criticised the Occupational Safety and Health Administration's (OSHA's) prescriptive emphasis on standards for work equipment on the grounds that it does not adequately reflect how workplace accidents occur. They contend that accidents are caused by a complex interaction between workers, equipment and the work environment, and that specification standards can address only part of the problem.

Little incentive to improve beyond minimum standard

Specification standards established in the early legislation have also been criticised for providing no incentive to employers to continuously improve OHS performance. Although these standards were intended to be minimum requirements beyond which employers should strive to improve, they came to be treated by employers as acceptable norms. Because the early legislation was drafted with little or no worker involvement, establishing minimum standards raises issues of equity. There was no way of being certain that the OHS standards required reduced risk to a level acceptable to those exposed. Indeed, inflexible standards were established at a time of rapid social change, with little consideration of the perspectives of people other than politicians and industrialists. Chapter 5 discusses issues of equity and communication in OHS risk management decision-making in more detail.

Lacking in flexibility

Early OHS law was not subject to systematic review and was updated infrequently. Thus the highly specific and detailed content of the legislation failed to keep pace with rapid technological change occurring in the workplace. Nor could the legislation respond to changing social expectations or judgements as to what was an acceptable level of risk. Gunningham (1996) argues that because specification standards are prescriptive, they do not allow companies to identify least-cost solutions to OHS issues and are therefore unlikely to be cost-effective in most instances. Rather than developing innovative solutions, industry is bound by prescriptive solutions, which may not reflect up-to-date knowledge of technology or industrial processes.

Thus, it is conceivable that compliance with rigid prescriptive legislation is more costly to employers than having the freedom to develop workplace-specific solutions to OHS problems.

The need for reform

By the middle of the twentieth century, a large number of Acts and sets of regulations dealing with OHS had been passed in the UK and other countries and territories of the Commonwealth, such as Australia and Hong Kong. The sheer volume of OHS legislation was itself a source of confusion to industry. Also, these Acts and regulations had been introduced on an *ad hoc* basis, and therefore to deal with new hazards as they became apparent meant that gaps in coverage remained. No all-encompassing provisions were made, and emerging hazards remained unregulated until such time as they were recognised and a new Act or set of regulations targeting the hazard was passed. Neither was there a regular review process through which the adequacy of existing legislation could be determined or discussed. Thus, certain groups of workers or hazardous processes were not covered by the OHS legislation, and workers and their unions had little say in determining to what extent known risks posed by hazardous industrial processes should be controlled.

In addition to these problems, the enforcement of the early OHS legislation was inadequate. Inspectorates were under-resourced and not in a position to monitor the number of operating workplaces. Furthermore, even when breaches of the provisions were observed, a non-prosecutorial culture, in which the emphasis was on education rather than punishment, prevailed.

The Robens Report

By the late 1960s, the limitations of traditional OHS legislation were apparent. The British government responded by commissioning a Committee of Inquiry into occupational health and safety. The Committee, under the charge of Lord Robens, produced a report in 1972. This report (often called the Robens Report) proposed dramatic modifications to the regulatory framework for OHS. These recommendations had a major influence on OHS regulation in Britain and other Commonwealth countries, including Australia, throughout the latter part of the twentieth century (Johnstone 1999b).

The Robens Report identified 'apathy' on the part of industry as being the cause of poor OHS performance, claiming that industry had come to regard OHS as something to be regulated by external agencies rather than something that should be proactively managed alongside other business objectives. The Robens Committee also suggested that the complexity and volume of detailed OHS legislation had contributed to this attitude. The Committee proposed that the law be reformed in such a way as to convey the message that the responsibility for managing OHS rested squarely on the shoulders of industry itself.

In attempting to achieve this, the recommendations of the Robens Committee were two-pronged. First, the Committee advised that a more unified and integrated system of OHS legislation should be created and, second, the Committee recommended the development of a self-regulating system.

The first objective was to be achieved through the replacement of existing statutory provisions and the creation of an 'umbrella' Act, stating in broad terms the responsibilities of employers, employees and others for OHS. This Act was to be supported by subsidiary regulations and non-statutory codes of practice, with the emphasis being on voluntary codes. The committee also recommended the creation of tripartite policy-making structures, through which regulations, standards and codes of practice could be developed and policies relating to the enforcement of administration of OHS legislation could be established. Such a tripartite decision-making structure would provide workers and their unions with a 'voice' in setting OHS standards and determining policy in OHS matters.

Legislative reform

Soon after the publication of the Robens Report, many of its recommendations were adopted in the UK in the enactment of the *Health and Safety at Work Act* (1974). This Act is all-encompassing in that its provisions cover every type of workplace hazard and every group of employee. Its provisions are expressed in general terms, and the Act provides for tripartite decision-making at both governmental and workplace levels. Thus, a key component of the requirements under the Act is that employers provide consultative processes at the workplace through which workers can participate in OHS decision-making. The Act is also an 'enabling' Act in that it provides that subsidiary OHS regulations can be implemented where necessary. The Act also gives evidentiary status to Approved Codes of Practice, which are non-mandatory sources of advice concerning how to comply with the Act or its subsidiary regulations. In the 1980s and early 1990s, all Australian jurisdictions adopted the Robens model of OHS legislation and Robens-inspired reform also occurred in other Commonwealth territories, such as Hong Kong. While the Robens model was tailored to the requirements of particular jurisdictions and therefore there are some differences in the exact wording and operation of provisions, there are also some common features of the existing Robens-style legislation. These are described below.

General duties provisions

A key feature of Robens-style legislation is the statement of 'general duties' of relevant parties including employers, employees, the self-employed and the designers and suppliers of plant and equipment. The 'general duties' contained in the Robens-style legislation embody the common law duty of care that had always existed (Gun 1992). These duties require employers to take practical steps to protect the health, safety and welfare of their employees and others who may be affected by their commercial endeavours. While the wording and detail of general duties requirements varies, some common requirements exist. Employers are usually required to provide a workplace that is safe and without risk to the health or well-being of workers and others who enter the workplace. In doing so, employers must comply with OHS regulations relating to specific hazards. Employers are also usually obliged to provide safe systems of work, appropriate OHS information and training and adequate supervision of the work under their control. Employers may also have general duties to monitor the health of employees and conditions at the workplace and maintain information and records relating to the health and safety of employees. Finally, a requirement of most laws is that employers report lost-time injuries and serious incidents to enforcement agents and keep standardised records of illnesses. Incidents must be reported within specified time frames and records of OHS incidents may have to be submitted at regular intervals to an inspectorate body.

Employees also have general duties under Robens-style OHS legislation. Normally, each employee is expected to comply with all occupational safety and health standards, rules, regulations, and orders issued under the law that apply to his or her own actions and conduct on the job. Employees are typically required to co-operate with their employers in OHS, and must not wilfully or recklessly interfere with any item provided in the interests of health, safety or welfare or wilfully place themselves or others at risk.

Principle-based standards

The 'general duties' provisions have frequently been labelled 'performance-based' because employers have discretion over how they are to comply with them (Gunningham 1996). However, in reality, they establish broad principles rather than performance outcomes. Performance-based standards could be said to include requirements, such as ensuring a maximum noise level of 85 decibels. Such a requirement specifies a required outcome without prescribing how this is to be achieved. By way of contrast, a prescriptive standard might specify that machinery of a certain type be insulated with a specified quality of soundproof material. The general duties provisions contained in Robens-style legislation are more accurately described as principle-based requirements in that they are broad statements of the responsibility of relevant parties (Industry Commission 1995).

The general duties are sufficiently broad to cover emerging hazards and ensure that no new hazards fall 'between the cracks'. The general duties are also adaptable and can accommodate technological changes occurring within industrial processes and are concerned with creating a positive attitude towards the prevention of work-related injury and ill-health. Prescriptive standards, by their nature, did not allow continuous improvement in accident or injury prevention to occur because innovative solutions were stifled. Employers have overwhelmingly supported general duties in OHS legislation, viewing them as a way to identify and implement solutions which best suit the operation of individual businesses.

Qualifying terms

The general duties are not absolute. The duties of care in Robens-style OHS Acts are limited by words like 'so far as is practicable' or 'reasonably practicable'. This means the degree of risk in a particular process or situation must be balanced against the measures by which it can be controlled. For example, the Victorian *Occupational Health and Safety Act* (1985) defines practicable as having regard to:

- the severity of the hazard or risk in question;
- the state of knowledge about that hazard or risk and ways of removing or mitigating that hazard or risk;
- the availability and suitability of ways to remove or mitigate that hazard or risk; and
- the cost of removing or mitigating that hazard or risk.

This definition of practicability underpins the cost-benefit analyses that are central to risk assessment and risk control decision-making processes described in Chapter 4.

Like the common law duty owed by employers to employees, an objective standard for this duty is 'what would a reasonable employer have done in the situation?' It is reasonable to expect that as the risk increases, the degree of effort exerted in controlling the risk should also increase (Industry Commission 1995). It is worth noting that while the 'general duties' requirements contained in Robens-style OHS Acts are subject to the practicability qualifier, the requirements in subsidiary regulations are absolute unless there is a statement to the contrary.

Consultation

Another key feature of Robens-style legislation is the implementation of consultative processes to ensure worker participation in OHS decision-making. Johnstone (1999b) points out that the self-regulatory emphasis of the Robens approach should not be confused with de-regulation because mandatory consultative processes are a key feature of what Robens meant by self-regulation. Consultative processes, including established OHS committees comprising of workers and management and elected employee OHS representatives with considerable powers are one of the great strengths of the Robens approach. The statutory requirement for employers to consult with employees or their representatives about OHS measures and the requirement for employee participation in the development of these measures is an extremely important departure from pre-Robens legislation. The participation of employees in organisational decision-making is consistent with the concept of industrial democracy. An important principle of industrial democracy is that people have a right to be involved in making decisions that affect them, particularly when those decisions can have an impact upon their health or safety (Industry Commission 1995).

OHS committees

Weil (1999) suggests that OHS committees can improve injury and illness outcomes in two ways:

1 they provide an ongoing forum for the correction of OHS problems; and
2 they augment government enforcement agencies in their regulatory role. Thus, workplace OHS committees improve the ability of regulators to enforce OHS law and ensure that employers comply with standards.

O'Toole (1999) also reports on the beneficial effects of OHS committees. In a study of manufacturing plants in the building products industry, he found that plants that had established and maintained OHS committees had lower accident frequency and severity rates than plants in which such committees did not exist. However, O'Toole also reports that plants in which such committees were established voluntarily experienced fewer and less serious injuries than those in which committees were mandatory. These findings suggest that the effectiveness of legal requirements to implement consultative processes may be limited by workplace cultures that do not support employee involvement in OHS decision-making. The difference between consultation and co-operation is discussed in Chapter 3.

Employee OHS representatives

The role of elected employee OHS representatives is also crucial. In the absence of a mechanism for ensuring employees' concerns are communicated and acted upon, it is unlikely that self-regulation could work. Employees' OHS representatives have certain rights and powers under Robens-style legislation. Employers must usually provide employee representatives with adequate resources to perform their role, and must release representatives for OHS training as necessary. OHS representatives have the right to view documents pertaining to the OHS of any member of the area or work-group they represent. They may also accompany inspectors on visits to the workplace and sit in on any interview between an inspector and the employer's representative or other members of the work group. OHS representatives have the right to file a complaint with the nearest OHS inspectorate requesting an inspection if they believe unsafe or unhealthy conditions exist in their workplace, and are also protected by anti-discrimination clauses providing that they may not be discharged or discriminated against in any way for filing safety and health complaints or otherwise exercising their rights under the law.

The role of the OHS representative is critical in ensuring that workers' OHS views can be freely expressed. The role of health and safety representatives should not be limited to a consultative one. Merely being consulted is insufficient. In order to ensure equitable decisions regarding OHS risk and its control, employees should also have some enforcement role. For example, in several Australian jurisdictions, employee OHS representatives are provided with considerable enforcement powers. In Victoria, elected employee OHS representatives have the power to issue provisional improvement notices and even order a cessation of work in the case of immediate danger. Despite employer concerns that the OHS representatives power might be abused, there is little evidence to suggest that this has happened. The legislation also makes it an offence for any representative to maliciously misuse his or her powers, thereby protecting employers from such abuse.

Emphasis on non-statutory codes of practice

Robens-style legislation emphasises the role of non-statutory codes of practice to provide guidance to industry about how to comply with the provisions of the Act. The reliance on non-statutory codes to provide practical guidance has been controversial. Some argue that giving approved codes evidentiary status has caused them to be treated as *de facto* regulations. While the law does not require compliance with codes, codes are admissible evidence in the event of a breach. Thus, an employer accused of failing to meet his or her general duty must prove that the OHS protective measures in place were at least as good or better than those prescribed in the relevant code. Employer groups argue that this reversal of the burden of proof has caused approved codes to be effectively 'mandatory'.

Gunningham (1996) warns against providing highly detailed and prescriptive codes of practice, saying these would foster a 'lowest common denominator' approach, in which employers would devote much energy towards implementing the detail of the codes rather than proactively seeking to improve their OHS performance over and above these standards. This is the same criticism levelled at the pre-Robens prescriptive OHS legislation.

Indeed, consistent with the criticism of pre-Robens legislation, employers groups have argued that detailed codes of practice inhibit creativity and innovation in the development of cost-effective OHS solutions (Industry Commission 1995).

Yet, while deploring a detailed prescriptive approach in codes of practice, employer groups are quick to criticise hazard-specific codes of practice, such as the code of practice for manual handling, as being too general to be useful. Employer groups suggest that industry-specific codes for certain hazards are needed. Gunningham (1996) argues against the development of industry-based codes on the grounds that generic hazard-specific codes are preferable because they provide guidance as to the correct risk management process to follow for a given type of hazard. For example, the hazard identification, risk assessment and risk control process is arguably generic, and applicable to workplaces in any industry. Hazard-specific codes of practice also contain checklists and guidance to duty-holders as to how to identify the types of hazards present, and the range of available risk controls for that hazard. Expressing a similar view, the Australian Department of Industrial Relations has commented on the role of codes that they 'are not intended to "translate" regulations into ready-made solutions for every workplace. This translation or adaptation has to occur at the workplace' (Industry Commission 1995).

However, these views do not address the needs of small businesses for technical advice. These businesses are unlikely to possess the specialist OHS expertise to apply generic hazard identification/risk assessment/risk control processes. The difficulties of small business in grappling with Robens-style legislation are particularly important in the construction industry and are discussed later in this chapter.

One possible means by which small business needs for practical technical advice can be addressed is through the provision of industry-specific guidance notes. Such an approach has been adopted by the Victorian WorkCover Authority. The Authority provides detailed technical advice in guidance notes published on many topics specific to the construction industry; for example, the use of earthmoving equipment as a crane, the prevention of breakage during the installation of frameless glass balustrades and many other topics. These guidance notes do not have formal legal status, and can be updated quickly as appropriate, but nonetheless retain a form of specification standard for small businesses to access.

Robens in construction

The implication of Robens-style OHS legislation for construction project managers is that they can exercise considerable managerial discretion in determining how to comply with the general duties provisions. No longer is there the assurance that, if they adopt a specified work method, they are in compliance with the legislation. There is evidence to suggest that the construction industry has struggled with the shift from prescriptive to principle-based OHS legislation. One possible reason for this difficulty is the intense competition that exists within the construction industry. Construction work is usually awarded on the basis of competitive tendering in which the pressures to cut costs to a minimum are acute. It has been argued that this competition ensures that persons with OHS obligations will only comply with these obligations in a certain way if it is known that all other duty-holders will also comply in this way (Department of Employment, Training and Industrial Relations 2000). Business survival dictates that costs will not be incurred until it is ascertained that all competitors are also incurring similar costs. This is most certain when regulations are expressed in 'black and white' terms, such that there is no discretion to reduce costs or ignore OHS requirements altogether. The structure of the industry and the implications of this structure on the operation of Robens-style legislation are discussed below.

Structure of the industry

The construction industry is characterised by a small number of large organisations and a plethora of micro-businesses (see also Chapter 1). Small businesses in the construction industry are unlikely to possess the OHS resources or know-how to effectively implement the self-regulatory principles of the Robens approach. This is likely to be exacerbated by the competitive nature of the industry mentioned above.

Research evidence supports this assertion. In the 1980s Dawson *et al.* (1988) undertook an analysis of the effect of the UK's Robens-style *Health and Safety at Work Act* (1974). They report that, of a number of industries, the construction industry had not embraced the self-regulatory approach to OHS. While the large construction firms had taken some initiative in establishing OHS management systems, small to medium-sized firms making up the vast majority of construction organisations had not. In a study of self-employed or small builders in the Australian state of Queensland, Mayhew (1995) found that there was a high level of confusion and uncertainty about the concepts underpinning the current OHS legislation, that is, the general duties and self-regulation. Furthermore, those builders who were aware of these aspects of the OHS legislation demonstrated only a very basic knowledge of the principles, which is likely to limit their ability to implement them. Another report by Mayhew adds to the evidence that small building firms in Australia and the UK do not cope well with Robens-style legislative requirements (Mayhew 1995).

Structural characteristics of the construction industry have been cited by trade unions as impediments to effective self-regulation in this industry. For example, the Australian Construction, Forestry, Mining and Energy Union

(CFMEU) expressed its opposition to Victoria's adoption of a performance-based approach in the Plant Regulations (1995). The CFMEU stated that:

> the structure of the construction industry is particularly ill-suited to the performance-based approaches now being adopted and that fundamental, regulatory prescription must remain an important foundation of OHS in this industry.
>
> (Industry Commission 1995, p. 361)

Need for prescription

A review of health and safety in the building and construction industry, undertaken by the Queensland Government, reports that industry participants are confused about their obligations under Robens-style legislation (Department of Employment, Training and Industrial Relations 2000). The final report of the industry task force reveals that, in all sectors of the building and construction industry, there is broad-based support for prescriptive regulations. The task force reports that the nature of the construction industry requires minimum standards with regard to OHS provisions. In the absence of these minimum standards, the competitive nature of the industry dictates that OHS standards will be compromised. The task force argues that it is essential that construction industry participants be told in specific terms what is required of them.

It is also the case that certain types of hazards might be best dealt with using a prescriptive legislative approach. In its submission to the Industry Commission (1995) review, the South Australian Government commented that there is a case for retaining legally enforceable specification standards where 'there is a known high degree of risk and specific controls which are applicable to all circumstances where the risk occurs' (Industry Commission 1995, p. 75).

As was noted above, the CFMEU opposed the adoption of performance standards relating to the design, manufacture, supply and use of industrial plant in the state of Victoria on the basis of the industry's structure. It can also be argued that the OHS aspects of working with dangerous items of construction plant, including scaffolds, cranes and hoists, should be regulated by means of detailed, prescriptive requirements because the degree of risk is known to be high. Clearly, there are specific controls applicable to these items of plant; for example, the erection of structurally sound scaffolding fitted with working platforms, ladders, guard-rails and toe-boards is applicable to all building construction projects. However, when the performance-based OHS (Plant) Regulations were enacted in the state of Victoria in 1995, 27 sets of prescriptive regulations were revoked. The revoked regulations included the Cranes Regulations (1989), the Cranes (Suspended Personnel) Regulations (1993) and the Scaffolding Regulations (1992). It is hard to imagine that managerial discretion could be exercised in deciding on the elements of a safe scaffold or the correct procedure for rigging a crane, and thus the decision for adopting performance-based standards in these areas should be questioned. The CFMEU argues that this 'de-regulation' has resulted in a lowering of the standard of care for employees in construction, and suggests that plant-related fatalities and serious injuries have increased as a consequence.

The need for prescription was echoed in the recommendation of the construction industry task force in Queensland. Following a review of the legislation, the task force recommended re-regulation in a number of key areas. These included: mobile cranes; requirements for the mandatory use of doggers; one plank exceptions for painters working on trestles and plant inspection records for use and maintenance (Department of Employment, Training and Industrial Relations 2000).

Employee representation

Another problem associated with implementing the Robens-style legislation in an industry in which small business predominates, is one of adequate employee representation. Meaningful employee representation is a key component of the Robens-based approach to OHS. In the absence of effective employee representation, the flexibility afforded to industry by the approach may result in the downgrading of standards of OHS protection for workers. Employee representation should provide a counter-balance to the exercise of managerial discretion in deciding on suitable methods for preventing occupational illnesses and/or injuries. Dawson *et al.* (1988) found that, in the UK, employee consultation on matters of OHS was very limited in construction and small businesses, and union activity in small firms is reported to be low.

The situation in Australia appears to be similar. For example, an Australian Workplace Industrial Relations Survey (Moorehead *et al.* 1997) revealed that 71 per cent of workplaces of between 5 and 19 employees had no union members. In contrast, 98 per cent of employees in workplaces of 500 or more are union members. While employee health and safety representatives do not have to be union members, Bohle and Quinlan (2000) suggest that unions play a critical role in promoting effective employer–employee consultation with regard to OHS. In particular, unions provide OHS training, advice and information to representatives and intervene in instances where OHS representatives suffer discrimination in workplaces. The occurrence of such discrimination appears to be increasing in some Australian states (Warren-Langford *et al.* 1993).

Biggins *et al.* (1991) present evidence to suggest that workplaces with active employee health and safety representatives have a more systematic approach to OHS. However, active employee representation in OHS matters is much weaker in small businesses than large ones. For example, the Australian Workplace Industrial Relations Survey found that only 25 per cent of businesses with between 20 and 49 employees had an OHS consultative committee (compared to 91 per cent of those with more than 500 employees), 55 per cent of businesses with between 20 and 49 employees did not have an elected OHS representative at their workplace (compared to 93 per cent in the 500 plus category) (Moorehead *et al.* 1997). The frequency of OHS audits, formal risk assessments and the implementation of an injury/disease reporting system all improved progressively with the size of organisations surveyed.

If it is accepted that meaningful employee consultation and input on OHS matters is a prerequisite for the effectiveness of Robens-style legislation, then these figures suggest that dramatic changes in employee representation in small businesses must be made before specification standards are revoked. Small businesses in the construction industry may be even less likely to have active OHS representatives because many tradesmen are self-employed or small businesses are family-run firms. Furthermore, all small construction businesses operate in an intensely competitive environment and business survival may be a more pressing issue than employee representation in OHS. One alternative approach might be that adopted in Sweden, in which small businesses are served by regional OHS representatives (Frick and Walters 1998).

Contractors and contingent workers

The construction industry is characterised by contingent forms of work, heavily reliant on contracting and subcontracting and the use of labour hire firms. Many operatives in construction are also self-employed or work for small businesses. These features can pose problems for allocating clear responsibility for OHS.

Contingent forms of work have been linked with undesirable OHS outcomes (Mayhew *et al.* 1996). It is argued they result in undue pressures to cut costs in an attempt to underbid other subcontractors. This is achieved by using fewer staff and cheaper equipment. Certainly, the use of contingent forms of work may result in such pressures and can also present difficulties for communicating OHS information or weaken chains of responsibility for OHS. However, legally speaking, contingent workers are covered by the general duties. For example, in all of the Australian OHS statutes, an employer's general duty is explicitly extended to cover workers in non-traditional employment situations. Johnstone (1999b) suggests that this is achieved by two types of provision, described below.

Contractors and subcontractors

The first type of provision extends the definition of *employee* to cover contractors and their employees. For example, in Victoria, section 21(3) of the *Occupational Health and Safety Act* (1985) states that the term 'employee' includes independent contractors engaged by the employer and the employees of independent contractors. The general duties of employers under section 21(1) and (2), including the duty to provide a working environment that is safe and without risks to health, are therefore extended to cover independent contractors and their employees.

In the construction industry context, section 21(3) operates such that a principal contractor has a responsibility to direct employees, but is also deemed to be the employer of subcontractors and their employees. More than one party at the same time can have an OHS responsibility. Thus, the principal contractor in control of a site has a general duty for the OHS of the subcontractor and the subcontractor's employees and the subcontractor also has a general duty for the OHS of its own employees.

If, however, a third party has control over the worksite (for example, in maintenance work being undertaken at a third-party premises), the contractor still has a duty towards his or her own employees, but the third party who is in control of the premises also has a general duty for the OHS of the contractor and the contractor's employees.

The courts have clearly held employers to be responsible for the OHS of the contractors they engage. Some examples of prosecutions are described in Box 2.1. Furthermore, imposing contractual OHS responsibilities on contractors or subcontractors does not remove an employer's duty of care. The clear message is that any employer engaging contractors to perform work at a site under the employer's control must manage that contractor's OHS as part of legal compliance. The management of subcontractors' OHS is considered further in Chapter 4.

Box 2.1 Example of prosecutions for contractor/subcontractor incidents

Case 1

When replacing broken glass on top of a 15-metre-high base frame canopy roof, an employee of a subcontractor engaged by Concrete Constructions Pty Ltd suffered a fatal fall. No scaffolding, platforms or fall protection had been provided. Concrete Constructions was convicted of breaching section 21 of the Occupational Health and Safety Act (Vic) on the grounds that it failed to provide for the OHS of a contractor and its employees. Concrete Constructions was fined $10,000 plus costs.

Case 2

Bestaburgh Pty Ltd hired a contractor to demolish a building. During the demolition work, the contractor fell 20 metres onto a concrete floor and was fatally injured. The contractor had not been provided with necessary instruction, information, training and supervision to undertake the work safely. Bestaburgh had also failed to provide a safe system of work and had ignored a Prohibition Notice requiring fall protection to be provided. Bestaburgh was fined for offences under section 21 for failing to fulfil its general duties for the OHS of the contractor and also for failing to comply with the Prohibition Notice. The company was fined $26,000 plus costs, and a director of the company was also convicted of the same offences and fined.

Case 3

An explosion at the Parkmore Shopping Centre resulted in Lend Lease company being fined $70,000. Lend Lease, and co-defendant P & O Services, each pleaded guilty in the Dandenong Magistrates Court to a WorkCover charge of failing to provide a safe system of work. P & O was fined $35,000. Two employees of P & O Services received serious flash burns in an explosion that occurred while they were undertaking repair work on live electrical equipment in the main switch room of a shopping centre owned by Lend Lease. A third employee received minor injuries. The incident investigation revealed that the three men were not qualified to carry out electrical work. P & O had been contracted by Lend Lease to carry out building maintenance but this contract did not include electrical maintenance, which should have been performed by a specialist contractor.

From the cases described in Box 2.1, it is clear that principal contractors have employers' obligations for the OHS of subcontracted workers and building managers, employers' obligations for the OHS of maintenance contractors. Moreover, these responsibilities cannot be delegated or contracted out.

However, in reality, there is sometimes confusion as to who is responsible for certain OHS-related activities. Perhaps this is most apparent in the responsibility for the provision of certain basic safety 'infrastructure'. For example, a general access scaffold, for use by all trades on site, may not be provided by a principal contractor who might regard the provision of access equipment to be the responsibility of the trades who would use it. This situation can present problems, and there have been industry calls to clarify matters by regulating that principal contractors have specific obligations. These obligations would include, but not be limited to:

- plant and equipment provided for common use by others including scaffold, trestles and ladders;
- provision of project-specific health and safety workplans;
- protection of the public from overhead lifting, falling objects and security from unauthorised entry;
- safe systems of working at heights;
- identification and marking of underground services and exclusion zones prior to excavation and trenching, drilling or boring work;
- good housekeeping and the provision of a system to collect and dispose of rubbish and unwanted materials;
- provision, maintenance and use of construction workplace amenities; and
- provision of common services (including power and water for use by others) (Department of Employment, Training and Industrial Relations 2000).

More confusing than whether a principal contractor should bear employer's responsibilities for the OHS of subcontractors and subcontractors' employees is the issue of identifying *who* is the principal contractor. The confusion is particularly difficult to resolve where there are multiple employers present at a workplace, as in the situation in which a contractor is engaged to build a dwelling and another contractor is engaged to build a swimming pool at the same time. Furthermore, there may be some circumstances in which the client/owner remains the person in control of building works. For example, definitions of the person in control of building work are usually designed to exclude owners of domestic premises but owner-builders are deemed to be principal contractors. An Australian industry task force has suggested that one way to clarify the confusion would be to have a default provision in the legislation that provides for an owner to become the principal contractor in the event that either an 'Instrument of Appointment' is not completed or more than one party is appointed principal contractor for the same geographical area of a project (Department of Employment, Training and Industrial Relations 2000).

Non-employees

The second type of provision providing protection to workers in non-traditional forms of employment is the duty of care imposed upon employers for the health and safety of non-employees, including members of the general public. This duty is also imposed on the self-employed. Thus, Victoria's *Occupational Health and Safety Act* (1985) states that:

> Every employer and every self-employed person shall ensure so far as is practicable that persons (other than the employees of the employer or self-employed person) are not exposed to risks to their health or safety arising from the conduct of the undertaking of the employer or self-employed person.
>
> (OHSA, p. 11)

This clause is very broad and covers independent contractors and their employees, salespeople, students visiting sites or members of the general public. Johnstone (1999b) notes that what is important in determining responsibility is not the employment relationship but whether the risk arises from the conduct of the undertaking of a duty-holder.

Another important aspect of this clause is that the duty is held for risks arising from the conduct of the undertaking of the employer or self-employed person irrespective of where the risk may actually operate. Therefore the duty is owed to members of the public passing by the site or occupants of neighbouring buildings.

Johnstone (1999b) suggests that the issue of responsibility for the OHS of workers engaged through labour hire companies is more complex. In some instances, the agency provides a placement service for a fee. Once workers are placed, a traditional contract of employment between the client of the labour hire company and the worker is formed. In this situation, the employing organisation has responsibility for the OHS of the worker as an employee. However, an alternative type of arrangement exists in which a company contracts with a labour hire firm to supply workers for a limited period of time. During this time, the company pays the labour hire firm for the workers' time, and the workers are paid directly by the labour hire company. In this circumstance, there is a separation between the organisation paying and controlling the workers (the labour hire company) and the organisation with temporary control over the workers' OHS. Johnstone (1999b) suggests that, in this event, it is likely that the labour hire company will retain a responsibility for the workers as employees and will also bear a responsibility to them as non-employees. The company engaging the services of the workers will also owe a duty of care to the workers as non-employees. These workers are therefore adequately covered by the general duties provisions.

Criticisms of Robens

We have already considered some barriers to compliance with Robens-style OHS legislation inherent in the structure of the construction industry. However, some more fundamental concerns exist as to the logic of the Robens arguments and validity of the assumptions upon which the Robens recommendations are based. A notable critic of Robens is Theo Nichols. Nichols (1997) suggests that the Robens recommendations are ill-founded because the Committee wrongly assumed the most basic cause of industrial accidents to be apathy. The Robens Committee argued that the mass of existing OHS legislation had created a belief among workers and managers that OHS was the responsibility of an external agent (the inspectorate) and, consequently, workers and management did not take responsibility for their own safety. What was needed, according to the Committee, was less OHS legislation and a more self-regulatory system. The Robens Committee assumed that self-regulatory emphasis would work because 'there is a greater natural identity of interest between the "two sides" in relation to safety and health problems than in most other matters' (Robens Report 1972, para. 66). Assuming this mutual interest, self-regulation would work because both workers and managers shared common interests with regard to OHS. In response to the Robens Report, Nichols co-wrote a pamphlet refuting this argument (Nichols and Armstrong 1973). In this pamphlet, Nichols and Armstrong argued against the 'apathy theory' and suggested that workers and employers shared no mutual interest in OHS. If such a mutual interest existed, they questioned, why is the occurrence of work-related injuries and illnesses so widespread? Nichols and Armstrong (1973) proposed an alternative theory concerning how and why industrial accidents happen. This theory was based upon incident investigations in a case-study organisation. All of the incidents investigated shared one common characteristic – they occurred when workers were trying to restore production after a temporary interruption. Nichols and Armstrong (1973) conclude that production and profit are the prime motivators for employers. While foremen and managers do not want to see workers being injured, production consistently takes priority over OHS. This focus on maintaining production at all costs is even evident in the terminology used to define accidents, referred to as 'lost time incidents'.

Adrian Brooks has similarly criticised Robens. She suggests that prescriptive specification standards were ineffective, not because they were inherently unsuitable, but because the enforcement of prescriptive legislation prior to Robens was so inadequate (Brooks 1993). Indeed, Nichols and Armstrong (1973) state that the UK Inspectorate's budget pre-Robens amounted to 40 pence a worker per year. Coupled with an explicit policy to persuade rather than prosecute and derisory fines, which averaged 40 pounds, the pre-Robens legislation could hardly be said to have been well-enforced in the UK. In the face of ineffective enforcement, Brooks suggests that non-compliance was probably deliberate, with employers knowing that a likelihood of getting caught in breach of specification standards was minimal. Even if employers were caught in breach of the pre-Robens legislation, penalties were so low that no real deterrent effect was felt. Thus, she rejects the argument that non-compliance was the result of industry's apathetic attitude towards OHS. Further, if industry failed to comply with statutory requirements for OHS, Brooks suggests that it is unlikely that industry will initiate procedures and effectively regulate itself (Brooks 1993).

Brooks also argues that it is wrong to suggest that prior to the Robens-inspired reform there were too many Acts and sets of regulations. She suggests that this argument is spurious because the number of Acts and sets of regulations in existence is not really the issue – what is more important is the number of Acts and regulations relevant to a particular workplace, a number she contends was relatively small. Thus it is likely that, contrary to the conventional wisdom of Robens-thinking, pre-Robens statutory OHS requirements would not have been too complex, overwhelming or difficult for industry to understand. Brooks argues that, rather than changing the whole basis on which OHS was regulated, better results could have been achieved by rationalising the existing legislation. This would have had the effect of reducing complexity while retaining unequivocal and legally enforceable specification standards.

The lack of precise and detailed requirements in Robens-style legislation makes it more difficult to enforce and makes it harder for employees and their unions to identify a breach, at least until something has gone wrong. Furthermore, given the de-regulation of the industrial relations environment that occurred in the UK in the 1980s and Australia in the 1990s, the power differential between workers and employers is more pronounced. In this climate, reliance on self-regulation may be even more questionable than when the Robens Committee published its recommendations in the 1970s.

Effective OHS legislation

Clearly, there is disagreement about whether OHS legislation should adopt specification standards or be principle-based. This has led Gunningham (1996) to suggest that, rather than focus on what should be regulated, a more fundamental question is what form this regulation should take or, alternatively put, what type of standards should be contained within OHS legislation? The Council of Australian Governments have identified the following principles for effective regulation.

- It should be kept to the minimum required to achieve desired objectives.
- It should minimise regulatory impact upon competition.
- It should, where possible, be focused on outcomes.
- It should be compatible with international standards.
- It should not restrict international trade.
- It should be regularly reviewed.
- It should be flexible and capable of amendment.
- It should seek to standardise bureaucratic discretion (Industry Commission 1995).

It seems that on most of these points, Robens-style OHS legislation would be preferable to its predecessor, though it is far from clear that Robens-based legislation yields satisfactory OHS outcomes, particularly in intensely competitive environments, such as the construction industry.

Unfortunately, research into the effectiveness of different types of OHS standards is inconclusive. Gun (1992) undertook a comparative study of the fatal injury rate in the United Kingdom and the United States since 1970, and found that the fatal injury rate had declined in both countries. The UK had adopted a self-regulatory Robens approach in 1974, while the US relied largely on the active enforcement of detailed prescriptive standards. Gun suggests that improvements in both countries may be due to social forces unrelated to the legislation and makes the point that specification standards should not be revoked until it is clear which form of OHS legislation is the most effective.

In 1994, a *Review of Health and Safety Regulation* was undertaken in the UK (HSC 1994). The review found that the Robens-style legislation in the UK was generally supported. However, areas in which improvements needed to be made were identified in the review. The review suggested that industry was confused as to the respective roles of legislation and codes of practice. In order to overcome this problem, employers and other parties with OHS responsibilities need to be educated as to the role of the legislation and codes. A more fundamental problem was the observation that, in the UK, there was a widespread misunderstanding of the standards contained in the legislation. In particular, the requirements to undertake risk assessment and the exercise of managerial discretion in the selection of appropriate controls were not well-understood. The Health and Safety Commission (HSC) concluded that this led employers to misdirect their OHS efforts. Again, this points to a need to educate duty-holders as to their obligations under the various types of OHS standards. Other problems identified during the UK review included the volume of legislation, many unnecessary sets of regulations, and the need to simplify the requirements for documenting OHS activities.

A more recent review of twenty years under the Robens-style, self-regulatory model suggests confusion still exists. The Institute of Employment Rights is critical of the system, highlighting that various shifts in labour market strategies have undermined the appropriateness of self-regulation, including the decline in union membership and density, the growth in small business, greater use of contingent forms of employment and a realisation that some employers are not able or willing to effectively regulate their own operations (Institute of Employment Rights 1999). The Institute also notes that work-related death and injury rates are still high and suggests that Robens-style legislation has produced 'unnecessary and confusing discretion' among employers (Institute of Employment Rights 1999, p. 44).

When considering the amount of freedom and discretion to permit people or organisations, in determining how to comply with any regulation, it is important to fully understand the context in which the regulation will apply (Coglianese *et al.* 2002). The recognition that a considerable difference exists between the context in which large businesses and small businesses, including many self-employed persons and subcontractors, operate has prompted the development of hybrid approaches to OHS legislation. The finding that some companies, notably SMEs, are ill-equipped to cope with the flexibility afforded under Robens-style legislation has led regulators to combine principle-based regulations with prescriptive voluntary codes of practice, and to include flexible equivalence standards in prescriptive regulations.

Process-based standards

A feature of recent OHS legislation and standards is the inclusion of process requirements. Process standards require that a duty-holder must follow a certain process in managing particular hazards or OHS generally (Johnstone 1999a). For example, manual handling regulations require that hazard identification, risk assessment and risk control processes be implemented in the management of manual handling risks. This type of standard goes further than establishing the general duties of obligation holders, and focuses attention on how OHS is being managed. Frick and Wren (2000) suggest that regulators in many countries are currently placing a great deal of emphasis on the requirement of duty-holders to adopt a systematic approach to the management of OHS. Since the end of the 1980s, Walters suggests that regulation of the *management of OHS* has been a prominent feature of legislation within the European Union (Walters 1998).

The legislative focus on the implementation of OHS management systems has increased the number of process requirements in OHS legislation. The rationale for focusing on OHS management systems is that it is more important to make sure that an effective OHS management system is in place than to inspect and enforce compliance with prescriptive standards because, if the management system is functioning as it should, compliance with detailed requirements will automatically follow (Walters 2001). The elements of an OHS management system are described in detail in Chapter 4 of this book. Johnstone (1999a) notes that the inclusion of process requirements is useful where regulators have difficulty specifying a required goal or outcome, but believe that the risk of injury or illness will be significantly reduced if a certain process is followed. Walters describes legislative provisions requiring the implementation of OHS management systems as 'regulating self-regulation' (Walters 2001, p. 3). Australian regulators have placed a similar focus on OHS management processes (Johnstone 1999a), although process-based OHS legislation is not as well developed in Australian jurisdictions as it is in Europe.

Process-based requirements often require that management activities be documented, and increasingly, compliance can be demonstrated only by documentation that activities, such as risk assessments, have taken place and their outcomes recorded.

Process requirements in construction

It seems likely that the properties of the construction industry, that limit its ability to respond to Robens-style OHS legislation, will also impact upon compliance with process requirements. The construction industry's ability and willingness to respond to process requirements in OHS legislation has not been rigorously evaluated. However, recent attempts have been made to examine the response of the building industry in the Australian State of Queensland to the process requirement to complete Workplace Health and Safety Plans prior to starting work on any construction site where total expenditure will exceed $40,000 (Johnstone 2000a). This requirement occurred as a result of a change to the *Workplace Health and Safety Regulations* 1995 (ss. 156–163, 166), which became effective on 1 January 1997. Work plans require principal contractors, employers and self-employed persons to identify potential hazards in their work, assess the risks associated with these hazards and communicate the control measures they intend to implement during the course of their work. Johnstone (2000a) reports that industry and the inspectorate have criticised the requirement for the following reasons:

- Industry participants think that by providing a work plan they are in compliance.
- Some industry participants are unable to understand risk management concepts and are ill-equipped to operate in a self-regulatory environment.
- Many industry participants buy off-the-shelf, tick/flick model work plans and thus do not analyse their own risks.
- Enforcement is inconsistent.

Despite these criticisms, a construction industry task force recently endorsed the value of work plans and OHS management systems stating that they were essential to strategic solutions to OHS problems in the construction industry (Department of Employment, Training and Industrial Relations 2000).

Ferris *et al.* (no date) report on a case-control investigation of the response of small builders to the requirement to prepare Workplace Health and Safety Plans. Their results suggest that the requirement was better accepted where the following conditions applied:

- builders were in geographical areas where industry associations had been actively discussing the requirement in seminars and in mail-outs;
- the industry association had developed standard contracts with the Workplace Health and Safety Plan requirement integrated;
- builders had developed links with the OHS inspectorate through long-term contact with the local office;
- clients included OHS clauses in their contracts; and
- the industry association had held promotional 'road shows' (in non-metropolitan areas).

In contrast, inconsistent compliance or non-compliance with the requirement to prepare Workplace Health and Safety Plans occurred where the following conditions applied:

- builders did not belong to an industry association;
- builders did not believe the requirement applied to housing sites;
- economic and time pressures were so acute that requirements were knowingly breached to enable business survival;
- ignorance of the Workplace Health and Safety Plan requirement was claimed;
- the benefits of the Workplace Health and Safety Plans were so obscure to the builders that nominal acquiescence rather than commitment was displayed;
- contractors worked solely on very small construction jobs or maintenance work which were excluded from the requirement; and
- builders were unaware of the requirement.

These findings suggest that construction firms are more likely to comply with process requirements when these requirements are actively promoted and where industry support for the requirements is clearly demonstrated. This support is critical if process-based legislation, designed to improve the management of OHS in construction projects, is to be effective.

OHS responsibilities in the construction supply chain

An important feature of Robens-style OHS Acts is the imposition of responsibilities 'upstream' in the supply chain by extending general duties for OHS to manufacturers and suppliers of plant, equipment and materials. For example, section 24 of Victoria's *Occupational Health and Safety Act* (1985) imposes duties on people who design, manufacture, import or supply plant for industrial use. The subsidiary *Occupational Health and Safety (Plant) Regulations* (1995) expand these duties to cover the conduct of risk assessments and implementation of suitable measures of risk control by plant designers, manufacturers, importers, suppliers and employers.

Duty-holders are also required to ensure that OHS risk information specific to an item of plant is transmitted through the supply chain to the end user. Similar duties are placed on the manufacturers of materials who are obliged to provide Material Safety Data Sheets (MSDSs) for their products.

Designers' OHS obligations

Risk management theory, described in detail in Chapter 5, holds that it is better to eliminate OHS risks at source than to try to control them once they are present. Within the construction industry, this 'source' is the design team (Martens 1998). Designers of construction projects, whether architects or engineers, have the opportunity to consider OHS in the project's design stage. It is widely accepted that design decisions can have an impact upon OHS during the construction phase of a project and during future occupation and maintenance of a facility. For example, designers make choices about the design, methods of construction and materials used, which could all impact upon the health and safety of those who build, occupy, maintain, clean, renovate, refurbish or demolish the structure.

The way that construction projects are organised, for example procurement method, contractor selection criteria, contractual relationships, communication arrangements etc., also have a bearing on OHS. In a report titled 'Safety and Health in the Construction Sector' the Commission of European Communities claims that 60 per cent of fatal accidents in construction can be attributed to decisions made before work commences on site. The Commission asserts that 35 per cent of fatal accidents can be attributed to design decisions, 28 per cent to work organisation and the remaining 37 per cent to site-related activities (Commission of the European Communities 1993).

Incorporating works contractors' experience and knowledge at the design stage can improve project 'buildability' and eliminate OHS problems at source. For example, the problem of insufficient means of anchorage for safety cables or lanyards can be overcome by designing columns with a hole above floor level to support guard-rail cables or provide an anchor point for lanyards. Similarly, manual-handling problems may be overcome by the specification of a maximum size of blocks to be used (Hinze and Gambatese 1994). Despite their influence on project OHS, the OHS responsibilities of construction design professionals were not explicitly addressed by the original Robens legislation. In some Australian jurisdictions (Western Australia, South Australia and Queensland), the obligations of the OHS designers of buildings and structures are established. However, the scope and nature of these obligations varies.

Bluff (2003) also suggests that legal action could still arise in several ways under the common law. First, as relationships between parties to a construction project involve contractual agreements, acts or omissions impacting on OHS, such as failure to consider OHS in design, could be initiated as alleged breaches of contract. Alternatively, in the event of injury, common law action could arise for the tort of negligence. This would require the plaintiff to prove that the defendant owed him or her a duty of care, that this duty was breached and that the breach caused the injury. Case law has established that owing a duty of care can include those who design buildings and structures. Lastly, in some jurisdictions, there may be the possibility of legal action for the tort of breach of statutory duty. This requires that:

- statute law establish obligations for the designers of buildings and structures;
- the plaintiff is a member of the class of people whose safety the statute law is designed to protect;
- the statutory obligation is aimed at preventing the kind of harm suffered by the plaintiff;
- the action is taken against a person on whom the statutory obligation is placed;
- on the balance of probabilities, the statutory obligation has been breached; and
- the plaintiff was injured as a result of the breach.

Slivak v *Lurgi (Australia) Pty Ltd* [2001] 205 CLR 304 is an example of a case in which a plaintiff sued a designer for injuries sustained during construction of a structure. The plaintiff was not successful because the court held that the statutory duty had not been breached but nonetheless the case sets a precedent for this type of action.

In June 1992, European member states adopted the Temporary or Mobile Construction Sites Directive (92/57/EEC). This required:

- that basic principles of risk identification and control be implemented at all project stages;
- that arrangements be made for the co-ordination of OHS during planning and execution of a construction project; and
- that better communication on OHS matters be achieved between parties involved in a project.

The UK responded to this Directive by implementing the *Construction (Design and Management) Regulations* in 1994. These *Regulations* identify key parties to a construction project, including the client, professional advisors, designers, the principal contractor and subcontractors or self-employed persons. Each of these parties has a defined set of statutory duties for ensuring that OHS risks are managed during the life of the project. In addition, the *Regulations* require that a planning supervisor be appointed, whose role it is to co-ordinate the activities of designers, collate OHS risk information relevant to the project into a health and safety file, and inform the client as to the competence and resource allocation of designers and contractors (Martens 1997).

The *CDM Regulations* encompass four stages as follows:

1 When commencing a project, a construction client is required to appoint competent persons to undertake the design and construction of the project. The client must also appoint a planning supervisor who ensures that designers co-ordinate their activity with other relevant professionals and that they also consider OHS in design decisions by identifying and reducing risks as far as practicable.
2 The planning supervisor prepares pre-tender OHS plans containing design decisions made with regard to OHS. These plans communicate that OHS has been considered in design and has been used in the tender documents.
3 When a contractor is appointed, a project OHS plan must be prepared before work can commence. This documents how the contractor proposes to manage OHS during the construction and commissioning phases of the project.
4 When construction commences, the principal contractor must monitor the implementation of the OHS plan and update it as necessary. The planning supervisor is also responsible for maintaining a safety file for the project containing all documentation, such as drawings, and OHS details related to the maintenance and demolition of the project. This is provided to the client once the commissioning is complete. The client is obliged to pass this file onto future owners in the event that the facility is sold.

The *CDM Regulations* adopt a life-cycle approach to OHS management in construction projects.

Effect of the CDM Regulations

Generally, industry participants perceive that the *CDM Regulations* have acted as a positive driving force for improved OHS in construction. There has been a perceived increase in co-operation between designers and constructors, which has led to improved buildability (Preece *et al.* 1999) and there is a general belief that the *Regulations* will lead to fewer accidents and better long-term health among construction workers.

The introduction of the *Regulations* has not, however, been an unqualified success, and a number of problems associated with their implementation have been identified. Research suggests that construction designers still do not treat OHS as a high priority (Entec 2000). A recent study undertaken by Construction Division inspectors in the Scotland and Northern England

Unit of the Health and Safety Executive's inspectorate revealed that many designers lacked knowledge of their responsibilities under the CDM Regulations and many had failed to consider the practical implications of their design for safety during the building of a structure (Rigby 2003). In addition, design risk-assessments were often of poor quality and in many instances simply stated 'contractor to develop a method statement' as the required risk control. Furthermore, Baxendale and Jones (2000) suggest that designers are slow to prepare risk assessments and pass this information onto others with OHS responsibility, including the Planning Supervisor. This amounts to little more than the wholesale passing of responsibility for OHS onto the contractor that the CDM Regulations sought to overcome.

In the HSE study, designers also showed a lack of understanding of the risk control hierarchy (discussed in Chapter 5) and relied heavily on the specification of personal protective equipment, especially safety harnesses as a control of the risk of falls from height (Rigby 2003). The practical issues involved in managing a harness-based fall arrest system were often not considered, including the provision of suitable anchorage points. Designers are also reported to have limited understanding of buildability issues and OHS issues relevant in different trades, or during different stages of the project's life cycle (Entec 2000). Designers need to be educated about their OHS responsibilities and in basic risk management principles. Less than 10 per cent of designers interviewed in the HSE study had received any training in the CDM Regulations (Rigby 2003) and there is clearly a need to incorporate an OHS component in tertiary and professional development courses aimed at construction design professionals. Gambatese and Hinze (1999) suggest a similar lack of awareness exists among construction design professionals in the USA.

The separation of functions in the traditional design-bid-build procurement approach is also likely to contribute to designers' lack of awareness of buildability issues (see also Chapter 3). Hecker and Gambatese (2003) report that construction workers' safety is considered more carefully in design-build firms, in which design and construction work are undertaken in the same company. This suggests that contracting strategies are also likely to impact upon design safety issues.

In addition to designers' lack of awareness, compliance with the CDM Regulations has also generated a massive amount of additional paperwork and excessive bureaucracy (Anderson 1998a) and the costs of compliance have proved to be higher than expected (Munro 1996). The assessment of competence of project participants is particularly costly. For example, it was reported that, in 1996, one large contractor received 5,360 prequalification questionnaires, which cost £589,600 to complete. The same contractor received 1,802 sets of tender documents, which generated costs of £495,550 for the preparation of OHS responses. The total of these costs for the year was £1.085 M. These costs were incurred even though only

10 per cent of the tender responses were successful (The Consultancy Company Ltd 1997).

Alternative regulatory models

The extension of OHS responsibilities to design professionals is acknowledged to be of importance in Australia (WorkCover NSW 1999; Gunningham *et al.* 2000). However, in Australia, at present, the OHS regulation for parties upstream of the construction process is piecemeal. Australia is in the fortunate position of being able to learn from the UK experience in formulating legislation. Bluff (2003) identifies several goals for this legislation. These are:

- It should enhance consideration of OHS in the design and planning of a wide range of construction works and improve OHS for anyone affected by these works.
- It should engage all parties with real control or influence in the design and/or planning of construction works.
- It should ensure that foreseeable risks are comprehensively identified and eliminated or minimised 'at source' i.e. as early as possible in the life of a projector.
- It should aim to ensure the OHS knowledge and capability of those involved in design and planning decisions.
- It should ensure that information is transferred from the design/planning phase to the principal contractor and other contractors engaged in the construction phase and those engaged in subsequent work on the structure.
- It should be readily enforceable and ensure the timely identification of construction works in the design/planning phase.
- It should be nationally uniform.

Bluff (2003) suggests that generic obligations, such as those contained in the UK's *CDM Regulations*, might be overlooked and recommends that duty-holders are required to address particular OHS issues, such as electrical safety, fire and emergency, scaffolding, movement of people and materials, plant, amenities and facilities, fall protection, structural safety, earthworks, manual handling and hazardous substances. The provision of a 'checklist' of OHS matters which have to be considered and addressed provides flexibility without specifying detailed preventive measures to be taken and could also have an educative function prompting designers to consider issues they otherwise may overlook.

Bluff also recommends the use of a detailed OHS file to transfer information from the design and planning stage to the construction stage. This file would contain all OHS information relating to the project, details of risks identified, how they have been controlled in design and planning and details of risks unable to be eliminated and required actions in the construction phase to minimise these. Bluff (2003) suggests that this file should be used as 'gateway' to construction in that, without it, construction work should not be permitted to proceed.

A problem experienced in the UK, is that design and planning work is often well advanced before the authorities are notified of a construction project. This makes enforcement of the *CDM Regulations*, as they apply to designers, difficult to effect. Bluff (2003) suggests that Australian legislation should stipulate who is responsible for notifying the authorities of the construction project and provide an incentive for early notification, for example to specify that construction work is not allowed to commence within 28 days of notification.

Enforcement and sanctions

Enforcement strategies

The primary objective of an enforcement strategy is to ensure compliance with OHS legislation and, in so doing, reduce exposure to the risk of work-related injury and illness to an acceptable level. Gunningham writes 'Legislation that is not enforced seldom fulfils its social objectives, and effective enforcement is vital to the successful implementation of OHS legislation' (Gunningham 1998, p. 213). The policies underlying the enforcement of OHS legislation, and the sanctions that are used to punish offenders and deter would-be offenders, are important determinants of industry's response to OHS legislation. Enforcement requires that duty-holders understand their obligations and know how to comply, suggesting that enforcement strategies should have some educational component. However, duty-holders are also likely to need some reasonable expectation that their work sites or practices will be inspected, and that, should breaches be found, prosecutions and meaningful sanctions will apply (Industry Commission 1995). Thus, enforcement strategies must also have a coercive component. It is not surprising, then, that much of the discussion about enforcement strategies hinges on whether, or in what circumstances, it is better to use a 'carrot' or a 'stick' approach.

Social control theory (Ellickson 1987) considers the rule of law in a free society and is a useful basis for examining how rules regulating the pursuit of different interests within society, for example governmental and corporate interests, are enforced. *Substantive rules* are defined as the core of a social control system in that they define what conduct is to be punished, rewarded or ignored. *Remedial rules* determine the nature and magnitude of sanctions to be applied in the event that substantive rules are broken.

Regulating agencies (or controllers) administer substantive and remedial rules. Controllers can adopt formal or informal control. *Informal control* is based on the development of social patterns between the regulator and the regulated. Thus, inspectors in the field use social interaction, providing advice, making requests and working with duty-holders to improve their knowledge of legislative requirements. A mutual understanding develops between the inspector and the duty-holder, such that the inspector might find cause to grant latitude due to extenuating circumstances. At the opposite extreme, *formal control* involves the precise and rigid administration of substantive and remedial rules. The relationship between the regulator and the duty-holder has little social meaning, and the only communication that occurs is of the rules prescribing appropriate behaviour. Little leeway is granted and sanctions are rigidly applied.

A study by Gilliland and Manning (2002) examined the relationship between field inspectors' use of informal and formal methods of control and compliance with health and safety-related regulations. The study found that formal control was associated with opportunistic behaviours, such as the use of deception and attempts to 'cheat the system'. In contrast, informal control methods were associated with lower reporting of opportunistic behaviours. Informal control was also associated with higher motivation to comply with health and safety rules, while formal control was associated with lower motivation to comply. These results seem to suggest that the 'carrot' approach to enforcement can be effective, particularly when the goal of current OHS legislation is self-regulation.

Despite this, research in America suggests that formal control methods may be very effective, and that the imposition of penalties significantly reduces occupational injuries. For example, Gray and Scholz (1991) report that when plants were inspected and firms were penalised in one year, their injury experience fell by 22 per cent in the ensuing years. Brown (1992) further suggests that a 10 per cent increase in the number of penalties would reduce the number of injuries by 1.61 per cent. In comparison, a 10 per cent increase in the size of penalties would reduce injuries by 0.93 per cent, leading Brown to conclude that, while the likelihood of being penalised and the magnitude of the penalty would have a deterrent effect, more injuries would be prevented by increasing the certainty that non-compliance will be penalised. Such a strategy would require more frequent inspections and a stronger tendency to penalise duty-holders for non-compliance – more readiness to use the 'stick'.

Ayres and Braithwaite (1992) called for more 'responsive' approaches to enforcing regulations, which would provide enforcement agents with a range of enforcement options. The available options are ordered according to their severity in an 'enforcement pyramid'. An example enforcement pyramid is presented in Figure 2.1. The enforcement pyramid allows minor issues to be dealt with using informal, persuasive means of control, while

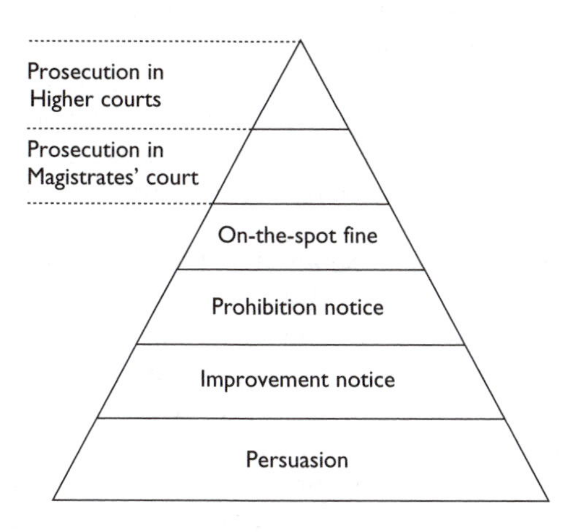

Figure 2.1 Enforcement pyramid (adapted from Ayres and Braithwaite 1991; Industry Commission 1995).

inspectors have at their disposal more formal means of control for graver offences. The enforcement methods are arranged according to the potential for social stigma and the magnitude of the penalty, with prosecution in the higher courts at the top of the pyramid.

As we have seen, prior to the Robens-influenced reform, government inspectorates were seriously under-resourced. There was also a pervasive non-prosecutorial culture and persuasion was strongly favoured. Only in the case of the most blatant and serious breaches of the OHS legislation were prosecutions brought and, even when defendants were convicted, fines were derisory. For example, in a historical analysis of OHS prosecutions occurring in the Australian State of Victoria, Johnstone (2000b) asserts that between 1900 and 1919 fines averaged 25 per cent of the maximum possible, and between 1920 and 1979 they averaged between 10 and 15 per cent. In 1979 the maximum fine was only AU$2,000. This situation led Brooks (1993) to suggest that if employers and employees were in fact apathetic towards OHS prior to Robens, then this apathy was the result of inadequate enforcement of the prescriptive legislation. Given that prescriptive legislation was never adequately enforced, it is not possible to draw the conclusion that it was ineffective.

Following the introduction of Robens-style legislation, a wider range of enforcement options was provided to inspectors, permitting more responsive enforcement. Thus, Robens legislation introduced *improvement notices* and *prohibition notices* (Hopkins 1994a,b). The former is a written direction requiring a person to remedy a breach of the legislation, and the latter is a direction to stop an activity that is posing an immediate risk to health and safety. Prohibition notices can be either in person or in writing, and are addressed to the person in control of the workplace.

In their enforcement of post-Robens OHS, inspectorates have continued to see their role, in part at least, as an advisory one. They use education and persuasion to encourage industry to comply with OHS legislation and prosecutions are still used as a last resort, being brought into play when persuasion fails.

In Victoria, the *Occupational Health and Safety Act* (1985) was the first Robens-style legislative instrument. Johnstone (2002) undertook a study of OHS prosecutions in Victoria between 1983 and 1999 and reports that, during these years, only a small proportion of visits made by government inspectors to industrial premises resulted in the issue of prohibition or improvement notices, and the number of prosecutions was much lower than the number of prohibition and improvement notices issued (Johnstone 2002). For example, in 1996/1997, 44,703 inspection visits were made, resulting in 3,219 improvement notices, 1,040 prohibition notices and only 57 cases prosecuted. Johnstone comments that this reflects the government's prosecution policy, which viewed prosecution as a last resort. It is also noteworthy that the vast majority of prosecutions – 87 per cent between 1983 and 1999 – arose as the result of a fatality or injury, and about 90 per cent of cases prosecuted in the 1990s involved machinery (Johnstone 2002). This finding substantiates Quinlan's concerns that 'very elementary and well-established forms of risk have remained depressingly resilient despite legislation' (Quinlan 1994, p. 14) and that prosecutorial activity does not reflect new and emerging hazards, such as manual handling, noise and hazardous substances.

At the time of its enactment, penalties under the OHS Act were AU$25,000 for a corporation and AU$5,000 for an individual. In 1990, these maximum fines were increased to AU$40,000 and AU$10,000 for a corporation and an individual respectively and in 1997, the maximum fine for a corporate offence was further increased to AU$250,000. Also, sentencing legislation allowed the courts to adjourn proceedings when cases had been proved without convicting defendants and to require defendants to commit to good behaviour or the fulfilment of other conditions for a specified period. Since 1991, courts could also impose fines without convicting defendants. This range of sanctions clearly illustrates that, in addition to a pyramidal enforcement strategy, there exists a hierarchy of sanctions, reflecting differing degrees of moral culpability. Many of these sanctions de-criminalise OHS offences.

Johnstone (2002) also examined sentencing outcomes for cases in which charges were proved. He reports that, in 1999, of 120 cases in which charges were proved, 101 defendants were convicted, 17 were fined without conviction and 2 received a good behaviour bond. The average fine was AU$14,673, which was 26.7 per cent of the maximum fine available. Given that prosecutions are only brought as a last resort and usually only when a death or serious injury has occurred, these sentencing outcomes reflect that what Carson (1979) termed the 'conventionalisation' of OHS crime continues. Thus, offences against the *OHS Act* are treated as 'quasi-crimes' and not assumed to have the same degree of gravity as other criminal offences. That the majority of OHS offences are heard in Magistrates Courts, which typically deal with minor offences, such as traffic violations, reinforces this notion.

On-the-spot fines

In some jurisdictions, infringement notices or on-the-spot fines have been adopted. For example, since July 1998, OHS inspectors in Queensland have had the power to issue infringement notices, or on-the-spot fines for certain breaches of the OHS legislation and for failing to comply with an improvement notice. On-the-spot fines were first introduced in New South Wales in 1991. Gunningham *et al.* (1998) surveyed the opinions of a cross section of stakeholders including employers, employees, safety representatives, government personnel and OHS inspectors to determine their views as to the impact of on-the-spot fines. They report that, in general, industry believes on-the-spot fines to be effective as a deterrent, and support the use of such fines in the case of recalcitrant companies. However, the construction industry was less supportive of the use of on-the-spot fines than other industries. Construction respondents, from both large and small companies, complained that fines were often issued for technical breaches that were not directly relevant to actual OHS performance, for example, in the case of inadequate OHS record-keeping. The survey revealed that relatively small fines were considered to be significant, particularly by small businesses and subcontractors. Given that on-the-spot fines are inexpensive and convenient to administer they can therefore provide a useful enforcement tool. Perhaps the most important advantage of on-the-spot fines is the immediacy of the penalty. The time it takes to initiate court proceedings means that, in many cases, construction projects may be complete before an OHS matter is heard in court. On-the-spot fines provide an immediate link between a breach and the punishment and, as such, can have an instantaneous rather than a delayed impact upon behaviour.

A construction industry task force in Queensland has proposed that OHS statistics be used to identify areas in which hazard-based, on-the-spot fines should be used. These include:

- working at heights;
- use of ladders;
- excavation and trenching work;
- installation, maintenance and use of electrical equipment;

- site housekeeping;
- the provision of construction workplace amenities;
- working in proximity to underground services; and
- protection of public safety (hoardings and gantries and so on) (Department of Employment, Training and Industrial Relations 2000).

In recent years, inspectorates have increased their enforcement capacity. For example, in 2000, the number of prosecutions by the UK HSE was 9 per cent higher than the number in 1999, while the number of identified breaches was 28 per cent greater (TSHP 2000). Similar emphasis on enforcement has been demonstrated in Australia; for example in Victoria, the number of inspectors has doubled, construction industry field teams have been created and on-the-spot fines are increasingly being used. In June 2002, Victoria's WorkCover Authority announced that it was going to take a tougher stance on OHS offenders, particularly in high-risk industries, including construction. In 2002, an estimated 60,000 inspections of Victorian workplaces were conducted – almost 30 per cent more than in 1999. This increase is the result of learning from the European and American experience, which has demonstrated that workplace inspections and enforcement have a significant impact on performance.

This enforcement effort is combined with increased fines for companies in breach of OHS legislation. For example, in July 2001, Esso Australia Pty Ltd were fined a record $2 million for 11 breaches of the *Occupational Health and Safety Act* (1985). On two of the charges Justice Cummins imposed the maximum fine of $250,000 and penalties were added to four of these fines under the provisions relating to companies with previous convictions. These fines were imposed for offences associated with the company's Longford Disaster, described in Chapter 4.

Enforcement authorities have also started to publish details of health and safety prosecutions in an attempt to pressurise companies that have been found to be in breach of the OHS legislation to improve. Thus, the Director General of the UK HSE has said, 'The convictions are there for everyone to see, including would-be customers, contractors, investors, employees and insurers. They all have a right to be aware of an organisation's health and safety record before they decide whether to invest their capital and labour' (THSP 2000, p. 9). Australian enforcement agencies have also adopted this strategy in an attempt to bring public attention to OHS offenders. For example, the Victorian WorkCover Authority produces an annual report summarising prosecutions brought under legislation administered by the Authority including:

- Occupational Health and Safety Act 1985
- Dangerous Goods Act 1985
- Equipment (Public Safety) Act 1994

- Accident Compensation Act 1985
- Accident Compensation (WorkCover Insurance) Act 1993.

This report provides details of each case, the defendant's details, penalties imposed and sentencing remarks of judges and magistrates.

The criminal law

The mainstream criminal law can also apply to OHS in certain circumstances. Thus when negligent conduct results in the death or serious injury of a worker, corporations or individual managers may be prosecuted for offences, such as manslaughter or criminal infliction of serious injury. Such charges may be brought in addition to, or as an alternative to, prosecutions for breaches of the preventive OHS legislation.

An argument often cited in support of the use of the mainstream criminal law is that the conviction of a 'real' criminal offence carries greater stigma than a conviction for breaching OHS legislation. It also captures public attention and, in doing so, is likely to have a greater deterrent effect. It is also argued that the principle of equal treatment means that if people in society can be convicted of manslaughter in the event of a death arising from their negligent conduct, so too should managers who, through their negligence, cause the death of a worker (Hopkins 1995). Neal (1996) agrees that the existing distinction between killings that occur in workplaces and those that occur outside workplaces is artificially created and cannot be justified.

Despite these arguments, as we have already noted, work deaths are widely regarded to be outside the realm of the mainstream criminal law and police and crown prosecutors have demonstrated a reluctance to bring criminal charges in respect of workplace death and serious injury (LRCV 1991; Lilley 1993; McColgan 1994; Ridley and Dunford 1997). There is evidence that magistrates do not perceive OHS issues as criminal. For example, a study, undertaken within the Department of Legal Studies at LaTrobe University in Melbourne, explored the opinions of Victorian magistrates. This study revealed that Victorian magistrates regard OHS offences as 'social' rather than criminal, and thus falling within a 'quasi-criminal' jurisdiction (La Trobe/Melbourne Occupational Health and Safety Project 1989).

The mental element

Gaining a successful conviction of a corporation in the mainstream criminal law is extremely difficult. Unlike offences created by the preventive OHS legislation, which are strict liability offences (that is, they do not require proof of a mental element), successful prosecution for a crime, such as manslaughter, requires proof of a guilty mind on the part of the defendant, or *mens rea*. The most likely basis for a manslaughter charge to be brought following a workplace death or serious injury is gross negligence (*Nydam* v *R* 1977). Manslaughter by negligence requires 'such a great falling short of the standard of care which a reasonable man would have exercised and which involved such a high risk that death or grievous bodily harm would follow, that the doing of the act merited criminal punishment'. Thus the Nydam test, as it is known, establishes the requisite *mens rea* for guilt in relation to manslaughter by gross negligence.

Establishing corporate guilt

In relation to a corporation, this *mens rea* is difficult to demonstrate because corporations do not have a state of mind (Wells 1989; Hopkins 1995). British and Australian criminal codes do not distinguish between natural persons and bodies corporate (Crabtree 1995). The basis of corporate criminal liability was established in the case of *Tesco Supermarkets Ltd* v *Nattrass*. Under *Tesco*, criminal conduct must have been committed by the board of directors, the managing director or another person to whom a function of the board has been fully delegated in order for it to be attributable to the company. This principle has been widely criticised because it limits corporate liability to acts and omissions performed at the top of the corporate ladder (Wells 1989; Field and Jorg 1991).

The application of the *Tesco* principle prevented a corporate manslaughter conviction following the *Herald of Free Enterprise* ferry disaster, which killed almost 200 people. The public inquiry into the disaster found that 'from top to bottom the body corporate [of the ferry company] was infected with the disease of sloppiness' (cited in Sheen 1995) yet the fault could not be attributed to any single controlling officer of the firm (McColgan 1994). Limiting the moral responsibility of a corporation to the acts of high-ranking company officers poses serious problems because priorities, set from above, determine the social context within which work is conducted on the shop floor (Field and Jorg 1991; Polk *et al.* 1993). In other words, the Tesco principle ignores the widely accepted notion of a corporate culture. The issue of corporate safety culture is discussed in more detail in Chapter 3.

Furthermore, under existing principles of attribution, it is not possible to aggregate the acts of two or more 'directing minds' to make one corporate offence. A British court of appeal upheld the direction of the Coroner in the inquiry into the *Herald of Free Enterprise* ferry disaster. The Coroner ruled that 'although it is possible for several persons to be guilty individually of manslaughter, it is not permissible to aggregate several acts of negligence by different persons, so as to have gross negligence by a process of aggregation'.

The view that corporations are clones of individuals, and therefore that the law should seek to identify a single corporate guilty mind at the head of the identity, is a fallacy, since corporate decision-making is typically diffused. Furthermore, it is easy, especially in large and complex corporations, for organisational structures to be set up to avoid such liability (Fisse 1994). This is especially true in decentralised, project-based industries, such as construction.

The circumstances of most industrial accidents, which occur as a result of multiple errors or omissions, make the conviction of a corporation difficult to sustain. These difficulties become apparent when examining the situation in the Australian state of Victoria. Since 1990, Victoria has maintained a policy of prosecuting individuals and corporations for manslaughter in cases of reckless or criminally negligent workplace deaths. However, since 1990 there has been only one successful prosecution. This case (*R* v *Denbo Pty Ltd*) is described in Box 2.2. In other cases, negligence on the part of individuals has been established but corporate guilt has not (Creighton and Rozen 1997).

Box 2.2 Case study: Corporate manslaughter case

The Queen v *Denbo Pty Ltd and Timothy Ian Nadenbousch* (Unreported, Supreme Court of Victoria, Teague J, 14 June 1994).

On 14th June 1994, the Victorian Supreme Court found Denbo Pty Ltd guilty of criminal negligence in causing the death of Anthony Krog. An AU$120,000 fine was imposed upon Denbo Pty Ltd. In his judgement, Justice Teague acknowledged that the amount of the fine ought to be substantial in order to achieve a 'generally deterring effect' but noted that owing to the company's liquidation, the imposed fine represented 'no burden on anyone as it will not be paid'.

Denbo Pty Ltd was owned by two shareholders, Ian Nadenbousch and his son, Timothy Nadenbousch. Timothy Nadenbousch, whose position was effectively that of director, was responsible for the running of the Western Ring Road project in Melbourne. Anthony Krog was an experienced plant operator employed to work on the construction of the Western Ring Road. He was fatally injured when the brakes of the dump truck he was driving failed while descending a steep track at the work site. The truck had been purchased by Denbo shortly before the incident. Timothy Nadenbousch knew the truck had defective brakes but put the truck into operation before ensuring the necessary maintenance work had been undertaken. Subsequent tests on the dump truck revealed that 'the braking defects were very obvious and very bad'.

Box 2.2 (Continued)

Both Ian and Timothy Nadenbousch and Denbo Pty Ltd were charged with breaches of the *Occupational Health and Safety Act* 1985 (Vic) and Timothy Nadenbousch and Denbo Pty Ltd were charged with manslaughter. Ian Nadenbousch was acquitted but Timothy Nadenbousch and Denbo Pty Ltd were committed for trial. At the trial, Denbo Pty Ltd pleaded guilty to the charge of manslaughter and Timothy Nadenbousch pleaded guilty to two charges under the *Occupational Health and Safety Act* 1985. The Crown dropped several other charges, including the charge of manslaughter against Timothy Nadenbousch. The only penalty Timothy Nadenbousch received was a $10,000 fine. The maximum possible fine at the time was $20,000.

Establishing individual guilt

The problems of prosecuting under the mainstream criminal law are significantly reduced in the context of individual defendants. It is much easier to prove the personal liability of individual managers compared to corporate liability. However, successful convictions are still relatively rare, possibly because in large, complex organisations, it is hard to identify with certainty the individuals whose conduct satisfies the gross negligence test. Nonetheless, there have been some notable examples in the USA, for example *Illinois* v *O'Neill, Film Recovery Systems Inc. and others* (1990) in which three company executives were convicted of murder following the cyanide poisoning of an employee.

Alternative models for attributing corporate guilt

The solution to the problem of establishing corporate guilt may lie in the adoption of an entirely new corporate liability law that establishes new methods for determining a corporation's *mens rea*. Field and Jorg (1991) contrast the English (and by extension Australian) principles of attribution with those adopted in Holland. Under Dutch law, an employee's actions can be regarded as those of his/her employer if the employee's actions belonged to a category of acts 'accepted' by the firm as being in the course of normal business operations. The notion of acceptance is arguably more suited to the evaluation of collective enterprises than the requirement to identify a guilty 'controlling mind'.

An alternative solution to the problem of attribution is that adopted in the Australian Model Criminal Code and widely adopted in Commonwealth legislation. Under this approach, corporate criminal liability is not based on the *Tesco* principle, but is based on the concept of 'organisational blameworthiness'. It requires that the external elements of the offence be committed by a person for whose conduct the corporate defendant is vicariously responsible. However, liability in relation to the mental element depends upon the corporation being at fault by having a policy that expressly or impliedly authorises or permits the commission of the offence or by failing to take reasonable precautions to prevent the commission of the offence. Such an approach is yet to be taken up by state legislators and thus the *Tesco* principle is still widely used.

The difficulties associated with establishing the requisite elements of the offence of manslaughter in the case of work-related deaths has led to repeated calls for reform of the criminal law. For example, in 2001, in the Australian state of Victoria, the *Crimes (Workplace Deaths and Serious Injuries) Bill* was introduced by the Labour government. This Bill created new criminal offences of corporate manslaughter and negligently caused serious injury by a body corporate and imposed criminal liability on senior officers of a body corporate in certain circumstances. Section 14A of the Bill overcame the problem of identifying a guilty 'controlling mind' by stating that the conduct of an employee, agent or senior officer of a body corporate, acting within the actual scope of their employment or within their actual authority, must be attributed to the body corporate. Furthermore, Section 14B allowed the conduct of any number of employees, agents or senior officers of the body corporate to be aggregated in determining whether a body corporate has been negligent. Section 14C identified the criteria required for senior officers to be guilty of manslaughter or negligently causing serious injury. First, it required proof that the body corporate committed the offence. If this were established, the prosecution would then have to prove all of the following:

- The senior officer was organisationally responsible for the conduct, or part of the conduct, of the body corporate in relation to the commission of the offence by the body corporate.
- The senior officer, in performing or failing to perform his or her organisational responsibilities, contributed materially to the commission of the offence by the body corporate.
- The senior officer knew that, as a consequence of his or her conduct, there was a substantial risk that the body corporate would engage in conduct that involved a high risk of death or serious injury to a person, and having regard to the circumstances known to the senior officer, it was unjustifiable to allow the substantial risk to exist.

These amendments to the criminal law would have changed the basis on which senior managers could be convicted of serious criminal offences in the event of workplace death or serious injury, and could have overcome the difficulties posed by diffused decision-making. In addition, the Bill allowed for senior officers of a business that was convicted of corporate manslaughter to be jailed for up to five years and fined up to $180,000, if they contributed to the commission of the crime. Companies convicted of corporate manslaughter would face fines up to AU$5 million.

The Bill, which was welcomed by trade unions, faced widespread opposition from employers' groups and was defeated in 2002. Following the re-election of the Labour Party, the government announced that it would not re-introduce the Bill in the present term of office. This decision caused outrage among trade unions and families of victims of fatal workplace accidents who argue that the reform had strong public support (*The Age*, 5 December 2002; *The Australian*, 4 December 2002). However, industrial law experts and unions caution that the push for reform will not simply go away due to a 'groundswell' of public sentiment in support of industrial manslaughter charges (*Australian Financial Review*, 29 January 2003, p. 8).

The investigation process

The process of investigating workplace deaths may account, in part, for the low prosecution rate. In cases of suspected homicide outside the workplace, a thorough police investigation is conducted with the express intention of determining whether the cause of death warrants a prosecution for manslaughter and, if so, to gather evidence necessary to pursue such a prosecution. In contrast, workplace deaths are investigated by state OHS inspectors with the primary purpose of identifying preventive measures for future implementation (Polk *et al.* 1993). Some jurisdictions, such as New South Wales, have established a special investigation unit for workplace deaths in an attempt to overcome this problem. In the UK, a protocol for liaison has been established between the HSE, the Crown Prosecution Service, the police and local government associations to establish the roles of these parties in relation to work-related deaths (HSE 2003a). This protocol makes it clear that where there is a suspicion that a serious criminal offence (other than a breach of OHS legislation) may have caused the death, the police will head the investigation but will work in partnership with the other parties. This protocol should therefore help to ensure that, where applicable, the investigation will be conducted with a view to collecting evidence to substantiate a criminal charge.

The decision to prosecute any serious criminal offence remains that of the Crown Prosecution Service. One difficulty may be that, even if sufficient evidence to warrant manslaughter charges is found, prosecutors may be reluctant to pursue manslaughter charges against corporations because they expect corporate offenders to defend their cases vigorously, resulting in lengthy and expensive proceedings and a reduced chance of conviction (Hopkins 1995).

Sanctions for corporate crime

In Australia, fines are the most commonly used sanctions against corporate offenders. Even disregarding the philosophical issue of the dollar value that should be placed on a human life, it is arguable that fines do not effectively influence the future conduct of corporations (Fisse 1994).

One difficulty with the imposition of fines is that the real burden may ultimately be borne by people who are entirely innocent. For example, shareholders may bear the cost or, if the company is forced to shut down, innocent employees may suffer. Fines do not necessarily lead companies to take internal disciplinary action against those responsible or change internal operating procedures (Fisse 1994). A company may simply opt to pay fines and regard them as 'purchasable commodities' or operating costs. Another difficulty with fines as a form of sanction is that they are prone to evasion through techniques such as asset stripping. In some instances, fines that reflect the seriousness of a criminal offence may be too great for a company to pay. This leaves a court with the choice of charging a lesser fine, or imposing an appropriate fine and forcing the company into liquidation. Once a company has gone into liquidation, it cannot be held liable for offences committed before dissolution.

Fisse and Braithwaite (1993) suggest there is a need for individual accountability in the maintenance of social control. The need to hold individuals responsible for their actions is overlooked or bypassed in the exercise of the criminal law in relation to corporations. Fisse and Braithwaite argue that the corporate form is used to obscure and deflect responsibility which, coupled with the growth of corporate activities in industrialised societies, poses a risk of the breakdown of social control. They suggest that corporations have the capacity but not the will to deliver clearly defined accountability for law-breaking, while courts have the will but not the capacity to deliver this. They recommend that a solution would be to combine the firm's capacity with the law's desire to achieve accountability. Thus, corporate offenders would be required to reform themselves and undertake internal disciplinary action, under the threat that, if they fail to do this to the satisfaction of the courts, a sanction such as forced liquidation or the withdrawal of a firm's licence or charter to operate may be incurred.

This model is somewhat similar to the corporate probation concept. The sanction of corporate probation may present a solution to some of the problems (Fisse 1994). Corporate probation provides for a judicially supervised period within which a company has to demonstrate that it has undertaken internal disciplinary action against those responsible for an offence or has undertaken actions to rectify defective operating procedures or physical aspects of the workplace. Probationary sanctions can be matched against the need for rectification and are not subject to limitations associated with the company's ability to pay a fine commensurate with the seriousness of the offence. Corporate probation is authorised under the United States' Criminal Code and has been recommended by the Australian Law Reform Commission in the context of the Trade Practices Act (ALRC 1994). The consideration of a similar approach to the punishment of OHS-related corporate offenders should also be considered.

The *Denbo* case represents the first and, to our knowledge, only Australian conviction of a corporation for manslaughter. Its significance lies in the important issues it raises relating to the application of the criminal law to workplace deaths and serious injuries. In the *Denbo* case, there was sufficient evidence to warrant a manslaughter charge. The facts of the case strongly suggested that the degree of negligence involved would satisfy the *Nydam* test. However, the decision to drop the manslaughter charges against Timothy Nadenbousch, and to proceed only in the charge against the company, highlights another problem with the current application of the mainstream criminal law to work-related deaths and serious injuries.

Although the Australian legal framework provides for both companies and individuals to be held responsible for corporate crime, the operation of prosecutorial discretion does not guarantee a well-balanced mix (Fisse 1994). Despite trade union calls to hold company officers personally liable for workplace death and injury, there have been few prosecutions of company directors or managers in Australia. As we have already noted, the demonstration of criminal liability on the part of an individual is often difficult to achieve, particularly in large, complex organisations and at higher echelons of the organisational hierarchy. In the *Denbo* case, these difficulties did not seem to exist. The circumstances did not render Timothy Nadenbousch's conviction for manslaughter unlikely. Timothy Nadenbousch was the sole representative of the company on the site and the criminal negligence attributed to the company related to his acts and omissions. Teague J stated that there was 'wilful neglect' on his part in relation to the maintenance and training on the site. Holding company officers liable for workplace deaths is believed to have a stronger motivational effect on managers to proactively manage OHS than pursuing corporate convictions. In re-allocating criminal liability from the individual to the corporation during pre-trial negotiations, the conclusions of the *Denbo* case could serve to seriously undermine the deterrent effect that Teague J indicated was necessary in his judgement.

Owing to the facts and circumstances of the *Denbo* case, the issue of when and how criminal liability can be attributed to a corporation was not discussed. The guilty plea offered by Denbo Pty Ltd meant that the attribution issue was not addressed during the trial and, as such, the principles of attribution remain unclear. In a company of the size and organisational structure of Denbo Pty Ltd, the issues of attribution were unlikely to have been problematic. However, because they were not addressed in the *Denbo* case, they were left open to future judicial interpretation.

The *Denbo* case also illustrates the inadequacy of fines as a sanction for corporate crime. Teague J highlighted the desirability of the general deterrent effect to be achieved through imposing a heavy fine on Denbo Pty Ltd. In the *Denbo* case, the fine imposed could never be paid owing to the company's liquidation less than a month before the trial. It is likely that public and industry recognition that the fine would never be paid devalued the general deterrent effect of the penalty.

Employees' compensation

We now turn our attention to legal mechanisms for compensating victims of occupational illness or injury. In most developed countries, workers who suffer a work-related injury or illness can look to two possible sources of financial assistance: They can bring a common law action for damages against the party responsible, usually their employer; and they can seek compensation from a statutory workers' compensation system. In the next section of this chapter, we discuss employees' compensation mechanisms and the advantages and disadvantages of these mechanisms as motivators for employers to improve OHS performance, and in meeting the needs of the victims of industrial accidents or illnesses.

Common law actions

If an employee can demonstrate that his or her injury or illness occurred as a result of negligence on the part of his or her employer, the common law is one means by which financial recompense may be obtained. A tort action for negligence can be brought against an employer in many jurisdictions. Such an action would be based on the theory expressed by Lord Atkin in *Donoghue* v *Stevenson*, that people are under a duty not to injure their 'neighbour' through careless behaviour. Employers are held to have a duty of care to establish safe systems of work and ensure that safety standards are enforced. Therefore, if a failure to meet this duty of care results in occupational injury or illness, the victims may seek damages. This duty of care is also implied in the employment contract, which gives injured or ill workers the right to sue employers for breach of contract (Creighton and Stewart 1994).

Historically, the scope for successfully bringing a tort action for negligence against an employer was limited by a number of legal principles which can be traced back to the very first reported English case of an injured worker suing for damages in 1837 (*Priestly* v *Fowler*) (Brooks 1988). These principles were:

- the doctrine of common employment, under which an employer could not be sued in respect of an injury caused by another worker;

- the doctrine of contributory negligence, under which any fault on the part of the injured or ill worker in the experience of the injury or illness precluded any recovery of damages; and
- the principle of *volenti non fit injuria*, whereby employees' voluntary assumption of risk was used to prevent successful claims against employers (Luntz 1981).

The principles of voluntary assumption of risk and common employment have been abolished, and the doctrine of contributory negligence has been modified, so that now an employer's liability may be limited by contributory negligence but is not extinguished altogether. Courts determine to what extent contributory negligence should reduce damages.

Injured or ill employees may also bring an action based on the tort of an employer's breach of a statutory duty. This is based upon the argument that, if an injury or illness was caused by an employer's breach of a statutory OHS provision, the employee may seek damages for loss arising from this injury or illness. The arguments on which actions for breach of statutory duty and negligence are based are often the same, and the two actions may be brought simultaneously (Creighton and Stewart 1994).

Common law actions arising from OHS issues are intended to serve two purposes:

1. They are designed to have a prevention or deterrent effect. Employers should be motivated to prevent occupational injuries and ill-health to avoid costly damages.
2. Common law damages are also intended to provide a source of financial support to injured or ill workers and their families.

The extent to which common law actions meet these objectives has been questioned. Common law actions are widely regarded to be an inefficient and expensive means of providing employees with compensation, and the length of time involved in processing and deciding on common law actions may have a negative impact on the rehabilitation and return-to-work of employees. Furthermore, the adversarial nature of court proceedings may jeopardise the prospect of employees' returning to work for the same employer, which is an objective of statutory occupational rehabilitation provisions.

A review of employees' compensation system in Victoria in 1984 reported that common law claims accounted for a small proportion of overall occupational injuries, yet consumed a large proportion of total compensation resources, due to wasteful dispute resolution procedures (Cooney Report 1984). Furthermore, the 'arbitrary' manner of determining the payouts made to successful plaintiffs in such actions has been criticised, and governments have expressed concern about the effect of unpredictable costs of common law claims on industry. Creighton and Stewart (1994) argue that the adversarial nature of common law claims means that the parties must argue their case in 'black and white' terms, when in reality, OHS issues are complex and caused by multiple factors. This, they argue, leads to a situation in which the outcome of these expensive legal proceedings is largely a matter of chance, undermining any deterrent effect that the award of damages might have.

In common law actions, damages are assessed on the basis of the principle of *restitutio in integrum*, which seeks to place the injured party back in the position they would have been had they not been injured. This is often impossible – for example, where serious physical disabilities have been suffered. It is also difficult to project the career path of young employees and to take into account projected changes in wage rates and inflation in the future. Arup (1993) suggests that the common law approach is problematic because, although organisations are vicariously liable for the actions of their employees, employers' liability at common law requires that an individual legal entity must first be found at fault. This is extremely difficult in modern firms, which are large and complex. It is likely to be particularly difficult in an industry, such as construction, where a high level of labour subcontracting is standard practice.

Total reliance on the common law in providing financial assistance to occupational injury or illness victims is not appropriate because the award of common law damages requires proof of fault on the part of the employer. The fact is that, when people are incapacitated and experience the physical and psychological damage of injury or illness and a loss of earnings, financial support is needed – irrespective of fault. Reliance on the common law would mean that employees would not be eligible for compensation in cases in which it could not be proved, on the balance of probabilities, that the injury or illness was an employer's fault. This would deprive some victims of much-needed financial assistance and force reliance on the social welfare system. As a result of these limitations, and in view of the inefficient operation of the common law as a means of compensating injured or ill employees, governments have created statutory 'no-fault' workers' compensation systems. At the same time, governments have attempted to limit or qualify employees' rights to bring common law actions against employers for work-related injuries or illnesses, in an attempt to restrict these actions. For example, in a recent case in Victoria, a man was awarded a record payout of AU$285,000 for injuries he sustained when he fell from a mobile work platform onto a concrete floor at the age of 15 (*The Age*, 21 February 2003). While the payout was substantial, the man's lawyer said it was considerably less than that which would have been awarded before employee rights to sue for negligence had been abolished in the state. Under current laws, the man could seek compensation for pain and suffering but not for loss of earnings. Pro-worker groups argue against limiting workers' rights to seek compensation via the common law on the grounds that, in requiring that fault be attributed, the common law provides an avenue for upholding the principle of individual responsibility for OHS and performing a valuable 'corrective justice' function (Veljanovski 1981).

Statutory 'no-fault' compensation schemes

In 1897, the *Workmans Compensation Act* introduced a statutory 'no-fault' system of workers' compensation in Britain. This legislation was based on a model pioneered in Germany, and required employers to take out a policy insuring against the risk that employees would suffer in a work-related injury, causing them to suffer a reduced earning capacity. An employee suffering a work-related injury and consequent diminished earning capacity was entitled to a level of income support, which would be funded by this insurance. This entitlement was not conditional on proof that the employer had been negligent, and it therefore covered workers who would be otherwise uncompensated (Bartrip 1985), although some consideration was still given to contributory negligence in assessing damages. This type of compensation system was introduced in all Australian jurisdictions in the first half of the twentieth century.

Under these new compensation systems, injured workers or, in the event of death, their dependants, were entitled to payments covering the cost of medical expenses arising from the injury, as well as weekly payments covering loss of earnings while absent from work. Lump sum payments were also available if employees suffered partial or total permanent incapacity. Payments for incapacity were assessed on the basis of the level of incapacity or body part affected, the amounts to be paid being listed in a 'table of maims'. Governing bodies also specified the hazards, which were accepted as having a work-related cause. Only specified injuries or hazards were compensable under these schemes. Initially, this list was heavily weighted towards acute injuries, and many work-related illnesses were not within the scope of the compensation systems. For example, in 1914, only six occupational diseases were listed under the Victorian Act (Quinlan and Bohle 1991). As knowledge of health effects of occupational hazards became better understood, the compensation systems have been reformed. Since their initial implementation, the list of compensable hazards covered by these systems has been extended, and now covers many illnesses for which a connection with work is presumed.

Also, the basis on which compensation can be awarded for these illnesses has been modified to include aggravation, acceleration or recurrence of the illness. Further, the standard of proof for establishing the relationship between occupational exposure and the illness has been amended, making it easier to demonstrate the connection between work and the illness. Despite these reforms, it is doubtful that the inclusion of additional occupational illnesses and diseases has kept pace with the introduction of toxic or hazardous substances into workplaces.

Traditionally, compensation boards have taken a conservative view of the occupational groups exposed to the risk of complex health problems, such as stress-related illnesses, mental health problems and heart disease. For example, Quinlan and Bohle (1991) argue that compensation for stress-related illnesses is most frequently awarded to white-collar workers, despite growing evidence that stress problems are more acutely experienced by blue-collar, manual workers. This issue has important implications for the construction industry.

The link between work, stress and death is receiving more attention and some legal recognition. For example a recent study by the Uniting Church's Urban Ministry Network in Australia revealed that work pressures were a significant factor in 109 suicides investigated by the Victorian coroner between 1989 and 2000. The report suggests this is likely to be an underestimation because of the lack of detail required by the coroner about work-related factors. The occupations most commonly affected were technical workers including trades (19 per cent), those in supervisory positions (18 per cent) and professionals (14 per cent). Significantly, in the UK, a widow is reported to have received a compensation settlement after a court ruled her husband's suicide was the result of work stress (*The Age*, 16 November 2002). A South Australian court also recently linked work stress to bowel cancer, awarding compensation to the widow of a man who died of the disease (*Sydney Morning Herald*, 29 March 2003). These rulings emphasise the need for employers to address work stress as an occupational health issue for both blue- and white-collar workers.

Quinlan and Bohle (1991) also suggest that groups of workers are excluded from existing workers' compensation systems, either directly or as a result of factors such as employment instability or labour market fluctuations. Thus, subcontracted workers, casual workers and the self-employed, groups which make up a large proportion of the construction industry's workforce, may enjoy only nominal coverage under compensation schemes and may be unable to access entitlements in the event of a work-related injury or illness.

The concept of work-relatedness

The concept of work-relatedness is a key determinant of the breadth of coverage of statutory workers' compensation schemes. The extent to which an injury or illness falls within a workers' compensation scheme is determined by external and internal boundaries (Clayton *et al.* 2002). External boundaries include such considerations as the form of the work relationship, that is, whether the injured or ill worker is employed under a contract of service or a contract for services (as in the case of independent contractors). The former employees are usually covered, while independent contractors are usually not. However, the situation is not clear-cut, and a range of different considerations may come into play in determining whether someone is a 'worker' for the purpose of workers' compensation eligibility. These may include:

- the degree of control exercised over the worker's activities;
- the level of the worker's integration into the primary business;
- whether the worker supplies his or her own tools;
- whether the worker bears the financial risks associated with the venture;
- whether the worker is free to perform work for other people; and
- whether the worker receives wages or is paid on submission of invoices (Clayton *et al.* 2002).

Other considerations are the degree of control an employer has over the work, and the extent to which the activities being undertaken at the time of the injury were for the benefit of the employer. Clayton *et al.* (2002) note a tension between the notion of control and employer benefit; for example, this tension arises in relation to the issue of whether injuries sustained on the way to work, so-called *journey* accidents, should be covered by workers' compensation. The travel to work is undertaken for the benefit of the employer, but it is unlikely to be within his or her control. At present, journey accidents are compensable in some Australian jurisdictions but not in others. This situation is inequitable and consistency is needed.

The concept of work-relatedness is also central to establishing internal boundaries of the coverage of workers' compensation schemes. These boundaries relate to the temporal and causal requirements for an event to be considered as work-related. In most jurisdictions, an injury or illness has to 'arise out of or in the course of employment'. Whether something arises out of the course of employment suggests that if a causal link can be established between work and the injury or illness, it is compensable. In Australia, the test for this causal link is known as the actual risk test. It requires the worker to demonstrate that the employment subjected him or her to the actual risk that caused the injury. This differs from the earliest test, which required that the risk of harm had to be peculiar to someone's occupation in order for the injury or illness to be deemed to have arisen out of the course of employment. The bizarre operation of this test was demonstrated in *Robinson's case*, in which a labourer who suffered a frozen foot while working all night in extremely cold temperatures was denied compensation, on the grounds that the risk of sustaining a frozen foot was not peculiar to his occupation.

In determining whether an event occurred in the course of employment involves consideration of how time and place elements relate to work-relatedness. Thus, it must be determined what the connection between work and an activity in time and space is, to determine whether an injury or illness has been sustained in the course of employment. In *Kavanagh* v *Commonwealth*, the Australian High Court ruled that this element was time-related, and that the worker had only to be engaged in something that was part of, or incidental to, his or her employment for an injury to be compensable. This was very broad and in the 1980s, the strain on workers' compensation schemes generated pressure for legislative reform. Reform of the Victorian compensation system under the Kennett government, in 1992, led to the removal of journey injuries from the compensation scheme's coverage and the introduction of a new concept determining work-relatedness. This was the degree of employment contribution to the injury. In Victoria, it is now necessary to demonstrate that employment must be a significant contributing factor to an injury or disease in order for it to be compensable. Other jurisdictions have adopted similar requirements.

The boundaries of state-managed compensation schemes have been the subject of considerable debate and change in relatively recent years. This change has in part been driven by the tension between breadth of coverage and the constraints on the costs of operating the schemes. It is also compounded by competition between jurisdictions to attract business and investment by promising lower operating costs, including workers' compensation premiums. For a detailed review of the concept of work-relatedness as it applies to Australian compensation schemes, see Clayton *et al.* (2002).

The debate about work-relatedness is certainly not resolved. For example, a recent decision by the Australian High Court about a Victorian worker who cut his finger while peeling an apple during a lunch break has the potential to dramatically widen the scope for workers' compensation claims once again (*Australian Financial Review*, 17 February 2003). The worker's employer and the Victorian WorkCover Authority sought leave to appeal the granting of $300 in medical costs to the worker under the scheme. The worker's lawyers argued that the case had implications for workers who are killed or injured at work due to major incidents not directly related to their job; for example, terrorist attacks or mass shootings.

Workers' compensation and injury/illness prevention

The extent to which compulsory workers' compensation premiums can provide a positive incentive for employers to improve OHS performance is often said to depend upon how closely premiums are linked to an organisation's OHS performance. Typically, premiums are set as a proportion of an organisation's wages bill. However, in most schemes, there are mechanisms to adjust premiums based on claims experience. The Industry Commission (1995) identifies four methods of setting workers' compensation premiums used in Australia: class ratings, experience rating, bonus and penalty schemes and upfront discounts.

Class ratings

Using class ratings, premiums are determined according to an industry category. High-risk injuries and occupations have relatively high class rates. The higher the class rate, the higher the premium for an employer of a certain size. In its basic form, employers' own experience does not affect the class rate, and therefore there is little incentive to employers to improve OHS performance in this system because improvements would not be reflected in lower premiums in the following year.

Experience rating

The experience-rating method usually takes a base rate, for example an industry class rate, and adjusts it to reflect an individual employer's recent claims experience. The size of the adjustment is the most important determinant of the incentive effect of this system. More weight is usually given to the claims experience of large compared to small firms therefore the adjustments are more significant and the incentive to improve is greater for large firms. This may limit the extent to which experience rating provides an incentive for many small to medium-sized construction firms to improve their OHS performance.

Bonus and penalty schemes

In some jurisdictions, bonuses and penalties are given to employers based on a comparison between the employer's claims experience and other businesses in its class. The size of the bonus or penalty is determined by the employer's size and maximum percentages of industry rates are usually set.

Upfront discounts

The provision of upfront discounts on employers' workers' compensation premiums has been used as an incentive to employers to implement preventive OHS measures; for example, to establish OHS management systems. These have the advantage that their effect is immediate. However, there is no guarantee that the OHS preventive effort will provide the expected reductions in claims and can present problems for scheme administrators who must then monitor the implementation of the promised preventive strategies.

Research in the USA, where experience-rating has been used for many years, suggests that experience-rating does have a positive reduction effect on the number of workers' compensation claims (Ruser 1991). Bruce and Aitkins (1993) also compared fatality rates in the construction and forestry industries in Ontario, before and after a shift from industry-rating to experience-rating methods of setting workers' compensation premiums. They report that, in both industries, fatalities reduced following the introduction of experience-rating. The Industry Commission (1995) reports that, since experience rating was adopted in New South Wales in 1987, there has been an improvement in the claims performance of large employers relative to small and medium-sized firms.

However, Clayton (2002) cautions that using a reduction in claims as an indicator of improved OHS management and performance is problematic because many work-related injuries and illnesses do not result in compensation claims. Clayton (2002) suggests that experience rating does have a significant behavioural effect but that this effect is in the better management of *claims*, rather than OHS. Linking compensation insurance premiums to recent claims experience is particularly ineffective when considering occupational illnesses, many of which have very long latency periods. This being the case, there is little economic incentive for organisations to take steps to prevent these illnesses.

Occupational rehabilitation

Rehabilitation is closely related to employees' compensation, yet there has been confusion and concern as to how the two interact. For example, Creighton and Stewart (1994) suggest that workers have been concerned that a focus on speedy rehabilitation and return to work will compromise their compensation entitlements. On the other hand, employers have been concerned that workers returning to work too soon could result in further injury and therefore increased costs.

The New South Wales WorkCover Authority defines occupational rehabilitation as:

> the restoration of the injured worker to the fullest physical, psychological, social, vocational and economic usefulness of which they are capable. It is all about getting injured workers back on the job as soon as possible.
>
> (Cited in Quinlan and Bohle 1991, p. 273)

In Australia, all compensation schemes require employers to provide a rehabilitation programme. On 1 July 2002, the law in Victoria changed regarding returning injured workers to work. Part VI of the *Accident Compensation Act* (1985) requires employers to establish and maintain risk management programmes and occupational rehabilitation programmes and prepare individual return-to-work plans for injured workers who have no work capacity for 20 or more calendar days (Victorian WorkCover Authority 2002). The law also requires that, where injured workers have current work capacity but still cannot return to their pre-injury jobs, employers must provide them with suitable employment while they recover from their injury, for up to 12 months following the workers' injury.[1] The suitable employment provisions apply unless an employer can demonstrate that it is not possible to provide suitable employment to returning workers.

The provision of suitable employment plays a critical role in the return-to-work process (Tate 1992). Krause *et al.* (1998) reviewed the return-to-work literature and concluded that the provision of suitable duties doubled the probability that an injured worker returned to work, halved the number of workdays lost due to the injury and yielded cost savings of between 80 and 90 per cent. However, the obligation to provide suitable employment poses a challenge in the construction industry due to the physically demanding nature of much of the work.

Creighton and Stewart (1994) argue that effective rehabilitation is a highly desirable objective. The human costs of work disability include impaired domestic and daily function, strained family relationships, negative psychological and behavioural responses, stress and loss of vocational function (Dembe 2001). Armstrong *et al.*[1] (2000) suggest that many people suffering work disability spiral into economic hardship and as many as 30 per cent live below the poverty line. It is understood that rehabilitation of injured workers back into the workforce prevents the devastating impact of employment loss for individual employees and the consequent loss of security, self-esteem and financial hardship.

However, rehabilitation is a complex process and many variables have an impact upon rehabilitation outcomes (Shaw *et al.* 2003). Some determinants of rehabilitation outcomes relate to the severity of the injury and characteristics of the individual worker (Kenny 1994). However, many determinants of rehabilitation outcomes relate to the organizational environment and management policies and practices (McLellan *et al.* 2001; Franche and Krause 2002). For example, proactive return-to-work programmes have been found to contribute to reduced disability duration, measured in lost workdays (Hunt and Habeck 1993; Amick *et al.* 2000).

James *et al.* (2003) suggest that formal rehabilitation programmes, including the formulation of a rehabilitation policy setting out what can

1 Or alternatively the sum of periods not more than 12 months in aggregate following the worker's injury, during which they have an incapacity to work.

and should be done to support the return to work of injured workers, the appointment of a rehabilitation co-ordinator and the establishment of a procedure for the management of the rehabilitation process, are associated with better rehabilitation outcomes.

Research indicates that employers' responses to workers on their return to work are also important determinants of rehabilitation outcomes. Strunin and Boden (2000) classify employer responses as follows: 'welcome back', 'business as usual' and 'you're out'. In the former response, employers actively encourage workers to return to pre-injury employment and provide accommodations to help this return. In the latter two responses, employers display either a benign neglect of workers needs or find reason to terminate them. Research suggests that where workers are accommodated, for example by being provided with alternate, light or modified work duties, they are significantly more likely to return permanently to work and significantly less likely to experience further periods of absence arising as a result of their disability (Butler *et al.* 1995; Crook *et al.* 1998; Krause *et al.* 1998). Unfortunately, Strunin and Boden (2000) report that employers' responses to work disability are often not positive, with the result that many workers who return to work, do not do so permanently.

Colella (2001) suggests that co-workers' reactions, in particular their perceptions of the fairness of workplace accommodation, may influence return-to-work outcomes. Modifications, such as restructuring work, changing schedules or trading tasks depend on co-workers' support for their implementation. How injured workers and supervisors believe co-workers will respond may also impact upon return-to-work outcomes. In a recent study of construction firms in Victoria, Australia, a large number of respondents expressed the belief that the provision of suitable duties was difficult because it places too much strain on other workers. This suggests that managers' beliefs about the impact of accommodation on co-workers may act as a barrier to the provision of suitable duties. It is critical that the importance of rehabilitation be communicated to everyone in the workplace to overcome such perceptions. The provision of education to all employees on the importance of rehabilitation was also emphasised by Bruyere and Shrey (1991).

Some argue that return-to-work programmes reduce the operating costs of businesses. For example, Ganora and Wright (1987) report that, in a company with 300 employees, the direct cost-to-benefit ratio of an injury management and rehabilitation programme was 1:12.6. Indeed, if implemented on an industry-wide basis, a reduction in the duration (and therefore the cost) of compensation claims would lower the rating that determines workers' compensation premium rates by industry categories. At present, some construction trades, such as roof tiling, have among the highest class ratings in Victoria, amounting to 7 per cent of the wage bill (Industry Commission 1995). At an individual enterprise level, a reduction in the duration of claims could also reduce costs because, in Victoria, the class or industry rate is taken as a base rate and then adjusted with reference to the firm's claims experience over the past three years. However, in a study of rehabilitation practices among construction firms in Victoria, the majority of construction firms indicated an increase, rather than a decrease, in operating costs as a result of rehabilitation requirements and the obligation to provide suitable duties. Either the cost benefits of rehabilitation are not realised or they are insufficient to act as an incentive.

Rehabilitation and effective return-to-work programmes also reduce the duration of compensation claims, a major determinant of compensation costs. Reductions in claims can therefore reduce the costs of government-administered compensation schemes, which are increasingly strained. Purse (2002) estimates that a rigorous policy to enforce statutory requirements for employers to provide suitable employment for injured workers who had recovered sufficiently to safely resume work saved the South Australian workers' compensation scheme between AU$60 and AU$106 million. However, this policy was short-lived. More recently, the South Australian workers' compensation scheme has suffered massive blowouts in its unfunded liability, attributed to the failure to resolve longer-term injury claims and promote effective return-to-work practices (*The Weekend Australian*, 22 March 2003).

In the current climate, both employers and employees remain hostile and suspicious of one another's motives. Kenny (1995) undertook a qualitative study of employers' and employees' experiences of occupational rehabilitation, and reported that employers adopted a 'victim-blaming' mentality, while employees adopted a 'system-blaming' mentality in many occupational rehabilitation cases. Consequently, the employer–employee relationship, which is central to effective rehabilitation, became polarised and quickly degenerated into open hostility, often resulting in court action, termination of the employees' employment, or both.

Workers' subjective expectations about the return-to-work experience have been linked to return-to-work outcomes (Foreman and Murphy 1996). Thus, workers' beliefs about the availability of suitable work, social acceptance of their co-workers and their capacity to retain work fitness are likely to be key determinants of successful return-to-work outcomes. The importance of maintaining communication between employers' representatives and injured workers during the occupational rehabilitation process is critical to making sure positive expectations of the return-to-work experience are fostered.

Trade unions and employees have also expressed concerns that workers' compensation entitlements will be compromised by return-to-work policies. Employers, on the other hand, have concerns that rehabilitation is costly and that workers who return to work too early may exacerbate their injury, resulting in greater cost to the employer.

Rehabilitation in construction

Working in the construction industry has been identified as a risk factor for chronic disability (Cheadle *et al.* 1994; Hogg-Johnson *et al.* 1994; McIntosh *et al.* 2000). There are several possible explanations for this.

First, the work arrangements commonly found in construction, such as short-term and occasional employment, are reported to be associated with longer periods of disability (Infante-Rivard and Lortie 1996). In construction, many workers have no direct long-term relationship with an employer and consequently employers are reported to be unwilling to accommodate workers with a disability (Welch *et al.* 1999).

Second, the majority of construction firms are small businesses with 97 per cent of general construction businesses employing less than 20 employees and 85 per cent employing less than five employees (ABS 1998). Working in a small firm is associated with disabilities of longer duration (Cheadle *et al.* 1994), possibly because small firms are less likely to have the resources to employ rehabilitation specialists or because they do not have the flexibility to offer alternate or light duties to returning workers (Kenny 1999).

Third, the type of injuries commonly experienced in construction, such as falls and manual handling, are associated with long duration disability (Hogg-Johnson *et al.* 1994; Tate *et al.* 1999). Also, the physical workload in construction jobs is likely to be a barrier to return to work and render the provision of modified duties difficult (Kenny 1999; Dasinger *et al.* 2000).

A survey of member companies of the Master Builders' Association of Victoria recently revealed that formal programmes for rehabilitation and return-to-work have not been universally implemented by construction companies in Victoria. Forty per cent of companies had no rehabilitation or return-to-work programmes and more than half of the companies had not appointed a rehabilitation co-ordinator. Even in companies in which a rehabilitation co-ordinator was appointed, more than one-third had issued no duty statement for the rehabilitation co-ordinator, suggesting that the role is not formalised and is performed in an *ad hoc* manner. In many companies, the rehabilitation co-ordinator was a clerk or a secretary and, in the majority of companies, the rehabilitation co-ordinator had not undergone specialised training. Together, these results indicate that, in many cases, the role of rehabilitation co-ordinator is performed by someone who lacks the requisite knowledge, skills or authority to fulfil this role.

The survey also revealed that many construction companies experience difficulty in the provision of suitable duties. The nature of the work itself is an often-cited source of this difficulty. While the provision of suitable duties is undoubtedly challenging in construction, Welch *et al.* (1999) identify a number of reported modifications made to the jobs of injured construction workers to help them return to work. For example, in their study, some workers were provided with temporary light duty assignments, involving no heavy lifting or climbing. Others, with sporadic symptoms, were allowed to change job assignments on days in which they experienced pain (Welch *et al.* 1999). The present lack of rehabilitation co-ordination in Victorian construction firms may mean that strategies for the provision of suitable duties are not adequately considered before it is decided to terminate a worker.

In relative terms, the OHS performance of the construction industry is known to be poor. While most injured workers will not require rehabilitation, those who have suffered a serious injury are likely to find returning to work difficult unless the rehabilitation process is effectively managed. Purse (2002) notes that the dismissal of injured workers is a deeply ingrained feature of the labour market in Australia and the construction industry, in particular, could perform a lot better in the area of occupational rehabilitation. However, some construction employers are taking occupational rehabilitation seriously. Box 2.3 outlines procedures followed at one large construction firm.

Box 2.3 Occupational rehabilitation at Baulderstone Hornibrook

Baulderstone Hornibrook has an occupational rehabilitation policy committing the company to the effective return-to-work of injured workers wherever possible. The implementation of this policy is managed by a Claims Manager whose responsibility it is to ensure that workers suffering a work-related injury or illness receive timely and appropriate medical treatment and ensure that these workers are supported in every way possible.

The Claims Manager is notified immediately in the event of an injury requiring off-site medical treatment. Thus, the occupational rehabilitation plan starts as soon as the injury has occurred. Communication between the Claims Manager, the line manager and the injured worker is recognised to be very important in determining outcomes and this communication is maintained throughout the rehabilitation process. While the Claims Manager recognises that insurance and claims management is essentially reactive, he strongly believes that ensuring injured workers receive excellent medical treatment and are able to get back to work at the earliest opportunity is the best outcome for everyone.

On notification of an injury, the Claims Manager completes the necessary paperwork and lodges a notification of the injury with the insurer. This ensures that medical treatment can occur without delay.

When injuries are severe, and it is clear at the early stages that a worker will not be able to get back to pre-injury duties, the Claims Manager works with the worker's line manager and the company's human resources department to arrange for a permanent long-term modified duties work programme. The Claims Manager emphasises that these modified duties have to be meaningful tasks because 'it is not sensible to give somebody meaningless employment because that has negative psychological impacts'.

The Claims Manager actively works to educate the company's management about return-to-work outcomes. He says that, at first, he experienced some resistance from construction managers. However, with education and provision of advice, he has managed to persuade managers that the company can make light duties and design new jobs where necessary.

The Claims Manager works closely with external service providers, such as occupational therapists, medical professionals and physiotherapists and says that this co-operation helps the return-to-work success rate.

In most cases, workers are keen to get back to work. The Claims Manager takes extra care to ensure that workers do not return to work too early and that they only do so when the conditions of work are suitable for their safe return.

If the worker is incapacitated, modification of the worker's pre-injury job is always the first option. The Claims Manager visits sites to investigate how jobs could be re-designed. He commented 'it is important that we go out on site. Many insurance managers don't get out on site and see the physical work environment. I'm led a lot by workers themselves. They know more about their jobs than I do.' The Claims Manager has to ask a lot of probing questions to identify the options. In some instances, there may be a risk associated with one part of a task. In this situation, the Claims Manager works with the injured worker and his or her line manager to identify a creative solution. One example was the provision of a lightweight aluminium ladder to enable a worker to get up into a mobile crane. Other solutions include the fitting of shock absorbers in equipment and the provision of appropriate footwear to reduce impacts to the knees, the hips and so on. The Claims Manager emphasises the importance of achieving the right fit between the person and his or her work equipment.

Box 2.3 (Continued)

If it is not possible to modify the worker's pre-injury job, another strategy adopted at Baulderstone Hornibrook is to take the injured person away from manual physical work tasks and train him or her for a supervisory role. Many go into OHS roles or re-train, for example learning computer skills. Workers who are re-trained in this way have been deployed in tasks related to contract administration.

The Claims Manager stresses the need to develop trust between himself and the injured workers and discusses options for return-to-work with them in an environment in which they are comfortable, for example in their own homes. The Claims Manager recognises that injured workers are often concerned about the impact of their injuries on family dependants and that it is necessary to manage the whole situation and to create a positive outlook. Failure to create a positive outlook can lead to depression and workers' early withdrawal from the occupational rehabilitation process. For added support, in some cases, injured workers are provided with a mobile phone and told to call the Claims Manager at any time that they have any questions or want to talk about their rehabilitation.

The Claims Manager also monitors the medical treatment that injured workers are receiving and ensures that the workers feel they are getting the support they need from external service providers. Where they feel that this support is lacking, workers are told that they can change service providers if they wish and some have even undergone alternative medical treatments, such as acupuncture, in the course of their rehabilitation. The emphasis is always on what is best for the individual.

When workers do return to work after an injury or illness, every aspect of their situation is considered, including how they get to work. Supervisors and managers, and others at site are made aware of the person's injury and, where relevant, are told that the worker should not be asked or instructed to perform certain tasks. The worker is also educated to refuse tasks that are unsuitable.

In some circumstances, when workers are incapacitated to such an extent that no suitable duties can be provided for them within the company, the Claims Manager seeks to locate these workers with other employers. This job placement is seen as the last resort. However, the Claims Manager identified several successful cases in which tradesmen had been placed with large hardware stores as retail employees.

The Claims Manager highlights the cost-incentives to effect the return-to-work of injured workers. He identifies cost-saving in dealings with the insurer, suggesting that return-to-work successes can significantly lower insurance premiums. Effective claims management can also reduce the administrative costs of work-related injury. For example, the Claims Manager suggests that quick access to medical treatment can reduce the costs of a claim from as much as AU$20,000 to AU$3,000.

While Baulderstone Hornibrook does not implement occupational rehabilitation processes for subcontractors' employees who are injured on their sites, the Claims Manager provides inductions, information and advice regarding occupational rehabilitation and return-to-work practices to subcontractors who seek this help.

Conclusions

This chapter has described the three objectives of OHS law. The shift from prescriptive to principle-based standards in preventive OHS legislation occurred in most Commonwealth countries following the publication of the Robens Report in 1972. There is little evidence that this shift yielded the dramatic improvements in OHS that were hoped for, and some theorists argue that the assumptions upon which the Robens recommendations were based were illogical and flawed. In particular, there is evidence that small businesses struggle to understand their obligations under principle-based OHS legislation. This poses a particular challenge to the construction industry in which the majority of businesses employ less than five people. Despite these criticisms, Robens-inspired legislation led to a significant improvement in mandating the establishment of consultative processes, through which employees and their representatives could play a part in OHS decision-making, and provided mechanisms for the resolution of OHS disputes through elected OHS representatives. More recently, preventive legislation has adopted a more process-based approach, requiring that organisations follow processes to identify and assess OHS risks in their operations and take steps to control these risks. The most wide-reaching process standards adopted in relation to construction operations are contained in the *UK's Construction Design and Management Regulations* 1994. These *Regulations* place statutory responsibilities on construction clients and designers for OHS during the entire life cycle of a structure. In meeting the requirements of this process-based legislation, construction firms must implement OHS management systems, the elements of which are described in Chapter 4, and adopt methods for the systematic identification, assessment and control of OHS risks, such as those described in Chapter 5. Rather than simply following the letter of the law, construction firms must now proactively manage OHS risk in their operations and there is considerable scope for discretion concerning how best to control OHS risks.

There is increasing public pressure to hold individual managers liable for OHS incidents and unions and pro-worker groups advocate the use of the mainstream criminal law. It is difficult to convict a corporation of a serious criminal offence, such as manslaughter; to our knowledge, the only Australian manslaughter conviction of a corporation was that of a construction company in Melbourne. Furthermore, in Anglo-Australian jurisdictions, there have been several attempts at reforming the criminal law to ensure that work-related deaths are dealt with in the same manner as other deaths. Public condemnation of workplace deaths imposes increasing pressure on prosecutors to use the mainstream criminal law where appropriate, and on legislators to reform the basis for attributing corporate criminal liability. These are issues that construction managers should not ignore, since use of the mainstream criminal law is likely to increase in the future.

Workers' compensation was also discussed in this chapter. We defend the use of the common law as a means of providing financial support for victims of occupational illness or injury, and believe that the civil claims process has a powerful corrective justice function. However, common law damages are unpredictable, and total reliance on the common law would mean that those workers who could not prove fault on the balance of probabilities would not receive the financial assistance they need. Thus, statutory no-fault compensation schemes are also essential. A critical feature of these schemes is the definition of work-relatedness and different jurisdictions adopt different approaches to this issue. This is problematic and inequitable.

Workers' compensation legislation also usually contains requirements pertaining to occupational rehabilitation. The effective return-to-work of workers who have been absent from work as a result of injury or illness is an important element of occupational rehabilitation schemes, and the formulation of return-to-work plans for individual workers are often required by law. Return-to-work process must be handled with sensitivity, and is more effective when there is communication and co-operation between external service providers, medical professionals, employer representatives and the employees themselves. Construction firms have typically performed poorly in occupational rehabilitation and return-to-work, and we argue that this is a key area in which the industry needs to improve. Wherever possible, employees should be returned to their pre-injury or illness duties, which may require that the workplace be physically modified. Where this is not possible, alternate or suitable duties should be identified wherever possible. Well-managed return-to-work processes can yield the best outcomes for all parties. It is important that line managers in the construction industry, both at site and office locations, recognise their obligations to workers who have suffered an occupational injury or illness, and work closely with medical service providers and the workers themselves to develop creative return-to-work plans for individual workers.

Discussion and review questions

1. Considering the nature and structure of the construction industry, to what extent is there a need to retain a prescriptive approach to OHS regulation?
2. Identify reasons for and against the use of the mainstream criminal law in punishing organisations and/or managers responsible for workplace deaths or serious injuries. In what circumstances, if any, should companies and/or individuals be charged with the offence of manslaughter following a work-related death?
3. Discuss the relative merits and disadvantages of no-fault schemes and the common law as a means of providing compensation for injured workers.

Chapter 3

Organisational issues

Introduction

Like many other industries, such as aerospace and automotive, the construction industry deals with projects. However, unlike most industries, the production run for the construction industry is often one unit rather than tens, hundreds or thousands. A good parallel is the shipbuilding industry, where a large investment is made in design and yet the product is reproduced, in the case of large vessels, in very limited numbers, with each new vessel different from the first. Consequently, although current management theory is relevant to construction, the industry is unusual due to its limited production runs, and some adaptation of current theory is required.

Organisation theory: Management and organisation

How do organisation and management theory apply to the construction industry? In many ways, construction organisations, whether they are companies or project teams, are very similar to the organic organisations described by Burns and Stalker (1961) and Lansley *et al.* (1974).

As such, they have to react very quickly to a turbulent and changing environment. The way that government has used the construction industry as a regulator of the economy is a prime reason for this need to be able to adapt and move quickly. Hence, although much of the manufacturing industry may be considered somewhat mechanistic in the way that it organises, the construction industry is very much typified by organic types of organisation.

As a consequence of this, the way that the industry is managed must be adapted: managers are expected to make decisions rapidly, based on incomplete information, and based on their own resources, rather than in a highly structured and planned manner. In the same way, construction operatives are often faced each day with new problems that are not planned for in the drawings and contracts under which they are employed. As such, they are expected to use their initiative and to continue working. Productivity is seen as paramount. Thus, the situation that the average construction worker faces is very different from that faced by the average factory worker in a highly structured environment. The construction worker must react rapidly to any change or new challenge, and be able to adapt and devise his or her own solutions to often difficult and far-reaching problems.

Construction workers tend to be given a high degree of licence and a high degree of self-control over their own work. When this situation is combined with the pressure to be productive and to reduce construction project times, there is a tendency to adopt non-standard solutions and to cut corners. Obviously, in such circumstances, OHS is one issue that is often not dealt with adequately by the site worker. Most construction organisations, whether their business occurs predominantly or on-site or in an office environment, are organic and actually encourage free thinking and problem-solving by their employees, and it can be very difficult to implement a structured and well-planned OHS management system in such an environment.

The major impact of the adoption of OHS management systems and performance-based OHS legislation has been in improving performance in the process and manufacturing industries where, largely, the production process is static and well defined. The construction industry is completely the opposite of this. The construction process is dynamic, moves around and across sites, is very unstructured, and is reliant upon the initiative of and opportunity for innovation afforded to the individual worker. As such, one could argue that performance-based legislation and OHS management systems are not so well designed for industries like construction. In fact, the view has been expressed recently that the construction industry would be far better off going back to the old prescriptive legislation, which offered a degree of control and structure on site, rather than moving further down the path of self regulation. Given the comparative weakness of trades unions in many countries, particularly in the construction industry (perhaps due to the high level of labour-only subcontracting), there is no safeguard to ensure that the self-regulatory framework described in Chapter 2 actually delivers. Organisational issues impacting upon the construction industry's ability and willingness to proactively manage OHS are described in the remaining sections of this chapter. This chapter focuses on the ways that the construction industry as a whole is organised, rather than examining the OHS management activities within a single company or construction project, which are described in Chapter 4.

Roles and relationships of the parties

All construction projects consist of a diverse group of stakeholders and participants. To name but a few, there are the clients, the design consultants, the contractor and subcontractors, the suppliers, the cost consultants, the architect and the project managers. These and their interactions have been described in a number of ways, but most of the descriptions focus on the issue of differentiation of organisations. The key issue then, in dealing with the parties to the construction project, is inter-organisational co-operation and the way that different companies act, the nature of the different cultures of both companies and organisations and of the professionals involved in the process. Cherns and Bryant (1984) described the construction project team as a *temporary multi-organisation* (TMO) consisting of organisations with competing and conflicting objectives. Newcombe (in 1990) describes the team as a *Construction Coalition*, in which each of the parties satisfies its needs rather than satisfying its organisational objectives. Given the nature of the organisation, it is not surprising that a number of different objectives surface during the course of the construction project. One of the problems with this is that OHS is often relegated to a low-level objective by many of the participants. As a consequence, the construction industry has a very poor record in terms of OHS performance, and this is undoubtedly due in part to the nature of the construction process, and the formation of these TMOs for each construction project.

Regulatory and legislative initiatives

In an attempt to rationalise this fragmentation in the industry, and to structure the way in which OHS is dealt with, the UK introduced the *Construction Design and Management Regulations* (HSE 1994). Similar regulations were introduced throughout Europe. The aim of these regulations was to focus attention at an early stage on the issues of OHS in use and OHS during construction, not forgetting the health issues in both instances. However, the regulations have been brought into place as a consequence of the nature and structure of the construction industry. The fact that the project team consists of essentially competing parties (in an organisational, economic and financial sense) has led to a lack of focus in terms of OHS issues. Implementation of such regulations was a way of forcing an OHS focus onto the industry.

Along with the regulations concerning design and management, there are also more general health and OHS regulations in many countries dealing with the setting up and maintenance of OHS management systems. Again, this has been enforced by legislation rather than by the natural reaction of construction organisations to put together such a system. This stems from the fragmentation that exists within the industry, in terms of both organisations and professions. It is unfortunate to have to say that, in many cases, architects do not see OHS as having primary importance to their role in the construction process and, often, the architect has no interest in how OHS should be managed throughout the process. As a consequence, many OHS issues that should be dealt with during the design process, relating to both construction and use, only surface during the construction, occupation and maintenance phases of the project. Yet, it is well known that the ability to improve processes is much greater at the early design stage than at the point of implementation. One could say, then, that the industry is characterised by a culture of carelessness or lack of care when it comes to designing OHS into a project.

Many OHS issues are dealt with in the construction contract and in the law of tort. One always has a duty of care to one's neighbour and fellow workers. However, much of the legislation passed has been produced in a prescriptive manner, and has brought about a culture whereby organisations attempt to minimise their input in order to satisfy, marginally, the prescriptive legislation and regulations. With the advent of process-based legislation, the avoidance of OHS matters should be lessened. There is no minimum standard, but there is the duty to act positively in order to invoke and enforce and manage a proper OHS management system. However, in practice, this has not worked very successfully today in many jurisdictions.

One problem that besets many newcomers to the construction industry is the plethora of procurement systems or contract strategies available. Unlike other industries, in construction, design is often separated from construction. There is no integration of the two processes, and there is often no continuity of the membership from one process to the next. Because of this, various contractual arrangements or contract strategies have developed. There has been little research in this area, but it is obvious, from a basic understanding of management theory and the concepts of integration, differentiation and specialisation, that the contract strategy chosen can have a significant effect upon OHS performance in construction projects. The following pages will deal with these issues in detail. The concept of procurement system or contract strategy will be introduced and explained briefly, and then the impact on OHS performance of the various types of contract strategy will be analysed.

Influence of procurement strategy on OHS

A large body of knowledge has been developed on the topic of procurement systems, and much of this is held within the International Building Research Council (CIB) Working Commission W092 on Procurement Systems. Best practice in procurement systems is based upon management principles such as differentiation, specialisation, integration, team working and the establishment of an over-arching set of objectives and management principles. The following section provides an overview of current thinking in procurement systems, and relates the theories embodied in the principles of good practice to the practical issues of incorporating OHS management into the construction process.

This section draws on the work of Rowlinson *et al.* (1999) in reviewing procurement systems. The fact that little has been written about the impact on OHS performance of choice of contract strategy indicates a major oversight in the way the issue has been dealt with thus far.

Procurement systems: An overview

The procurement concept in construction has been defined in many ways. Hibberd (1991) began with the general definition of the term procurement offered by the Oxford English Dictionary: *the act of obtaining by care or effort, acquiring or bringing about*, and then argued that thinking about the concept of procurement can raise awareness of the issues involved both in challenging generally accepted practices and in establishing strategies.

Others have attempted more focused definitions, including this one: 'the acquisition of new buildings, or space within buildings, either by directly buying, renting or leasing from the open market, or by designing and building the facility to meet a specific need' (Mohsini and Davidson 1989).

A meaningful definition for our purposes here is as follows: 'Procurement is a strategy to satisfy client's development and/or operational needs with respect to the provision of constructed facilities for a discrete life-cycle' (Lenard and Mohsini 1997, p. 84).

This sought to emphasise that the procurement strategy must cover all of the processes in which the client has an interest, the whole gestation and life span of the building, from planning, design and construction to use and facility management, including OHS issues associated with those who occupy, use, maintain and ultimately demolish the facility.

However, McDermott and Jaggar (1991) have argued that for some research purposes the usefulness of definitions such as this is limited. For example, as a means of comparing projects or project performance across national boundaries, these definitions are limited to developed market economies. This criticism is supported by Sharif and Morledge (1994), who have drawn attention to the inadequacy of the common classification criteria for procurement systems (for example, *traditional*, *management*, *design* and *build*) in enabling useful global comparisons. Thus, to develop a fully informed discussion of OHS issues, one must take into account the context in which the project is developed.

Even comparisons between developed economies are fraught with difficulties. Latham (1994), in a review of procurement and contractual arrangements in the UK, has noted the difficulty of drawing conclusions from existing studies: 'some international comparisons reflect differences of culture or of domestic legislative structures which cannot easily be transplanted to the UK'. Davenport (1994) has reported that the French do not recognise the British/North American concept of procurement. National differences also exist in understandings of OHS. The way in which OHS is dealt with in different countries is a function not only of legislation, but also of the underlying culture, the attitudes of those involved and, more importantly, the social, economic and political environment. Also, in construction there are many different professions and organisations, and the way in which these come together is markedly different from country to country. Hence, the approach to the management of OHS is very much an international issue, but is also highly dependent on the country in which the construction is taking place. Systems that may work well in, say, the United Kingdom, may not work well at all in Australia or Hong Kong.

For example, behaviour-based safety management systems that use motivational techniques to improve workers' OHS behaviour are reported to work well in the construction industries of Britain and other European countries. (Chapter 8 presents a description of such programmes.) However, a study carried out in Hong Kong revealed that the behaviour-based programmes were ineffective, because even basic OHS infrastructure for provision of adequate equipment, worker training and attention to OHS planning and monitoring were often absent. In this context, the ability to work safely was not within the control of workers, and motivational techniques focusing on individual behaviours were of limited effectiveness.

The extent of union membership and activity is also likely to determine the effectiveness of self-regulatory legislative frameworks for OHS (see also Chapter 2). While trades union membership in construction is traditionally high in countries such as the UK, and particularly Australia, and unions there play a key role in negotiating for improved working conditions and OHS in collective bargaining processes, worker involvement in trade unions is much lower in Hong Kong. There, the proportion of unionised construction workers fell from 15 per cent in 1976 to only 8 per cent in 1986 (Turner *et al.* 1991), and unions are not in a strong position to push for improved OHS standards. The employer–employee consultative processes that are a key part in the Robens-inspired regulatory model are likely to suffer from a serious power imbalance.

Procurement – theoretical foundations

Procurement is a social science, which implies that the disciplines of history, sociology, economics, psychology, law and politics can all contribute to furthering understanding. The same is true of OHS. Rarely, however, do researchers or practitioners make their approach explicit. Green (1994) argues that research has reflected the positivism of functionalist sociology, and has largely ignored the validity of naturalistic inquiry. This implies that the adopted research methodologies tend to establish causal relations from a distance, having assumed that there is an objective reality existing independently of human perception, rather than to engage the researcher in what is being researched.

Green attempts to identify and characterise paradigm shifts in procurement practitioners during the 1980s and early 1990s in the UK. Although very few practitioners consider their operative paradigm, Green maintains that a 'default paradigm of practice' can be identified from behaviour in the field. Green draws upon the work of Morgan (1985), who identified eight different organisational metaphors, amongst which practitioners might recognise their own views of the world. These metaphors liken organisations to:

- machines
- biological organisms
- brains
- competing cultures
- political systems
- psychic prisons
- states of flux and transformation and
- instruments of domination.

An important issue is raised here. The way in which we view our world colours how we actually organise and maintain that world. Much of the work relating to OHS management systems is drawn from the systems theory viewpoint. However, there are many other ways of looking at how OHS can be managed, and a move towards a more individual centred and personal approach may well draw significant dividends. Hence, the metaphors we use to describe OHS management styles are very important in analysing responses to OHS problems.

Much of the work of practitioners and researchers in CIB W92 can be associated with the machine or biological metaphors. For example, the systems and contingency approaches are used extensively (both explicitly and implicitly). Hughes (1990), drawing upon a tradition of using the systems approach to analyse the organisation of construction projects, starts with the premise that buildings are procured through organisational systems. He describes various means for designing flexible procurement systems that are appropriate to each project. Naoum (1990), Naoum and Coles (1991) and Naoum and Mustapha (1994) used systems-based models in comparing the performance of alternative procurement systems. In addition, Carter (1990) outlines the use of data-flow diagrams and activity profiles as a means of improving the management information systems within designers' and contractors' organisations. Such approaches, whilst being useful in terms of illustrating how procurement systems work, are focused on strategic decision-making and do not deal with the implications for construction-site practice, to date at least.

McDermott takes this argument further: 'Many researchers have conducted their work within a socio-technical framework' (Rowlinson and McDermott 1999, p. 7). McDermott and Jaggar (1991), Newcombe (1994), Jennings and

Kenley (1996), and McDermott (1996) have all emphasised the relevance to procurement research of the 1960s work of the Tavistock Institute. Jennings and Kenley (1996), for example, have argued that the dominant functionalist paradigms of the 1980s failed to extend beyond the technical logistics of procurement and into consideration of the social aspects of organising for procurement. There are parallels here for OHS management. Psychological and social aspects of OHS management are very important. In this book these are discussed in more detail in Chapters 5–8, but it is important to note here that the nature of the approach to OHS management is very important in determining how individuals react to their role and position in the OHS management system.

The contingency approach is evident in the work of Swanston (1989) and Singh (1990), amongst others. Many researchers draw upon the models and insights provided by the *Transaction Cost* or *Markets and Hierarchies* approach developed by Coase (1937) and expanded by Williamson (1975). Doree (1991) notes the drift towards the outsourcing of design services by public clients and asks the question: *When should unified governance structures (in-house production) and when should market governance structures (contracting-out) be applied?* He concludes that although the move towards contracting-out can be justified based on design production efficiencies, the equation is not complete if the transaction costs of operating in the market and the opportunity costs (of, for example, increased life-cycle costs) are not considered.

Chau and Walker (1994) investigated the nature of subcontracting in the Hong Kong construction industry and concluded that the decision to subcontract is not random, but is predicated on the attempt to minimise transaction costs. A similar methodological approach is adopted by Alsagoff and McDermott (1994) in investigating the true level of infiltration of relational contracting in the UK construction industry. Cheung (1997) uses a transactional analysis to construct a model for determining the most appropriate form of dispute resolution procedure. Transaction costs can be an important element in terms of OHS management. If work is outsourced, considerable transaction costs can be incurred if contractors do not implement adequate OHS policies and procedures. Indeed, Lingard *et al.* (1998) suggest that the contractor selection decision is critical to minimising costs, including those arising from contractors' and subcontractors' poor OHS performance.

Dahlman categorises transaction costs as *search and information* costs, *bargaining and decision* costs and *policing and enforcement* costs (Dahlman 1979). Transaction costs can also be categorised as either *ex-ante* or *ex-post. Ex-ante* costs include the costs of tendering, negotiating and writing the contract, while *ex-post* costs may be incurred during the execution and policing of the contract, or of resolving disputes arising from the contracted work (Williamson 1975).

Time spent determining the OHS requirements of the project and evaluating contractors' and subcontractors' past OHS records and their ability to deliver adequate OHS performance can minimise the need to closely monitor their OHS performance once they commence work. It can also minimise the costly occurrence of occupational injuries and incidences of ill-health, which can impact upon the principal's business reputation.

OHS is seen as an additional cost by many contracting organisations, and even by the consulting organisations that oversee the contractors. However, it is an essential cost for ensuring a safe and healthy working environment. Given the extent of contracting, the issue of transaction costs is highly relevant to construction sites. Transaction costs have not been thoroughly investigated in relation to the costs of managing OHS when work is outsourced compared to when it is undertaken in-house, and is one area that needs further research.

The BPR (business process re-engineering) paradigm and the so-called new (lean) production philosophies that have penetrated other industries have been investigated in construction. For example, Baxendale *et al.* (1996) investigate the implications of concurrent engineering for roles and relationships within procurement systems, and Lahdenpera (1996) argues for re-engineering procurement in a fundamental way rather than continuing to study incremental and non-fundamental issues.

Green (1997) argues for a soft systems interpretation of BPR for application in construction. BPR, he suggests, as currently practised, is relevant only where the problems are easily identifiable. In complex circumstances, as in construction, these techniques will not lead anywhere.

Business process re-engineering (Mohamed 1997) and lean production (Alarcon 1997) are both the subject of significant research activity elsewhere. BPR is predicated upon the assumption that efficiency and effectiveness are the main objectives of any organisation. This is a difficult argument to counter, but OHS management comes within its remit. If a site is unsafe or unhealthy, then a series of other costs will occur: for example, the cost of rescuing and removing injured workers, and the cost of medical treatment.

However, there are sound counter-arguments to this viewpoint and these are dealt with in detail at the end of this chapter. With this in mind, a view is presented as to how BPR could have positive impacts on OHS but the alternative view and evidence cannot be ignored. Management theory is, after all, theory and how theories are implemented, often in part rather than in whole, is a key determinant of success. The concept behind BPR is to strip the process and organisation down to its essentials, review and re-model how the process takes place and rebuild a new organisation based on a rational process that allows for greater efficiency and effectiveness. BPR deals with a whole range of management issues, and has as its objectives improved quality, reduced costs and more timely performance.

Business process re-engineering may have a role to play in defining where OHS management systems fit within the whole organisation, and within a particular construction site. In fact, BPR could provide the mechanism by which the whole of the construction process becomes more holistic, and OHS becomes an integral part of the whole system.

Once the whole system and process are analysed, then new objectives, or objectives that have been previously overlooked, can be incorporated into the system. The most obvious objective to be included in such a BPR process is to make OHS – in fact, *zero accidents* – a prime objective. By dismantling and rebuilding the whole organisation, and remodelling the process, the OHS management system can be built into the project as a whole, rather than as an add-on, as is often the case at present. Hence, if one is a director of a construction company or a client body who aims to improve construction OHS, and is determined to improve efficiency through BPR, then there is great scope and opportunity for major OHS improvements. Examples of organisations that have achieved such improvements are commonly seen only in large organisations: take for example the petrochemical industry, or the gas production industry, or the British Airports Authority. All these organisations have built the objective of zero accidents into their business processes.

Jennings and Kenley (1996) also emphasise the importance of recognising the theoretical underpinnings of procurement research. They argue that procurement systems go beyond technical logistics, and it is the perception and response to project objectives by organisations that is a key determinant of procurement system suitability. Liu (1994) takes this further, and discusses a cognitive model of the procurement system in which the goal–performance relationship is paramount.

Kumaraswamy (1994) discusses the appropriateness of developed countries' procurement systems when applied to less-developed countries, and argues that a sustainable and synergistic procurement strategy must be developed in such situations. The power paradigm (Newcombe 1994) suggests that selection criteria for procurement systems are less important than the realisation that 'procurement paths create power structures which dramatically affect the ultimate success of the project'. Using this paradigm, Newcombe criticises the fragmentation and friction evident in the traditional system. This is further supported by Walker (1994b) who draws the conclusion that project construction speed is strongly determined by how well clients relate to the project team.

Each of these findings has an implication for OHS. For example, if the largest share of responsibility for OHS during construction continues to fall upon contractors, and the power structure inherent in competitive tendering continues to force down prices alongside the imposition of ever-decreasing construction times, OHS will continue to suffer. This is a central theme in the 'industry reports' of Egan in the UK and Tang in Hong Kong; the lowest price is not the cheapest (in all sorts of ways). Integration of OHS objectives into the procurement process is the key issue.

The importance of conceptualising the procurement problem is that lessons can be learnt in one context and transferred to another. While practitioners clearly need to focus on solutions, the role of the research community is to provide the conceptualisation and theory that will lead to the development of best practice, and to establish and satisfy appropriate project objectives, including OHS.

Definitions

One of the problems that have beset this particular area of construction research is a lack of clear definitions for terms such as *procurement systems*, *contract strategy* or *design-build*. The definitions used in this book are briefly discussed below; a more detailed discussion of the definition of *procurement systems* is found in Chapter 12 of Rowlinson and McDermott (1999).

The first issue to clearly distinguish is what is meant by *construction*. In this book, we are looking at the whole project life cycle, from initial inception to realisation and use; this accords with Walker's (1996) systems view of the process. First, for reasons of simplicity of presentation, the project process is divided into three distinct processes: *design*, *construction* and *use*.

Within the concept of *design*, there is the whole range of planning, funding, structural and architectural design and documentation – in short, all of those activities that are necessary in order to be able to break ground on a new site.

The *construction* process is seen as involving all of those activities, be they technical, managerial or strategic, which make up the realisation phase of the project, where the physical facility actually appears.

On completion of this phase, the facility is actually *used*, and this is an important part of the whole process; the *use* phase of a project has a major impact on the client's perception of whether the process has been successful or not.

Each of these phases poses its own OHS problems; additionally, these problems are compounded and passed from one phase to the next. Thus, problems in design influence OHS during construction, which, in turn, lead to post-occupancy health and safety issues and OHS issues in the maintenance and demolition of the facility.

The perspective adopted in this discussion is multi-faceted, in that a client's view of the process is presented, as well as a contractor's view and a workers' perspective (including occupiers and users), particularly with reference to such issues as contract strategy and OHS management and concepts such as partnering. Thus, the overall perspective adopted is one of simplification of the phases of the construction process, and the polarisation of the procurement system into two actors: the *client* and the *construction industry*, those who make the building grow.

The type of construction referred to is generic, in the sense that we are dealing with all types of projects in the building, civil engineering and process industries. However, it has to be noted here that many of the examples used come from the building industry, and so the insights offered are particularly focused on this sector.

Contract strategy: Procurement systems

Procurement is about the acquisition of project resources for the realisation of a constructed facility. This is illustrated conceptually in Figure 3.1, which was produced by the International Labour Office (Austen and Neale 1984). The figure clearly illustrates the construction project as the focal point at which a whole series of resources coalesce. Central to this model is the client's own resources that are supplemented by the construction industry participants, that is the consultants and the contractors along with the suppliers and subcontractors. The model clearly illustrates the need for the acquisition of resources in order to realise the project. This acquisition of resources is part (and *only* a part) of the procurement system. This part of the system can be referred to as the contract strategy – that is, the process of combining these necessary resources together. The contract strategy is not the procurement system but only a part of it; the rationale behind this definition is that the procurement system involves other features, such as culture, management, economics, environment and political issues.

Conventionally, contract strategies have been described as, for example, the *traditional approach*, *construction management* or *build-operate-transfer*. However, writers such as Ireland (1984) and Walker (1994a) indicate that

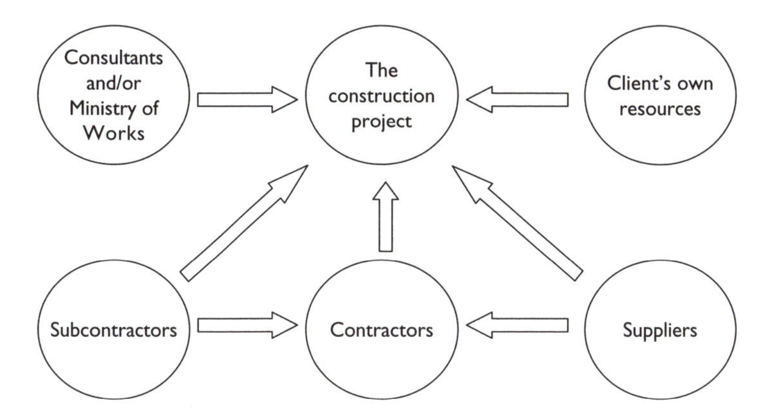

Figure 3.1 Procurement (after Austen and Neale 1984).

there is little difference between many of these supposedly alternative strategies. This situation has occurred due to reluctance by many writers to clearly delineate the variables that make up a contract strategy. The list in Figure 3.2 indicates an increasing integration of design and construction expertise within an organisation as one works down the list. However, to assume that these labels will uniquely define a contract strategy is a false supposition. What is needed is a set of key variables that can uniquely define a contract strategy, rather than the arbitrary list of definitions given in Figure 3.2.

Why has this range of contract strategies developed within the construction industry? One of the reasons is illustrated in Figure 3.3. The logic behind the model is that each participant wishes to obtain as much financial advantage as possible from the process. Because the process takes place in a competitive market, there is a cycle of fluctuating pressure on prices at all times, and commensurate downgrading in the priority of OHS issues. A downward pressure thus forces the contracting participant to look to alternative means to recoup its profit. This leads to claims-conscious behav-

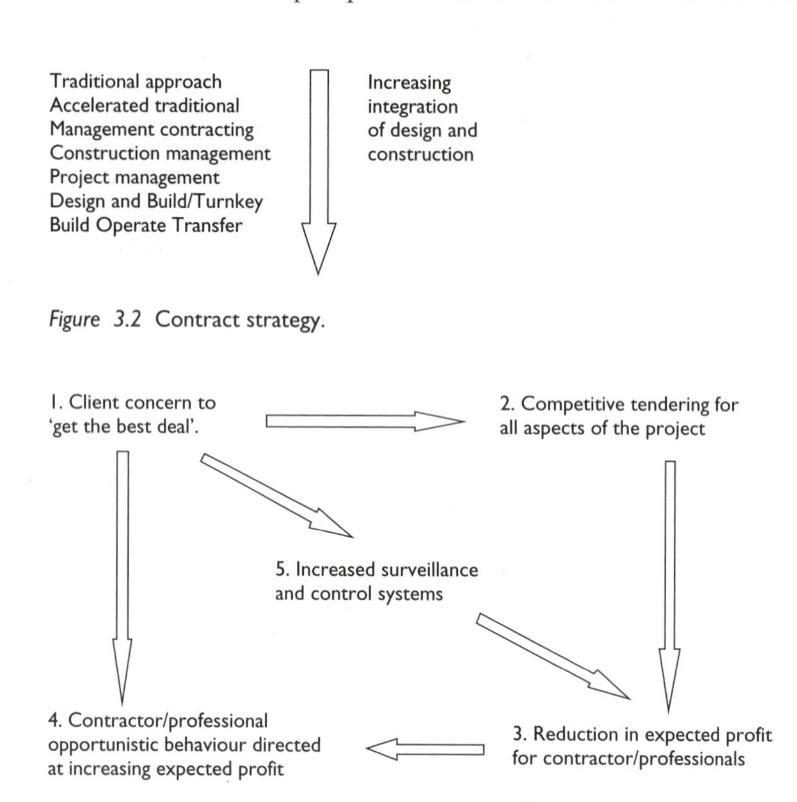

Figure 3.2 Contract strategy.

Figure 3.3 Vicious circles in construction procurement (adapted from Curtis *et al.* 1991; reproduced by kind permission of CIRIA).

iour, and can also stimulate reductions in quality, functionality and OHS performance. Consequently, the client and its advisers are forced to exert greater surveillance over the contractor in order to minimise the effects of this behaviour. As shown in the Figure 3.3, this results in a vicious cycle of negative behaviours. The result may be poor OHS performance, 'claimsmanship' and poor quality.

One way of avoiding this model, which is based upon the traditional contract strategy, is to adopt alternative contract strategies. Other forces are at work in bringing new contract strategies to the marketplace. The competitive nature of the market forces organisations to innovate if they wish to grow and secure market share. Hence, as the construction industry typically has been a conservative and somewhat traditional industry, there has been, in the past, ample scope for the introduction of new and innovative strategies. However, until the late 1960s such attempts at innovation were quite rare, especially in the UK construction industry. With the advent of more experienced and sophisticated clients, there has been an opportunity for industry leading-contractors to explore new routes. Also, as buildings have become technically more complex, and clients managerially more sophisticated, there has been an increasing recognition that the conventional (traditional) approach to procurement is inadequate. In recent years, modern management concepts, such as BPR and partnering, have taken root in client organisations, and the construction industry has experienced these concepts both at second-hand, when working with a major client, and first-hand in the rush to reorganise in the face of declining markets. Consequently, factors have combined to force the construction industry into the position where it has to change to survive.

Much of the literature in this area uses terminology such as the *traditional approach*, *design and build*, *build-operate-transfer*, *management contracting*, etc. In order to clearly define the types of arrangements in which the OHS issues and strategies described in this book arise and are implemented, a generic taxonomy of organisational forms is given below. The function of this taxonomy is to provide a clear and simple description of construction project organisation forms which, when taken with other contract strategy variables, uniquely define a strategy that is further clarified when put in the context of the overall procurement system.

Traditional

The traditional (or conventional) approach is shown diagrammatically in Figure 3.4, and its key characteristics are the separation of the design and construction processes, the lack of integration across this boundary and the employment of a whole series of separate consultants to design the project, with an independent contractor to take charge of the construction process. Typically, the project team is led by an architect charged with the responsibility

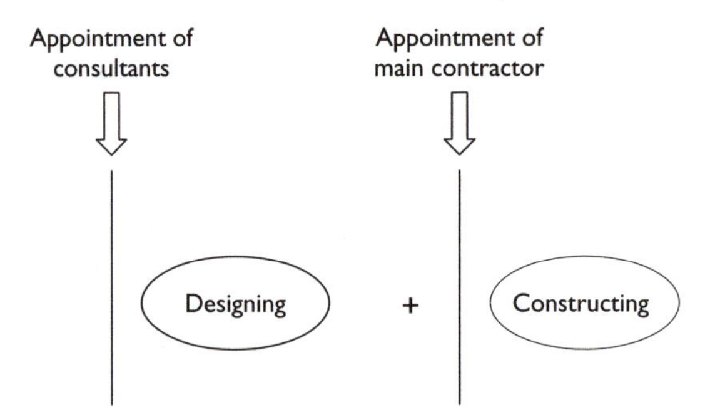

Figure 3.4 The traditional approach (after Austen and Neale 1984).

for both designing and project managing the project. Other consultants (such as structural engineers and quantity surveyors) will join the design and administration team through the life of the project, and the contractor will be selected by competitive tendering on a fixed price bid. The contractor's input to the design process will be minimal, often nil, and in many countries most of the production work on site will be subcontracted to other organisations. The design and construction processes and their sub-tasks are seen as sequential and independent.

Important characteristics of the traditional approach are the level of differentiation among process participants, and the fragmentation in the tasks accomplished in order to complete the project. What effect does this have on the OHS management system implemented on the project? The obvious answer is that the high level of differentiation and specialisation leads to a situation where OHS are not considered during the early phases of the project. As a consequence, OHS are not built into the design process, nor are OHS considered in the in-use phase of the building or structure. Thus, it would seem, from a purely theoretical viewpoint, that the traditional process will be very weak in providing for OHS, both in-use and during the construction phase.

Typically, the designers of a project have very little influence on the construction process, methods and materials used. The contractor is asked to tender during a very short period just before construction starts. The tender process is one in which a large amount of data have to be assimilated, sorted, analysed and costed in order to prepare a bid for the project. During this process, it is inevitable that shortcuts will be taken, and that certain issues will receive less attention than others. Traditionally in the construction industry, OHS has been the issue that has not been addressed at this stage. Hence, when construction actually starts, there is a need to consider the OHS plan for the project in a very rushed and hurried manner. Many projects in Hong Kong are prime examples of this approach, whereby the tendering period has been so compressed that when work starts on site, there is neither a method statement nor has any type of risk assessment been undertaken.

In the traditional process, there is a heavier reliance on legislation and contract documentation to ensure that the risk assessment actually takes place. Inevitably, what happens is that generic risk assessments are produced and presented to the client or the client's representative, and no very detailed analysis takes place based on construction methods or the risk in context. The problem stems from the divorce between design and construction in this particular approach, and, perhaps more importantly, from the hurried, highly competitive environment in which the tendering process takes place. There is a tendency to ignore OHS, or to put in a simple prime-cost item for OHS that in most instances is inadequate. Given the nature of the competitive process, and the fact that the lowest bid usually wins, most contractors take a rather cynical view and determine that expenditure on OHS can actually lose them the contract. Hence, unless clients take a very positive attitude and insist that OHS be included in the documentation and costed in a bid (and costed realistically), it is almost inevitable that the traditional process will deliver a project in which OHS have not been thought through fully. Such attitudes instigated the passing of the CDM Regulations in the UK, legislation requiring for pre-construction, on-paper consideration of OHS by clients, designers and contractors.

Design-build

The design-build approach is a unitary approach characterised by single-point responsibility offered to the client by the contractor, and the opportunity for overlapping the design and construction phases. As with the traditional approach, there are many variants on the basic theme of design and build. Of particular interest are the variants which include project financing, and which go under the headings *build-operate-transfer*, *build-own-operate-transfer*, *build-own-manage* and the like. The organisation of a design-and-build project is more complex than the traditional project at the tender stage, as differently priced bids with different design solutions are often competing for the same project. The adjudication of such bids is a complex process, and requires, to be fair to all bidders, some assessment scheme in place before bids are submitted.

Design-build projects go under many names: for example *design-build*, *design and build*, *design manage construct*, *design and manage*, *build operate transfer* (BOT), *build own operate transfer* (BOOT), *build own operate* (BOO), *turnkey*, etc. The underlying principles of all of these systems are that the client body contracts with one organisation for the whole of the design and construction process, the overlapping of the design and construction phases, and the concept of single-point responsibility. Hence, design-build can be considered as one of the three generic types of organisations in contract strategy, and each form of design-build can be uniquely defined by addressing the status of the other contract strategy variables, such as leadership, selection process, payment systems and so on.

The use of design-build methods opens up the process of competition in more than price alone. The form of the finished project is also a part of the competition, and therefore the process is not only more open but more costly for the bidders. In such a situation, it is necessary that the bidders be given some guarantee that the competition they are entering will be both fair and limited to qualified bidders. Pre-qualification of bidders is a virtual necessity for design-build projects (Figure 3.5).

Design-build organisations can be categorised into three main forms. This categorisation is based on the differentiation which each of these forms exhibits in terms of time, space and profession.

The pure design-build organisation strives for a complete and self-contained construction system. All necessary design and construction expertise resides within one organisation, and this is sufficient to complete any task that arises. This organisation must specialise in a particular market sector due to the complexity of today's building projects. All aspects of design and construction can be highly integrated, and much of the experience gained in design and construction is fed back into the organisation. Thus, the potential for organisational learning in the design-build organisation is far greater than with other procurement systems.

There is a tendency, because of the need for market specialisation, for pure design-build organisations to stay within the medium-size range of contractors. Additionally, pure design-build organisations are unlikely to develop in countries with small construction markets; currently the United States has a large number of recently established design-build organisations that can take advantage of the size and scope of the US market.

The second form of design-build organisation is the integrated design-builder. Such an organisation takes a less holistic approach to the design and construction team, and buys in design or construction expertise

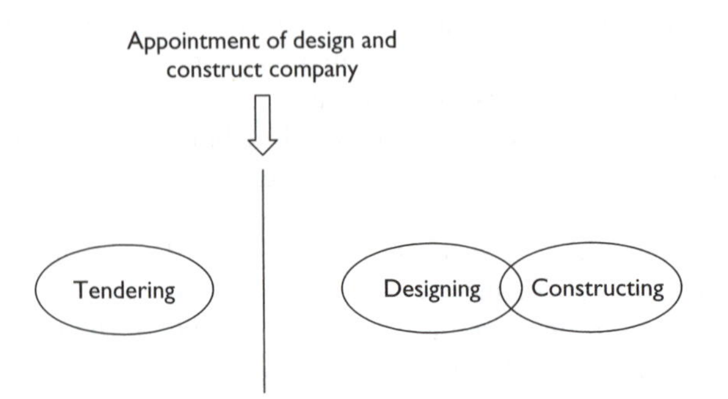

Figure 3.5 The design-and-construct approach (after Austen and Neale 1984).

whenever necessary. This may take the form of architectural or other consultancy services, but a core of designers, engineers and project managers exist who are experienced in their own speciality and the workings of the organisation. These permanent staff members provide the linking pin between the internal and external organisations, and exert an integrative influence on the team. The design and construction teams may be separate organisations within a business group, and this group may participate in the whole range of contract strategies. This more general approach to design-and-build tends to be a development from a general contracting background, and these organisations tend to be larger, more mature companies seeking particular market niches.

Finally, the simplest way for a construction organisation to enter the design-build market is to operate in a fragmented design-build mode. To perform in such a way, the design group can be quite small, perhaps consisting solely of project managers whose task is to liaise with the client and appoint consultants to develop designs. Major companies have the ability to expand such units quite rapidly if required but, as with the traditional contract strategy, a major effort is required to integrate the work of the various consultants. This type of organisation is characterised by a lack of sense of identity and absence of feedback loops between the design and construction processes. The integration and co-ordination problems inherent in the traditional approach are likely to manifest themselves along with role ambiguity amongst the professions, as they attempt to come to terms with working for a construction organisation acting as design-team leader. Such fragmented design-build organisations have the capacity to take on large projects, and such an approach is regularly used with BOT projects. The design-build forms are shown in Figure 3.6, and a more detailed discussion can be found in Rowlinson (1987).

A fundamental concept in the design-build system is single-point responsibility. Theoretically, given this situation, the management of OHS throughout the whole design, construction and in-use process should be more readily possible than with other contract strategies. If we take into account the BOOT and turnkey projects, then design-build offers the unique possibility of building OHS into the whole of the facility management process. In order for a design-and-build organisation to function effectively, it must display the attributes of high levels of integration and co-ordination. The design and construction organisation, and in the case of BOT projects, the whole life cycle of the facility is in the hands of one integrated organisation. As such, this type of organisation has been identified as one in which transaction costs are minimised, efficiency is paramount, and the organisation can, in general, provide a highly buildable and efficient solution to many design, construction and in-use projects.

Thus, from a theoretical standpoint, a design-build approach should be ideal for building OHS into the whole of the design and construction process.

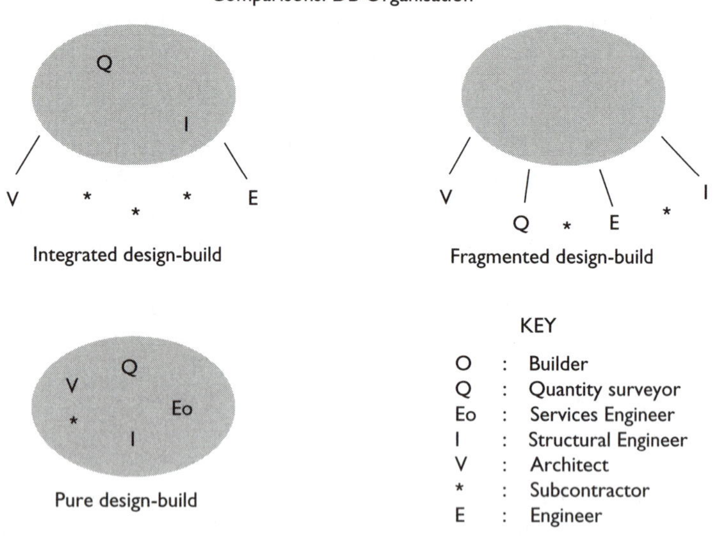

Figure 3.6 The design-build organisation.

However, little research has been undertaken in this area, so there is no strong evidence to show that design-build organisations actually perform better in OHS on construction sites. However, the Building Research Establishment in the UK has strong links with a group of design-build contractors within the UK, and they claim that design-build produces safer construction sites. This is anecdotal evidence, but there is a fundamental theoretical basis for assuming that design-build will produce a safer facility and a safer construction process. Fundamentally, the design-build organisation looks at the whole process in a holistic manner. This incorporates a value-engineering approach into design-build projects and offers the opportunity to build in risk analysis and risk reduction planning at an early stage of design.

Indeed, it would appear almost inevitable that any construction process that adopts a value-engineering approach should be able to build in a much safer method of construction, and to produce a much safer facility with respect to maintenance. However, to achieve this it is important that OHS be incorporated into the concept of value and that *value* be understood not just in terms of cost reduction and functionality.

If OHS is properly integrated into the value-engineering process, the only factor that militates against improved OHS is the tendering process. As indicated previously, this can be a very difficult and tedious process in the design-build field, as the ability to compare dissimilar designs at different prices is a serious problem. In this instance, the design process and tendering process are combined, and may be part of a very hectic schedule. This can impede, at least in outline design, the incorporation of OHS considerations. Carefully considered time and organisational perspectives are necessary in order to achieve the theoretically possible holistic approach that design-build offers.

One cannot say definitively that the design-build approach will always produce safer construction, but one can expect that to be the case. There are many types and forms of design-build organisation (see Rowlinson 1987) and the integrated design-build organisation offers the OHS advantages mentioned above, while the fragmented traditional organisation offers few of these advantages, and may actually *stop* the incorporation of good OHS practice in the overall design and construction through the conflict of vested interests.

Divided-contract approach

The divided-contract approach is illustrated diagrammatically in Figure 3.7. The key principle in this form is the separation of the managing and operating systems. It can be clearly seen that the project organisation is overarched by a managing system. This managing system is generally provided by a management contractor, a construction manager or a project manager. The tasks of design and construction are undertaken by separate organisations that specialise in the technical aspects of the process, and their inputs are integrated and co-ordinated by the management organisation. The high degree of specialisation allows for the fast-tracking of the project, which is the fundamental characteristic of this type of organisation form.

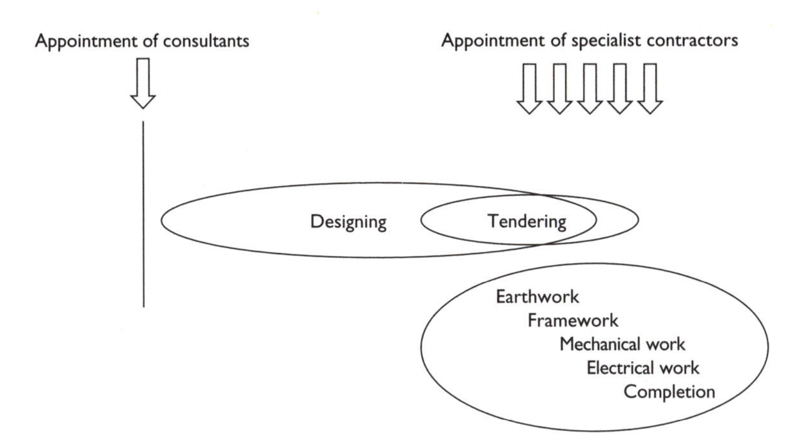

Figure 3.7 The divided-contract approach (after Austen and Neale 1984).

The nature of this type of organisation form, and the underlying goal of providing a specialist management role, requires that the managing organisation be appointed at the outset of the project, and that the role of the contractor be one of consultant builder rather than constructor. This change in role leads to a reshaping of the roles that all of the other professionals play, and there is scope for role ambiguity and conflict to arise. Hence, it is essential that roles and reporting relationships are clearly defined at the outset, and that those consultants providing technical services are fully aware of the limits of their authority and responsibility. The management role is increasingly being undertaken by management consultants, and this is a threat to the traditional construction industry participants, who may find themselves distanced from their client base if this trend continues.

In essence, these representations are idealisations of a complex TMO, and each project will be organised in a different manner from the next. The nature of the payment system, legal documents and selection system will define more clearly the roles of each participant and the exact nature of the contract strategy. In addition, the personalities involved, particularly of the different organisation leaders and their power bases, will have a significant effect on how the project organisation acts in practice (see, for example, Walker and Newcombe 1998).

Two of the main characteristics of the divided-contract approach are

1 the employment of multiple specialists during both the design and construction process; and
2 the overlapping of project phases to a significant extent.

The former should give the opportunity for improvement in OHS, both in construction and in use, by the early involvement of OHS specialists in the team. The latter, however, is likely to present problems because the project is fast-tracked and some OHS issues may be omitted from consideration.

In general, however, the divided-contract approach adopts a value-engineering stance, as all elements in the project are subject to review and critique, with the aim of improving the overall solution. This should give ample scope for consideration of OHS issues throughout the whole of the process, and should allow for better incorporation of OHS provisions and a more thorough risk assessment. However, one must bear in mind the countervailing situation, since the project process is very complex and iterative, and many changes are likely to take place when this contract strategy is used. Also, the constant change may lead to a situation where OHS issues that had been considered and dealt with arise again in another form.

For instance, if we 'value engineer' a steel beam, which can be lifted into place by a crane, onto prepared bearings, from ground level and determine that a reinforced concrete beam cast in-situ is a better alternative due to steel supply and transportation problems, we then have to provide falsework, formwork, working platforms, access ladders, fall protection and a host of other precautions that were not required for the steel beam.

An iterative approach to OHS management must be adopted when the divided-contract approach is in use. Indeed, the divided-contract approach is characterised by a series of organisations working together in this TMO. Although this can lead to conflict, the nature of the conflict can be constructive, in that all of the contributing organisations will have a different perspective on OHS. Hence, it is possible that better and more innovative solutions to OHS issues would be developed due to the synergistic nature of the divided-contract approach.

Another characteristic of the divided-contract approach is the intimate involvement of the client in all of the processes. If the client has a strong desire and need to produce a safe and healthy project, they will drive the project team to deal with these issues in a substantive manner. This can be a very powerful method for improving OHS in construction projects. However, the negative element in this is that the teams put together in the divided-contract approach are temporary and will disband and move on, and their commitment and ongoing organisational learning will be limited compared to, say, a design-build organisation. Hence, mechanisms for storing and saving the lessons learned and the solutions developed must be devised for such organisations and for such client types. The issue of organisational learning is paramount in both the management of OHS and in the production of effective and sensible risk assessments. This issue is discussed in Chapter 10 and an example is cited by Walker and Hampson (2003) in relation to Australian National Museum project. Business strategies, such as those embodied in relational contracting, also assist in this process.

The use of trades contractors in this approach allows the fast-tracking of the project, since works packages are being let on a continuous, rolling programme as the design is developed. Although this leads to fast construction, a price has to be paid for this in terms of abortive work and re-work. For example, foundations may well be over-designed to enable a rapid start on site (see Moss and Rowlinson 1996 for an example). This strategy does encourage the use of value-management techniques to improve design solutions, and on large projects, the opportunity to build a large integrated team at the site exists.

The use of works packages also enables competitive tendering to take place for all elements of the construction process, so that, although an extra management fee has to be paid, a level of competition is maintained. Given the fast-track characteristics of this strategy, and the extra costs associated with an overarching layer of management, this approach has, conventionally, not been recommended for non-complex projects.

The use of multiple layers of subcontractors has been highlighted in a number of industry reports as a main cause of problems in both control of quality and control of OHS. The inevitable consequence of subcontracting is an expanded chain of command, and the probability that issues will be neglected or forgotten during the process of production. The issues that are most often neglected are those of the OHS management system. One of the problems with the continued use of labour-only subcontracting is the lack of training and education provided to operatives. The fact that they, and their gangers and foremen, have little knowledge of OHS management and OHS management systems leads to a situation where productivity is paramount and OHS is often neglected. The multi-layered subcontracting system leads to the situation where control by the main contractor is very difficult and, when one is working on a large, multi-storey building, the ability to control each individual operative is strictly limited. Hence, without the ability to observe continuously operations that are taking place, there is a strong possibility that dangerous occurrences and incidents will occur on a regular basis. If the workers involved were full-time employees of the main contractor then a high degree of control could be implemented in training, education and observation on site. However, the nature of the multi-layered subcontracting system is that little, if anything, is spent on training and education and the discipline of the subcontracted worker is very poor. Often the workers arrive on site, leave and return at odd times. Many operations undertaken using subcontractors are typified by a lack of control and discipline. For an OHS management system to work effectively, control and discipline are essential ingredients. For more on the role and use of subcontractors, and their effect on OHS, see Chapter 4.

Contract strategy variables

It should now be clear from the foregoing discussion that the organisation form adopted cannot uniquely define the contract strategy being used. This takes us back to Ireland's concept of 'virtually meaningless distinctions between nominally different procurement forms' (Ireland 1984). In order to more clearly define contract strategy, a minimum of seven separate variables must be considered:

1 organisation form
2 payment methods
3 overlap of project phases
4 tendering and selection process
5 source of project finance
6 contract documents and
7 leadership.

These are described below.

Organisation form

Organisation form has been discussed in detail above. It defines the responsibilities of each of the disciplines in the project life cycle: whether they are directly responsible for making decisions or only expected to give advice, and at what stage in the project life cycle they should be involved. These generic models allow only a partial evaluation of the construction process, but do give a starting point for further definition.

The effect of organisation form on construction OHS has been discussed in the previous section. It is obvious, when applying the principles of management theory to the organisation forms, that they each have characteristics of differentiation, specialisation and integration, and co-ordination that will determine how well-suited they are to incorporating OHS considerations into the earliest phase of the project process. It is undoubtedly essential that consideration of OHS issues takes place at the outset of a project rather than at the stage at which the contractor has tendered.

Payment methods

Payment methods, particularly for contractors, are either cost-based or price-based. The latter places more financial risk on the contractor, whilst the former places more financial risk on the client. Veld and Peeters (1989), in their paper 'Keeping large projects under control: the importance of contract type selection', discuss seven different payment methods and indicate the advantages and disadvantages of each. These, when combined with organisation form, give a reasonable feel for the risks involved in a particular contract strategy. Few standard contracts contain direct incentive provisions for the contractor to complete the contract earlier or for a cheaper price. If completion of sections of the work and overall completion are critical, a payment system linked to milestones within the agreed construction programme can be used. Cost reimbursement and target contracts can also provide incentives to the contractor for reducing costs.

The way in which the contractor, and indeed the consultant, is paid during the project has a serious impact on the way OHS is treated. Take, for example, the traditional construction process, whereby a contractor is selected as the lowest bidder under a competitive tendering system. There is no incentive in the system to incorporate adequate resources for OHS as this will put the bidder at a disadvantage in the tender adjudication process. Although this attitude cannot be condoned in an ethical sense, it is in fact a reality of the commercial business world. Another example might be the adoption of cost-based systems whereby the contractor is reimbursed the full cost of the construction work plus a percentage for overhead and profits. In this approach, particularly where an open-book approach to costs is used, there is an incentive for the contractor to perform well in regard to OHS without any risk of being underbid by other contractors less conscientious in this area. In an attempt to overcome the limitations of the lowest bid method, the Hong Kong Government Works Bureau instigated a 'pay for OHS' scheme whereby it attempted to take costs associated with OHS systems out of the bidding process. Although this was a novel and well-meaning attempt to improve the situation, there is no strong evidence to show that the system has actually worked, and there is anecdotal evidence to indicate that the system has been open to abuse.

Overlap of project phases

Overlap of project phases determines the degree of acceleration or fast-tracking within the construction process. In most instances, the traditional process is difficult to fast track and little overlap occurs with this organisation form, but varying degrees of overlap occur in both the design-build and the divided-contract approaches. The decision to overlap project phases also affects the selection process (for both contractors and consultants). Conventional wisdom has it that the divided-contract approaches can achieve the maximum overlap, with design-build also being able to achieve relatively high degrees of overlap and, consequently, fast project times.

The more project phases overlap, the more difficult and complex management of the process becomes. This has an obvious implication for OHS management, in that the participating contractors and organisations will often be under quite serious time-pressure in order to deal not just with the progress of the works, but also the changes that inevitably come about – for example, whilst using the divided-contract type approach. Hence, the successful management of OHS in such a situation requires a well-developed administrative and management system to ensure that changes are logged, and their implications for method statements and risk assessments are noted and acted upon. It is often said that time is the enemy of OHS, and the work of Peckitt *et al.* (2002) in the Caribbean indicates that the slow pace of work and life in the Caribbean actually helps in improving OHS performance.

Tendering and selection process

The way in which contractors' pricing methods often fail to account for OHS requirements, and it is common for the unit-rate estimated for an activity to ignore safety issues (Brook 1993). Despite the importance of providing resources for OHS, research suggests that estimators have little or no involvement in pre-construction OHS planning and OHS advisers are similarly excluded from the tendering process (Brown 1996). Research suggests, however, that the inclusion of safety costs in a tender can reduce the lost-time accident frequency rate from a range of 2.5–6.0 per 100,000 man hours worked to a range of 0.2–1.0 per 100,000 man hours worked (King and Hudson 1985). To ensure that tenderers price OHS appropriately, and to facilitate fair comparisons, construction clients inviting tenders could specify the way in which prospective contractors should allocate OHS costs in their bids. The Hong Kong Government Works Bureau 'Pay for Safety' scheme being one example of this.

Contractors may be selected by *open competition, select competition* among a limited number of pre-qualified contractors or by *negotiation* in one or more stages. Similar methods may be adopted for the selection of consultants. This variable is linked to both payment methods and the overlap of project phases, in that open competition is normally used with price-based bids, and projects that need to be fast-tracked might use select competition. Any of these processes can be used with any of the organisation forms, but conventionally a set pattern has been implemented, with the traditional system adopting open or select competition, and design-build moving towards select competition or negotiation in the selection process. There is no necessity to believe that conventional wisdom is correct.

The nature of the selection process has implications for OHS management. For example, in an open competitive tendering system, the considerations in terms of OHS are quite limited. However, when a contractor is involved in any negotiation, or even a select-tender process, the playing field is rather more level. As such, the worries of being undercut by an unqualified and 'dangerous' competitor are much reduced. In situations where negotiation takes place, the contractor can clearly make the client know of the OHS costs and needs, and have these adequately budgeted.

Source of project finance

The source of project finance can have a significant impact on the contract strategy and procurement system chosen. If the client body provides finance, then it essentially has a free hand in the choice of strategy, but if third parties or the contractor organisation provide part or all of the finance, then strings will be attached. With aid agencies operating in non-industrialised countries, there is a tendency to specify use of funding-country contractors and products, and this can have a negative effect upon the local construction industry (Kumaraswamy and Dissanayaka 1997). However, these arrangements also have the potential to allow important OHS knowledge and technology transfer to these countries. If funding-country contractors are to be specified, it is important that these requirements be written into financing agreements: it is all too common for contractors to move into a new location and adopt lower OHS standards than they do in their home country, often citing local practice, inadequate training and competitive bidding as excuses for an inexcusable practice. With contractor finance there is a tendency for these to be provided only when a franchise or other similar BOT-type agreement is incorporated into the project.

The source of project finance has many implications for the construction project. One of the major points to bear in mind is that the financier has a responsibility to ensure that the finance provided will actually be adequate for the whole process. This means that the financier should be prepared to consider issues of OHS, and ensure that adequate provision is made for these during the design and construction period and that adequate budgets are actually provided. Many financiers will make it clear that OHS is a key issue in terms of adjudicating tender bids, but this needs to be clearly stated and documented in the financial agreement.

Contract documents

The contract documents used in construction projects are, in the main, drafted by the clients or industry bodies with representatives from all parties. The drafter will obviously weight the conditions towards their own interests but this is not necessarily a problem as long as the contract is well understood and disputes adequately documented. More important is the appropriateness of the documents to the type of contract strategy being used. Issues such as the degree of completion of drawings at commencement of construction and the use or not of bills quantities or schedules of rates are important considerations. The contract documents should match the strategy and procurement system adopted. Most standard contracts are suitable only for a particular contract strategy but documents, such as the Engineering and Construction Contract (previously known as the New Engineering Contract, NEC) produced by the Institution of Civil Engineers, UK, is flexible enough to be used with all types of contract strategy.

Most construction projects are run using standard construction contract documents. These documents have traditionally been rather weak in providing for construction-site OHS. The provisions are often general and certainly not detailed. Legislation is in place in many countries to add to the provisions in the contract documents, and many sophisticated and conscientious clients have built OHS performance goals, with incentives and penalties, into their contract documents.

Leadership

An important strategic decision is the choice of project team leader. Any of the project participants, including the client, can take up this role. The choice of leader should be based on a number of factors including personality, expertise, experience and an analysis of the roles and responsibilities to be allocated to the participants. In the past, the role of leader went to the architect by default in the traditional system. If a contingency view of contract strategy is adopted, then this choice-by-default must be questioned, and the leader fitted to the strategy adopted.

Leadership is vitally important in all occupations and for all processes. Leadership in OHS management is essential. Countless studies have shown that where a leader is recognised and where that leader makes it known that OHS is an important issue, performance is improved (for a review of the literature on this topic, see O'Dea and Flin 2003). A champion is needed in all organisations, and indeed on all construction sites. By having a strong leader committed to OHS, a strong OHS culture can be developed. Leadership must come not just from OHS professionals, but also from the construction professionals and administrators who deal with the whole process on the construction site and within the construction company, and within the clients' and consultants' organisations. Many examples of this type of approach are shown throughout this book.

Authority and responsibility

The distribution of authority and responsibility is an important issue in any organisation, and of paramount importance in project organisations, which have the characteristics of TMOs. Walker (1996) discusses the nature of authority and responsibility in detail, and suggests a methodology for devising an appropriate structure of authority levels within project organisations. Walker sees this as the key to success in project organisation.

All organisations must have clearly defined lines of authority, and with authority goes responsibility. In an organisation that takes OHS seriously, it is important to delegate authority for the efficient and effective running of an OHS management system to a senior member of the management team. This person must be a champion for OHS, and must ensure that his responsibilities are undertaken diligently, effectively and on a continuing and consistent basis. Hence, at the outset of a contract, when the initial concept has been devised but no development has taken place, it is essential that a leader within the project team be identified who has the authority and responsibility to ensure an effective health and safety management system. OHS management systems in themselves are discussed in Chapter 4. These systems are neither technically complex nor difficult to put into place. The real issue lies in maintaining functioning of such a system, and ensuring that the outputs of the OHS management system are effectively dealt with and are implemented on a continuing basis.

Supervision on site

Safety responsibility and safety leadership are particularly important at the level of site supervisor/foreman. In a recent piece of research, undertaken on

Hong Kong construction sites, Rowlinson *et al.* (2003) found that although foremen knew their responsibilities in areas such as worker orientation, explanation of safe ways of operation, holding OHS meetings and coaching workers, they were less aware of their responsibilities in areas such as accident investigation, inspection for hazards and discipline issues with workers. In those areas where they had responsibility, there was often a mismatch in the authority they were given. This reflects a failure in the management system, and foremen are really the interface between the management and the operational systems. As such, foremen play a key role in ensuring that the OHS management system operates effectively. It appears from the results of this study that this role is not being performed properly, and that the interface associated with it is an area requiring urgent attention in most Hong Kong construction companies. Recent research shows that the more OHS-aware supervisors are, the more positive the OHS climate on construction sites (Mohamed 2002).

As far as the great majority of the foremen were concerned, management did not measure their OHS performance in any area. All six areas investigated returned low scores, including handling new workers, training workers, OHS practice, discipline, co-ordination and motivation. The underlying impression given by the foremen was that they were not considered part of the OHS management team. This is an important finding, and reflects a serious shortcoming in the way OHS management systems are operated on Hong Kong construction sites. By neglecting the role of the foreman, who is the main interface between worker and management, the most potent resource for the promotion of OHS improvement is being under-utilised. Effective implementation of any OHS management system largely depends upon the ability of supervisory personnel (Agrilla 1999). Moreover, the system should provide the means for controlling and monitoring performance (Smith *et al.* 1998). Lack of performance measurement by management indicates little interest in benchmarking and continuous improvement – which could be interpreted, in some cases, as having a minimum level of OHS management commitment. Research shows a strong association between the latter and relatively poor OHS performance record (Mohamed 2000).

From this research, one might conclude that, overall, the OHS supervisory performance of foremen in the Hong Kong construction industry is poor. However, this conclusion must be carefully considered. Although the results indicate that performance is not as good as it could be, this is not necessarily a failing of the foremen themselves. The research shows that foremen are unclear as to what their responsibilities are or should be. This stems in part from a lack of formal authority in many areas of site supervision related to OHS. In addition, foremen generally believe that their OHS knowledge and experience is limited, and this seriously impairs their ability to perform to high levels when it comes to site OHS. All of these factors together point to a failure of the OHS management systems, as implemented by their companies, to properly train and educate foremen. They neither clearly define the foremen's roles and responsibilities nor ensure adequate formal authority is given to foremen when it comes to OHS matters. This is a failure on the part of senior management to adequately address the nature of the company's OHS management system, and, in particular, to address the problem of the interface between management and worker (which is most often the position in which the foreman find him- or herself).

Recent work in small-world theory postulates that information is disseminated through an organisation by means of 'well-connected' nodes in a network (Barabassi 2002). Networks may take many forms, such as the Internet, the biotechnology network, a company's supply chain and so on. Each is a network, but with differing properties. In the situation discussed here, the network is an information network, and the node, the foreman, is not being used effectively. However, the potential of the foreman in terms of instilling OHS awareness and a positive OHS culture is great. By addressing the role of the foreman as a key node for OHS promotion, a great potential influencing network can be realised.

Performance

The performance of different organisation forms, and indeed procurement systems, has been the subject of much research over the past twenty years. No definitive outcome has stemmed from this research, but a series of commonly held beliefs is presented in the Table 3.1. The reader should be aware that comparison of the performance of the different organisation forms is fraught with difficulty, and the opinion expressed in the table cannot be relied upon in all circumstances.

One of the main arguments for better OHS performance of the design-build and the divided-contract approaches is that the opportunity for integrating the design and construction processes and bringing the different professions together (*at an early*) stage exists, whereas this is highly unlikely in the traditional approach. However, choice of organisation form is only a strategy –

Table 3.1 Hypothesised performance of 'organisation forms'

	Traditional	*Design-build*	*Divided-contract*
Speed	L	H	H
Cost	M	L	H
Potential for incorporating variations	H	L	H
Cost certainty	M	M	L
OHS management	L	H	H

(L = low, M = medium, H = high)

no matter how good the strategy, it must be implemented and this is the crux of the matter, implementing good opportunities.

Construction industry issues

A number of basic issues, identified in reports worldwide, have a negative effect on construction project performance. These are a recurring set of issues based upon the way the industry organises itself. They fundamentally stem from the sentient differentiation between the various professions within the industry. The client, a key contributor and stakeholder in the process, also has a major effect on these issues. For instance, the insistence on competitive tendering and the acceptance of the lowest price have been seen to lead to a vicious cycle of surveillance, opportunistic behaviour and more surveillance: this has a detrimental effect on time, cost and quality performance.

As these are failings of the overall management system, it can be assumed that OHS will also suffer in the same way. Hence, the key elements identified as having a negative effect on performance (that is the competitive tendering system, the acceptance of the lowest price, the proliferation of multi-layer subcontracting and the adversarial relationships that exist) can all be seen to work negatively in construction-site OHS performance.

Another issue that should be considered in less developed countries is the appropriateness of the systems and technology being employed. If technology transfer is to take place effectively, its impact on OHS is a key issue that must be addressed at the outset of the project. In many instances, the use of novel procurement systems, or the use of a new technology can lead to an increase in OHS problems.

Managing contractors and subcontractors

There has already been discussion of the role of subcontracting in the construction industry. The divided-contract approach makes use of trades contractors as subcontractors to the main contractor. Yet, even in the traditional and the design-build approaches, it is quite common for a high percentage of the construction work, up to 95 per cent in Hong Kong for example, to be subcontracted to a series of trades and labour-only contractors. The problem with this approach is that many of the subcontractors are small and medium-sized enterprises that do not have the spare capacity to provide adequate training and education and to implement OHS management systems. Consequently, they regularly fail to meet their OHS obligations, and become a major source of problems to the main contractor. It is unreasonable to expect that responsibility for OHS can be subcontracted to these organisations. The main contractor must bear the risk that the use of such multiple layer subcontracting brings with it.

In an unpublished piece of Hong Kong-based research (Lam 2003) it was found that the biggest contractors, measured by turnover, perform badly on items such as OHS training, job hazard analysis, working at height, manual handling and mechanical plant and equipment maintenance and procedures – the same elements that contribute to the majority of accidents on Hong Kong construction sites. Across all contractors, large and small, surveyed in the research, a systematic 'fatal' failure in the safety management system (SMS) was found. This has implications for the implementation of the SMS as a whole across the construction industry, and brings into question the suitability of the self-regulatory approach to OHS, described in Chapter 2.

Some small contractors were found to perform very well. This good performance relied to a great extent on a number of factors: among others, the existence of a senior person within the organisation who was prepared to champion OHS, the degree of partnering between main and subcontractor, and the education and training received by the 'champion' and instilled in his foremen. These findings draw together the findings reported by Lingard and Rowlinson (1991), Rowlinson and Matthews (1999), and Rowlinson and Lam (2002) in the specific context of the smaller construction firm in Hong Kong.

In an attempt to deal with this issue, the University of New South Wales produced its 'six pack' or 'subby pack' (www.constructionalliance.org), which provides a series of tools the main contractor and subcontractors can use to monitor and control performance. Such systems can help in monitoring and guiding subcontractor performance, but the bottom line is that all systems implemented on a construction site to manage OHS must be maintained on a regular basis. If the subcontractor cannot do this, it is incumbent on the main contractor to ensure that this happens. In recent years, there has been a move towards alternative project supply arrangements in order to deal with issues including OHS as well as time, cost and quality.

Subcontractors are a critical part of the modern construction process. This aspect of modern construction has enabled main contractors to become the managers of the construction process, rather than the builders. As competition has become greater, the reliance on subcontracting has increased. Main contractors now seek to become more competitive when tendering by reducing subcontractor estimates, employing onerous contract conditions, and generally adopting unfair management practices. This is not to say that subcontractors have not caused problems as well. With ease of entry into the construction industry, subcontractors are quickly established, many do not survive, and main contractors subsequently find it difficult to meet client expectations when subcontractors default.

Discipline and powerlessness

At the site supervisory level also subcontractors provide problems. In research reported by Rowlinson *et al.* (2003), a number of foremen said that they had inadequate authority to take disciplinary action against problem workers. Workers, especially the more experienced, would often resist instructions because they knew that foremen could not fire them. In fact, most subcontract matters are negotiated at contract manager level, and there is no emphasis on person-to-person communication or supervision. This is a structural issue that the industry needs to address.

A number of foremen said that under many circumstances, they had the authority to stop unsafe acts on site, but they had no authority to transfer a worker out of their division or fire them, though this authority under many circumstances was, in their opinion, necessary. Undoubtedly, this would lead some foremen to feel powerless. Research studies have shown that powerlessness is positively associated with lack of coping with job control (Ross and Reynolds 1996). The foremen's only recourse was to report such cases to higher management or to the subcontractors' representative on site, and ask them to take action. This is an example of the negative influence that the economic basis of the subcontracting system can have on OHS performance.

Other subcontracting issues

Several other characteristics of subcontracting give rise to negative OHS outcomes; this includes the 'payment by results' system which is based on the amount of work not the time required, thereby encouraging subcontractors to minimise time and maximise profit. Furthermore, not all contractors expect to be proactive in including subcontractors in OHS discussions; research conducted in Australia revealed that almost 40 per cent of surveyed contractors rate their proactiveness in this matter as average or below (Mohamed 2000).

The contribution of specialist and trade subcontractors to the total construction process can account for as much as 90 per cent of the total value of the project (Gray and Flannagan 1989). The result of this, and other factors, is that main contractors are concentrating their efforts on managing site operations rather than employing direct labour to undertake construction work. The increase in the use of subcontractors can partially be attributed to the increased complexity of the construction of buildings. However, this increase in complexity, the oversupply of subcontractors and the declining construction output, such as have been experienced, for example, in Hong Kong, have cultivated adversarial tendencies that have a negative effect on the main contractor–subcontractor relationship.

Main contractors have realised that one of the greatest potential areas for cost savings lies with subcontractors. However, subcontractors have caused problems too. With easy entry into the construction market place, subcontractors have been established with very little capital investment. Many of these companies do not have the necessary skills to undertake work satisfactorily and, consequently, are unable to give their clients the service they require. Moreover, many of the bad traits common to the main contractor–sub contractor relationship are also common to the subcontractor–sub-subcontractor relationship.

Partnering

Partnering is seen by many as a way of improving the relationship between main contractors and subcontractors. Research in the UK and USA has shown that partnering can facilitate better relationships, with the overall aim of benefiting everyone within the construction process. Although partnering is employed in Hong Kong, its current application is limited to public sector works only.

There are no fixed definitions for partnering. Partnering is a relationship that occurs at a particular time to meet the needs of all parties concerned. Matthews (1996) identified that one of the most commonly cited definitions for partnering was that proposed by the Construction Industry Institute (CII 1991).

Construction Industry Institute defines partnering as:

> a long-term commitment between two or more organisations for the purpose of achieving specific business objectives by maximising the effectiveness of each participant's resources. This requires changing traditional relationships to a shared culture without regard to organisational boundaries. The relationship is based on trust, dedication to common goals, and an understanding of each other's individual expectations and values. Expected benefits include improved efficiency and cost effectiveness, increased opportunity for innovation, and the continuous improvement of quality products and services.

Two types of partnering exist. *Project Partnering* is partnering undertaken on a single project. At the end of the project the partnering relationship is terminated and another relationship is commenced on the next project. *Strategic Partnering* takes place when two or more firms use partnering on a long-term basis to undertake more than one construction project, or some continuing construction activity (RCF 1998). Matthews *et al.* (1996) note that companies are likely to commence their partnering relationships by adopting the short-term project partnering. Only once the practicalities of applying the partnering philosophy are fully understood will strategic partnering relationships be developed.

Partnering and strategic alliances have been seen as mechanisms by which technology, both hard and soft, can be transferred from the contracting organisation to the subcontracting organisation. Rowlinson and Matthews discuss how partnering can help to improve the subcontractors' OHS management system. However, this improvement requires a culture change within the industry. Both main contractor and subcontractors must be aware of their responsibilities and be willing to take on these responsibilities and improve their performance in a synergistic manner.

Term contracts, maintenance and facilities management

Having dealt with the main issues affecting contract strategy, it should be apparent that a whole life-cycle approach to OHS management is important in the construction industry. As such, a series of additional issues should be considered. If it is found that relational contracting and partnering are beneficial to the performance of the construction industry then consideration should be given to extended contractual relationships, term contracts and other issues. These would include maintenance and facility management at the earliest stages of a project. By addressing these issues, OHS can be built into the project and the constructed facility, so that the overall level of OHS can be incrementally moved up a notch or two. These elements can be further reinforced by repeated use in term contracts.

This infers a whole life-cycle view of construction and development and it may well be that a logical extension of this is for facilities managers to become the drivers in terms of specifying the final product, the constructed facility. There is a parallel here with green construction and sustainability. Basically, the underlying parameters for value in a project are not least initial cost of the facility or its components but the whole life-cycle performance of the facility and its components. This does not necessarily equate to lowest life-cycle costs; rather it reflects least environmental regret. In the same way, building OHS into the whole lifecycle of a facility has the ultimate goal of least OHS regret, which necessarily equates to zero accidents and no facility-related ill health.

Lean construction and OHS

One management fad widely embraced by policy makers and industry spokespeople is the concept of lean construction. This concept was wholeheartedly endorsed in a government-initiated task force report titled 'Rethinking Construction'. The concept of lean construction is based upon the ideas of lean production, which are purported to have led to enhanced efficiency and competitiveness in parts of the manufacturing sector. In *Rethinking Construction*, the car industry is highlighted as a success story from which the construction industry could learn to eliminate waste and use its resources more efficiently. 'Lean Construction' was developed from the working philosophy of 'Lean Production' developed by Japanese car manufacturers. The fundamental principles of lean construction are: reduce the share of non-value-adding activities; increase output through systematic consideration of customer requirements; reduce variability; reduce life-cycle times; increase process transparency and build continuous improvement into processes. The most important 'instruments' of lean production are: multifunctional groups; simultaneous engineering; just-in-time production (JIT); long-term relationships and customer orientation.

However, as Green (2002) points out, the construction industry has traditionally adopted a regressive 'command and control' model of human resource management, in which the human resource is seen primarily as a cost. In this context, the notions of 'becoming leaner' and 'cutting out the waste' are likely to result in increased pressure on employees to work harder and longer, with less room for error. Green (2002) argues that lean production has serious implications for human resource management and warns of the link between lean production and longer working hours and OHS complaints. Yates *et al.* (2001) conducted a survey of Canadian employees of motor vehicle manufacturers and report that employees working at plants that had embraced lean production processes reported the lowest degree of empowerment, a diminished quality of life and poorer health and safety conditions. Also, in an ambitious multi-industry time series study examining the introduction of lean production management approaches and occupational injury rates, Askenazy (2001) concludes that the introduction of such techniques coincided with the reversal of a downward trend in injury rates in most industries. Though construction did not experience such a downward trend, the introduction of these techniques was not yet widespread in construction during the period studies. These studies cannot demonstrate cause and effect with certainty, but they provide enough evidence to warn against the blind adoption of these techniques.

There is also evidence to suggest that lean production not only leads to a higher exposure to OHS risks, but that it also contributes to an inability to respond to emergency situations in an appropriate matter. Consequently, emergencies can escalate into crises resulting in significant organisational losses. For example, Nichol (2001) cites cost-cutting which resulted in the reduction of on-site engineering capability, as one of the key factors in the inappropriate response to an abnormal situation that arose at the Esso Longford gas plant in September 1998, which left two people dead and eight injured. Nichol suggests that Major Hazard Facilities (MHFs) in Australia have failed to learn from the Longford disaster in that engineering is still treated as a cost centre rather than a contributor to profit. The reduction in engineering capability prohibits the ability for experienced engineers to impart their knowledge of applied engineering principles to a younger generation of employees who will take over the management of the existing plant. This has serious implications for OHS and risk management, because it results in a chasm in the corporate memory and thus limits the extent to which organisations learn from their mistakes.

Thus to adapt innovative management techniques like lean production to the construction industry without thought of the consequences could expose construction workers to greater risk of occupational injury and illness, worsen the industry's OHS record and cause harm to the industry's reputation. Nonetheless, there may be an opportunity to adapt some of these measures to work in favour of rather than against OHS performance. For example, an important aspect of the lean production approach is the incorporation of a *DFX* approach, where *x* may be *buildability*, *flexibility* or *economy*. From an OHS point of view, *design for OHS* is a paradigm that may be adopted. An example of how this approach and information technology (IT) can be combined is given by Hadikusumo and Rowlinson (2002) and discussed in Chapter 9. The incorporation of long-term relations into the business process, such as is favoured by lean-production advocates, also reflects the objectives embodied in the partnering approach, and Rowlinson and Matthews (1999) indicate how this approach to contracting can assist smaller organisations to develop and maintain their OHS management systems. However, given the empirical evidence suggesting that lean production may be harmful to OHS, any initiatives must be designed and implemented with great care and cautiously evaluated.

Summary

It is obvious that the influence of procurement systems and contract strategy on occupational OHS is far-reaching. It should also be obvious that OHS is a strategic issue for projects and that it should be considered at the outset of a project. Certain contract strategies, and particularly payment systems and business relationships, such as partnering, can have a major influence on the effectiveness of OHS management systems. Hence, the underlying lesson from the discussion presented above is that key strategic decisions made at the outset of a project have a major influence on the effectiveness of the management system. However, there is emerging evidence that widespread adoption of some modern management concepts, such as those classified under the umbrella term *lean production*, can actually be harmful to workers' health and safety. It is therefore imperative that the effect of different procurement strategies and management methods on OHS performance in construction be carefully evaluated.

Discussion and review questions

1. Describe how contractual relationships in traditional, design-build and the divided-contract approach to construction contracting could impact upon a project's OHS performance.
2. What role should construction project clients and financiers play in the management of OHS? How could these parties improve the OHS performance of the construction industry?
3. In what ways, if any, could project or strategic partnering improve OHS performance if undertaken between:

   (a) clients and main contractors;
   (b) main contractors and subcontractors; and
   (c) contractors and suppliers?

Chapter 4

Systems management of OHS

Introduction

The earliest efforts to prevent undesirable OHS outcomes focused on the provision of a safe physical environment, and addressed issues such as the provision of machinery guarding and safe mechanical equipment. As OHS practitioners and researchers became aware that individual behaviour played a role in the occurrence of occupational injury and illness, the focus shifted to the individual. Within this tradition, OHS programmes focused on individuals' risk behaviours were commonplace. While these traditional approaches led to enormous improvements in OHS performance in the twentieth century, there is compelling evidence that workplace organisational factors play a key role in OHS. For example, Shannon *et al.* (2001) present a detailed review of the literature linking workplace organisational factors to occupational injuries, occupational ill health and musculoskeletal disorders. Workplace organisational factors impacting upon OHS include the organisation's OHS management activities, in particular the extent to which an OHS management system is in place, and the effectiveness with which this management system operates. There is also a growing recognition that less tangible features of an organisation also contribute to OHS performance. These features include the organisation's philosophy, culture and employee relations situation. Contemporary OHS theory and best-practice models focus on both the OHS management system and the organisational culture (Hale and Hovden 1998).

In this chapter, we describe the process of OHS management and consider how components of an OHS management system, such as OHS planning and OHS training, are undertaken in the construction industry. We consider how the management system must address challenges inherent in the organisation of construction work, such as the management of OHS in construction design and the management of subcontractors' OHS. We explore the relationship between a corporate OHS management system and the management of OHS within projects, and consider the roles and responsibilities of different levels of management within a firm. We also explore the related issues of safety, leadership and culture, and identify characteristics of a strong OHS culture.

OHS management

Implementing an OHS management system is an important first step in ensuring that OHS is systematically managed within an organisation. The key elements in OHS management are shown in Figure 4.1. The aim of an OHS management system is to ensure that the productive work of a company is designed and performed with workers' OHS in mind, that managers make decisions following a systematic evaluation of OHS risks, and that work is adequately planned, resourced and controlled so as to prevent occupational injuries or illnesses. The management activities required to achieve good OHS performance are essentially the same as those required to achieve success in any area of business activity. They require a clearly defined policy, well-defined plans incorporating specific objectives, strong management commitment, the provision of sufficient resources, a systematic training programme, effective monitoring and reporting of performance and a process for reviewing performance and making improvements.

OHS policy

Defining a corporate OHS policy is the first step in the OHS management process. Policy should be established following a detailed analysis of an organisation's current situation with regard to OHS. Thus, OHS processes in place and performance to date should be carefully considered to determine how satisfactorily the organisation is performing. The external

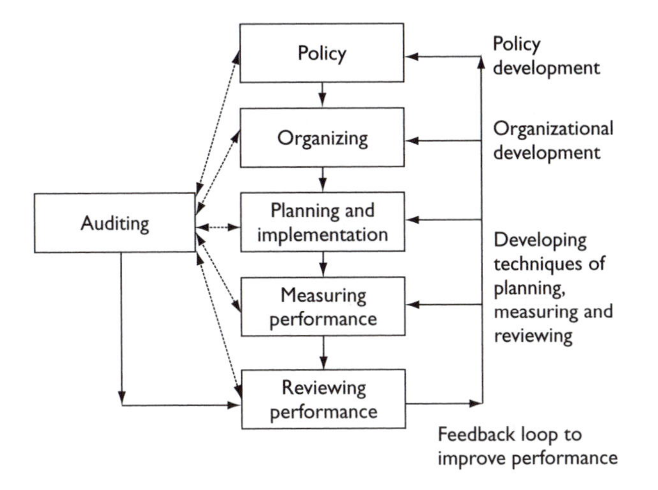

Figure 4.1 Key elements of successful OHS management (adapted from HSE 2000, p. 4).

environment in which the organisation conducts business should also be considered. Questions to ask include:

- How well is the organisation performing in comparison with competitors?
- What external pressures exist to improve OHS?
- Are these pressures likely to change in the future?
- What impact does OHS have on the firm's business?
- How is OHS performance affecting employee morale and project performance?

Once the organisation has analysed its own performance and its business environment, it can formulate an appropriate policy statement. Typically an OHS policy statement contains a statement of the organisation's commitment to OHS, and identifies the OHS responsibilities of all employees, including senior managers in both corporate and project management roles and specialist OHS or legal advisers. Policy statements also usually contain statements of principles in which the organisation believes. For example, it may state that 'all occupational injuries and illnesses are preventable' or that 'OHS is as important as productivity or quality'. Policy statements are usually signed by the company's most senior manager, and are freely available to all employees within a company. The OHS policy is therefore an important document defining the company's overall approach to OHS. As the organisation's external business environment will change over time, policy statements must be responsive and therefore should be reviewed at regular intervals and updated as necessary.

Organising

The next step in the management of OHS is creating an organisation in which roles, responsibilities and relationships support the systematic planning and control of OHS. The organisation design is important to the effectiveness of OHS management and comprises the way that OHS is controlled, through allocating responsibility for OHS functions, identifying OHS objectives and monitoring progress towards meeting these objectives. It is important that OHS responsibilities are allocated to managers and that a senior manager, at the top of the organisation, is charged with co-ordinating and monitoring the OHS management activities.

Responsibilities for OHS activities must be clearly allocated so that no important tasks 'fall between the cracks'. This is particularly true in construction projects, in which several different organisations are often involved. Thus, for example, contractors' and subcontractors' OHS responsibilities must be clearly understood.

Performance standards should be written to specify, in detail, who should do what, to what standard and when. Performance standards specify what people in various roles need to do, in relation to the management of OHS. For example, supervisors or foremen might be expected to conduct weekly toolbox meetings with work teams, a site manager might be expected to induct all new workers and visitors who enter the site and a project director might be expected to examine the past OHS performance of all subcontractors prior to engaging them.

Holding people accountable for their OHS responsibilities might involve including performance standards in job descriptions and using them as the basis of measuring how effectively employees are performing their jobs. Thus, meeting performance standards with regard to OHS management activities should be incorporated into formal performance appraisal systems, which are often linked to bonus schemes and career opportunities.

Another important aspect of organisation design is the extent and nature of supervision. OHS legislation typically requires employers to provide adequate supervision in partial fulfilment of their general duties (see Chapter 2 for more information on OHS legislation). It is very important that supervision be carefully considered in designing systems of work, because different tasks and teams will require different levels of supervision. New workers or younger workers require more supervision than experienced workers, and high risk jobs typically warrant more supervision than low risk jobs. However, a certain level of supervision is necessary, even for experienced workers, to ensure that OHS standards are consistently met.

In the decentralised construction industry, in which many subcontractors or tradesmen operate as semi-autonomous crews, it is essential that management control of OHS be maintained by the principal contractor, who cannot delegate his or her responsibility to provide a safe system of work with adequate supervision. Thus, although the authority to act will be delegated to subcontractors and supervisors, it is essential to make sure that those exercising discretion are competent to do so and do not take, or direct others to take, any unacceptable risks.

Consultation

Worker involvement in OHS is an important requirement of the Robens-style legislation described in Chapter 2. It is also a characteristic feature of organisations with good safety performance. The establishment of consultative mechanisms is therefore an important part in organising for OHS. Joint employer–employee OHS committees and the appointment of employee OHS representatives are means by which worker participation in OHS is often elicited. The role and function of these formal consultative mechanisms is usually prescribed by the law and there is often provision for paid-time off for employee health and safety representatives to undertake OHS training and other facilities required to support employee representatives in their role.

However, the quality of consultation and the effectiveness of OHS committees and employee representatives often depends upon the extent to which they are encouraged to play a role in OHS planning and monitoring performance.

Also, a healthy communication climate should exist, such that employees at all levels feel free to express their views about OHS, and have a role in setting OHS performance standards and writing procedures for controlling OHS risk relevant to their work. The involvement of supervisors and others in risk assessment and decisions about how to control OHS risk is particularly important because these people have a close knowledge of the practical aspects of the work. In order for risk controls to be relevant, it is essential to incorporate this knowledge in decision-making and development of controls, including safe work procedures. Not only does the participation of those who will supervise and perform the work help to ensure that risk controls are practical, but it also helps to develop a sense of 'ownership' of OHS solutions. This makes it more likely that these solutions will be voluntarily adopted.

In the context of a construction project, the operation of consultative mechanisms may be difficult to achieve, but it is important that subcontractors and their workers are involved in consultation with decision-makers wherever practicable.

While consultation is beneficial and contributes to the development of good working relationships between management and employees, in the short term, disagreements about risks and their control often arise. Consultative processes therefore require that a clear and effective issue resolution procedure exists. This procedure should establish when and how independent, specialist help can be sought to decide upon issues that cannot be resolved between management and the workforce.

In Victoria, the *Occupational Health and Safety (Issue Resolution) Regulations* (1999) require that employers nominate management representatives who must deal with OHS issues. The Regulations specify a procedure for resolving OHS issues that must be used unless an alternative issue resolution procedure has been agreed. Figure 4.2 depicts an issue resolution procedure that might be used in a construction project.

Communicating OHS information

Establishing an organisational environment in which OHS information can be communicated effectively is an important aspect of organising for OHS. OHS information needs to flow within the organisation and be communicated between members of the organisation and people outside it.

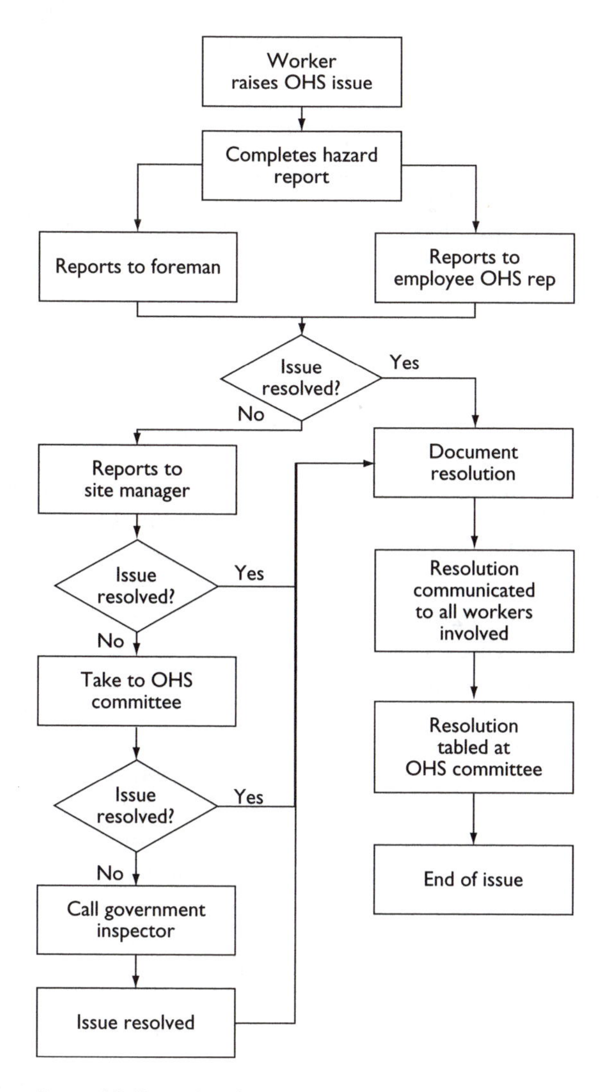

Figure 4.2 Example of an issue resolution procedure.

Information about OHS legislation, technological developments and OHS risk and risk control methods needs to be brought into the organisation. Health and safety information also needs to be communicated outside the organisation. For example, statutory reporting requirements for OHS incidents usually exist. In some instances health and safety risk information may need to be communicated to local residents or planning authorities. In the event of major incidents, communicating with the media may become an issue. The importance of managing this information flow is highlighted by the Esso Longford Disaster case study presented later in this chapter. In order that this communication is handled effectively, processes for information flow need to be established and relevant employees trained in appropriate communication methods. In certain situations, professional advice may be needed as to how to communicate information to the target audience, for example the public or the media.

Information about OHS also has to be communicated internally and organisations should establish systems to ensure that important information is understood. For example, the OHS policy, plans and objectives need to be communicated throughout the organisation to people who will manage their implementation. Formal written communication is one means of achieving this but, in many instances, important information may be better communicated face to face, through information sessions, presentations or meetings.

As well as formal communication channels, informal communication can also occur. One important means by which OHS information is communicated informally is the visible behaviour of managers. For example, the organisation's commitment to OHS can be communicated by managers regularly undertaking 'safety walks' on site, through including OHS on project meeting agendas, chairing OHS committee meetings or being actively involved in investigating OHS incidents.

One important source of information that needs to be communicated to the workforce is OHS risk information about building products, plant and equipment and work processes. Communication in the form of warnings and risk information plays an important role in shaping people's hazard-risk judgements and safety behaviour. The format of information provided is related to the accuracy with which people interpret and the speed with which they comprehend safety messages. For example, the provision of too much information on chemical warning labels is reported to have the negative impact of discouraging people from seeking more detailed safety information from MSDSs (Lehto 1998). The provision of MSDSs documenting chemical hazard information is an important responsibility incumbent upon managers in workplaces in which chemical products are used. However, MSDSs are written in technical language and may be difficult for workers to understand (Phillips *et al.* 1999). It is therefore insufficient to provide access to MSDSs without providing training in how to interpret their contents.

Research also indicates that people perceive a risk to be more serious and are more likely to comply with warnings about an item or activity when the risk warning information emphasises the likely severity of an injury, as compared to the probability (Wogalter *et al.* 1999). Thus, it is recommended that OHS warnings focus attention on how badly a person might get hurt if the warning is not heeded. The provision of accurate information about the magnitude of potential injuries is likely to prompt rational safety-related responses.

The use of pictorial warning signs and safety instructions are recommended as a means of conveying safety-related information (Young and Wogalter 1990). As the construction industry globalises and becomes more culturally diverse, the use of pictorials as a means to communicate health and safety information is likely to be increasingly effective.

ISO 3864 1984, *Safety colours and safety signs*, was developed in an attempt to standardise colours and symbols used in safety signage around the world. Safety signage on site should comply with this standard. The use of pictorials is based upon the proven theory that visual stimuli are maintained in the memory longer than verbal messages. However, the beneficial effects of displaying safety information in pictorial form should not be taken as a given. Recent research suggests that the meaning of some pictorials is not readily apparent and that some industrial safety pictorials have a comprehension rate of 50 per cent or less (Wogalter *et al.* 1997). Thus, many safety pictorials may fail to convey their intended messages and it is recommended that pictorials be explained in training and accompanied with a short written explanation. This may be particularly important in construction as the educational level of many construction workers is relatively low and ensuring that risk information is clearly communicated and accurately comprehended is an important management responsibility.

OHS competencies

Ensuring that the workforce is competent to fulfil their OHS responsibilities is another aspect of establishing organisational OHS capability. All employees need to be able to perform their work safely, and managerial and supervisory personnel need to understand their legal obligations as well as the principles and practices of OHS management. Training plays a key role in the development of these competencies.

Despite the importance of OHS training, the construction industry does not have a good record of investing in training its employees, investing less in training its employees than many other industries (Loosemore *et al.* 2003). A study undertaken in the UK in the late 1980s revealed that the construction industry provided less training to its employees than any other industry sector, including comparable industries in which casual employment is common, such as retail or catering (Training Agency 1989). In construction, training is often regarded as a cost rather than an investment in the human resources of an organisation and training programmes are the first activities to be axed in lean times. However, a strategic approach to employees' training views training as an important driver in the achievement of the organisation's business objectives, including OHS objectives. It has been estimated that less than a quarter of the construction workers in the European Union receive any training in OHS (EC 1993).

In Australia, there is evidence that the situation is beginning to improve. Australian Bureau of Statistics figures indicate that between 1997 and 2001, training course completions among construction industry employees rose by 54 per cent. Also, in the same period, OHS courses as a percentage of all training completed rose from 12 to 17 per cent (ABS 2003). However, closer examination of these figures indicates that these improvements might have only occurred in larger businesses with little change in the delivery of training to employees in small businesses, suggesting that a large number of workers in the construction industry may not benefit from the increased provision of training (ABS 2003). Also, the low baseline means that further improvement is needed if the construction industry is to build up the OHS knowledge, skills and abilities of its workforce.

A strong OHS training programme is reported to be a feature of organisations with good OHS performance (Smith *et al.* 1978). The objective of OHS training is to provide workers, at all levels, with the knowledge skills and abilities they need to perform their jobs safely as well as to effect positive change in workers' behaviour, and consequently, organisational performance with regard to OHS.

The provision of training is an explicit requirement of the Robens-style OHS legislation, described in Chapter 2 and is an essential component of OHS risk control strategies described in guidance materials on the implementation of health and safety management systems (HSE 1997; Cooper 1998). However, there is also a moral dimension to the case for OHS training, based on the premise that everyone who is involved in an industrial process has a 'right to know' about the hazards of their work. Training is therefore applicable to everyone whose health or safety could be impacted by the activities of the organisation and should be freely provided in the exercise of the common law duty of care.

Training needs analysis

While much research and effort has focused on OHS training design and delivery, just these two stages in the training cycle, alone, are unlikely to meet the needs of an organisation. A more systematic approach to OHS training is needed. An essential first step in the training cycle is a *training needs analysis*. Conducting a training needs analysis involves performing a thorough analysis of each job to identify the distinct tasks involved and to determine the OHS risk inherent in these tasks. Risk can be present due to hazards in the work environment, complex performance sequences, infrequent or difficult tasks, the need to use sophisticated equipment or complex or subtle cues to which workers must respond in appropriate ways (Lindell 1994).

In an organisation, the training needs of different groups of employees must be considered. These groups include:

- senior management;
- project managers;
- plant managers, mechanics and fitters;
- plant and equipment operators;
- site staff, including foremen and supervisors;
- design professionals, including architects and engineers;
- specialist staff advisers, including OHS professionals;
- tradesmen and labourers; and
- contractors and subcontractors.

In analysing the OHS requirements of particular jobs, reference can be made to incident reports and investigation data, and workers themselves can be asked about how they perform their jobs, including the sequences of work and materials, tools or equipment they use. When analysing the training needs of managerial or supervisory employees, the analysis should cover the jobs of these employees as well as the OHS aspects of work performed by their subordinates.

Following the training needs analysis, shortfalls in OHS knowledge, skills and abilities are identified and priority groups are identified for training. These shortfalls represent a deficit which OHS training programmes must make up. Particularly 'at risk' workers may include new employees, those who have transferred into new positions, temporary employees and those who have been recently promoted (Dawson *et al.* 1988). Studies indicate that most construction accidents happen during a worker's first four weeks on a construction site. Inexperienced workers are more likely to be involved in an accident. Laukkanen (1999) suggests that, because construction sites are constantly changing and new hazards emerge on an almost daily basis, on-the-job training in hazard recognition is required for new workers.

In the UK, the HSE (1991) classify training needs into three categories:

1. organisational needs
2. job needs and
3. individual needs.

Organisational training needs are common to everyone in the organisation, and include knowledge of the OHS policy, arrangements for consultation and communication with regard to OHS issues, the organisational structure, and OHS responsibilities. *Job* training needs include management needs, for example knowledge of OHS law, leadership skills, risk management methods and knowledge of processes for OHS planning, measurement, auditing and achieving continuous improvement. Non-management job training needs include a knowledge of OHS performance standards relating to a particular job, OHS procedures and emergency procedures. *Individual* training needs are those that are specific to individual workers, or groups of workers; for example, new workers need to receive project-specific induction training, and re-training may be required when an individual moves from one job to another or when new equipment or technology is introduced.

In some circumstances, project-specific training programmes may be required. Each construction project is different, and consequently assessing the training needs in different construction projects should be undertaken in the project's pre-planning stage. For example, in work in remote areas, emergency evacuation training may need to focus on removing injured or ill workers to the nearest medical centre. In such projects, all workers may need to be trained in first aid. Appropriate training provisions should be determined at the outset of large or unusual projects and be documented in the project OHS plan.

The output of any task analysis should be expressed in terms of specific desired OHS behaviours so that trainees understand the objectives of a training programme and so that training outcomes can be easily evaluated.

Training design and delivery

Once specific training objectives have been decided upon, the next step is to decide how to deliver the training. Training can be carried out in-house, or delivered by external professional trainers. In certain high risk jobs, training may have to occur off-site using simulation methods. Training delivery will depend upon a number of criteria including:

- the nature of the subject matter and desired learning objectives;
- the number of trainees;
- trainees' preferred learning styles;
- resources available to support the training; and
- logistical issues, such as geographic location of workforce.

Where OHS training is designed to prevent human error, training design and delivery methods must reflect the type of error behaviour in question. Training course developers must recognise that different learned error control mechanisms will be effective for different error types. Cooper (1998) suggests that training can instil safe practices into habituated work routines, thereby reducing the incidence of skill-based errors, such as slips or lapses. Such errors occur in familiar circumstances, such as when workers omit to carry out a step in a procedure, or perform a routine series of actions out of sequence. OHS training designed to enable workers to perform a repetitive task with skill should involve the repetitive performance of the task until it becomes automatic. When routine tasks are not performed frequently, they may be 'overlearned' or practised using periodic drills, to make sure they are not forgotten. Rehearsal of emergency drills is a good example of this method.

Training can also be particularly helpful in the prevention of rule-based or knowledge-based errors. Rule-based errors occur when problem-solving rules are wrongly applied. Training can provide a set of safe rules that can then be applied to particular situations. For example, if a crane strikes live overhead power lines then remain in the cabin and attempt to move the crane away from the lines without anyone else approaching the crane. The use of rules is particularly important in circumstances in which counter-intuitive actions are safest, such as remaining in the cabin of the crane when one's first instinct might be to try to get out of the cabin. Training based upon the learning of a correct sequence of steps to follow in a particular situation can also be best achieved through practising sequences. The use of checklists and flowcharts to guide rule-based behaviour can be useful prompts to this method of learning, the objective of which is to raise workers' awareness and prompt appropriate responses to features in the work environment.

However, some OHS training is not based on a set of pre-determined correct steps but requires that workers respond to new and unfamiliar problems. Knowledge-based errors occur when new situations are encountered and errors of judgement are made. Rule-based actions are not available and individuals must 'think on their feet'. Training can assist in the selection of appropriate solutions in previously un-encountered situations. In these circumstances workers need problem-solving skills, and techniques such as brainstorming and lateral thinking will be useful. Training programmes designed to provide these abilities would benefit from the analysis of accident or disaster case studies to examine appropriate responses to non-routine OHS problems. Without such training, people would have to rely upon a 'trial and error' approach to build up this knowledge.

Decisions about training delivery methods must also take into account the important fact that the recipients of work-related training are adults. Training delivery methods must therefore be informed by theories of adult learning, as distinct from those of learning in general. Cheetham and Chivers (2001) summarise the principles underpinning the andragogy theory of adult learning as follows:

- that mature adults are self-directed and autonomous in their approach to learning;
- that they learn best through experiential methods;
- that they are aware of their individual learning needs;
- that they have a need to apply newly acquired knowledge or skills to their circumstances;

- that learning should be seen as a partnership between teachers and learners; and
- that learners' experiences should be used as resources in the learning process.

In accordance with this theory, training materials should not utilise a formal lecture delivery mode, but should draw upon the experiences of training participants and incorporate other mechanisms for participants' learning, for example class exercises, syndicate work and group discussions.

Choice of instructor is also an important decision when deciding on training methods. Instructor performance is one of the most important determinants of trainees' responses to training (Weidner *et al.* 1998). In particular, it is important that trainees trust the instructor (Merrill 1994). Worker-trainers are reported to communicate better with trainees and are able to share their own experiences so that trainees identify with them to a greater extent than they would with professional trainers, safety or human resource specialists (Fernandez *et al.* 2000). Kurtz *et al.* (1997) reviewed the performance of peer and professional trainers and found that trainees who were trained by peer-trainers had greater confidence in their ability to perform recommended actions and were more likely to change their behaviour after receiving training. This suggests that providing 'train the trainer' programmes for workers, for example employee health and safety representatives, who could then train others, may be a beneficial strategy.

Training workers in communicating OHS issues and expressing their OHS concerns has been referred to as 'empowerment-based training', and may also benefit the functioning of employer–employee consultative processes. The underlying premise of empowerment-based OHS training is that training workers to take action on health and safety issues is essential to improving OHS in work environments. Such programmes are aimed at promoting workplace change by training workers to raise OHS concerns and to negotiate solutions to these issues in their workplaces. A learner-centred training approach is adopted in empowerment-based programmes. In this approach, participants' direct experiences in the workplace and their OHS knowledge are drawn out. Easy-to-read fact-sheets containing information about specific OHS issues are provided, and participants work in small groups to solve a series of case-based problems. Findings are then reported by each group to the full class, enabling learning to be shared. Trainers, who are workers themselves, support the groups as they work on the problems and facilitate the report-back session. These training strategies are designed to legitimise workers' collective knowledge of the health and safety issues relevant to their workplaces, and aim to develop an understanding that workers' expertise in OHS issues relating to their work equips them to act as change agents in the workplace. The problems used are often designed such that inadequate OHS conditions occur alongside threats of discipline or dismissal. Worker participants have to learn the skills to advocate for health and safety changes in the context of difficult organisational contexts. Lippin *et al.* (2000) report that empowerment-based training effectively increases participants' willingness to raise health and safety concerns, and also that these actions lead to actual workplace changes.

Training evaluation

It is important that the outcomes of OHS training are measured against the training objectives. Too often training is simply assumed to have achieved the expected result and there is little rigorous assessment of actual outcomes of training. Training evaluation is important for five reasons, summarised below:

1 It indicates whether objectives have been met and identifies remaining training needs.
2 It provides feedback to trainers about their performance.
3 It is an intrinsic part of the trainee's learning cycle.
4 It enables the costs and benefits of investment in training to be appraised.
5 It can be used as a benchmark against which to compare other OHS interventions (Cooper and Cotton 2000).

Much training evaluation is limited to an assessment of immediate impressions. However, it is important to consider the long-term effects of training, particularly when anticipated learning outcomes include changes to organisational culture and the implementation of workplace health and safety changes. Learning takes time to diffuse within organisations, and the effects of some of the arguably more important outcomes of OHS training may not be observable in the short term.

Any OHS training evaluation exercise must be carefully designed to ensure that outcomes are measured and attributed to the training intervention appropriately. Chapter 10 contains further information on the importance of using rigorous evaluation methods.

Training transfer

Research suggests that delivering training that meets specified learning objectives is not sufficient because learning and behaviour are linked in a complex way. Goldstein (1993) has observed a low correlation between learning how to do something and actual job behaviour, and suggests that learning, of itself, does not automatically translate into behaviour change at the workplace. An important element in the effectiveness of a training programme is therefore the extent to which learning is put into practice when employees return to the workplace. This is known as training transfer. The extent of training transfer has a key bearing on the ability of training to impact upon organisational performance in OHS, or any other aspect of business performance that is the target of training. It is therefore critical to understand the organisational factors that impact upon training transfer. Holton (1996) developed a model of training transfer that has received some empirical support (Figure 4.3). Holton's model has been used to predict training outcomes and probably explains why some training programmes do not achieve the desired results (Donovan *et al.* 2001).

Holton's model identifies three outcomes of a training programme: learning, individual performance and organisational performance. Assuming that learning occurs effectively, the transfer of learning is determined by three factors: motivation to transfer, the transfer climate and transfer design.

Motivation to transfer may be influenced by employees' expectations of the benefits to be gained from the transfer of learning, for example the prospect of promotion or higher pay. The inclusion of OHS in managers' and supervisors' performance appraisal processes is one way to enhance the motivation to transfer learning. Supervisors should also be trained to recognise and appropriately reward OHS behaviour demonstrated by the workers they supervise.

Holton (1996) identifies what he terms the transfer design as a cause of employees' failure to transfer learning to the workplace. This refers to the possibility that intellectual learning may occur but that trainees are not provided with the opportunity to practice the training in the work context or taught how to transfer this learning. One solution to this would be to ensure the training programme closely reflects the work environment to ease the transfer (Yamnill and McLean 2001). The use of practical exercises in training is another means of facilitating training transfer.

Finally, Holton's model suggests that offering a suitable transfer climate is an important determinant of transfer. Empirical work by Smith-Crowe *et al.* (2003) confirms that, in high risk environments, the organisational climate for the transfer of safety training moderates the relationship between safety knowledge and performance. Thus it is therefore important

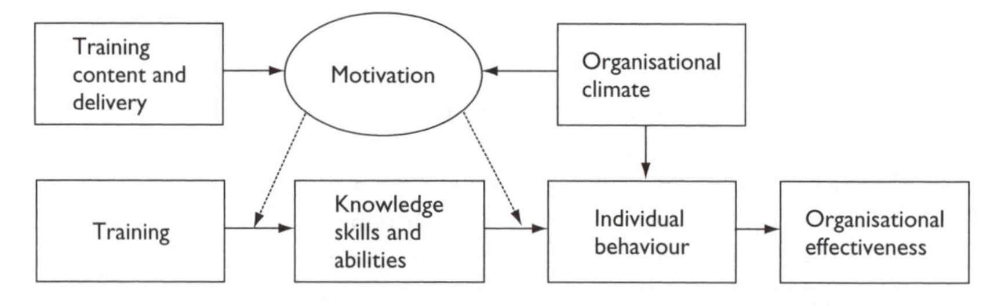

Figure 4.3 Training transfer model (adapted from Holton 1996).

that if an organisation's climate is strategically focused on the transfer of safety training, the relationship between safety knowledge and performance will be strengthened. Training programmes should not be developed and delivered without consideration of the work environment because, on completion of a training course, trainees return to the workplace and respond to cues in the environment. Cues that remind trainees of their training can facilitate transfer of that learning. Also, when the learning is put into practice it should be reinforced with praise or positive feedback to ensure that the desired behaviour is maintained. Thus, in the case of OHS training, it is important that safe behaviours are recognised and positively reinforced by others in the workplace. Failure to commend safe behaviours that have been learned through training may prevent the effective transfer of learning to the workplace. For more on organisational safety climates, see Chapter 10.

In addition to the implementation of OHS training programme, other management activities can support the development of employee OHS competencies. For example, OHS competencies can be assessed during the recruitment and placement of employees to ensure that workers have the capabilities they need to perform the work allocated to them. In the construction industry, this function should be extended to the consideration of subcontractors' OHS competencies when making selection decisions, in much the same way as construction clients might pre-qualify principal contractors on the basis of OHS criteria.

Planning and implementing

Planning is a critical part of OHS management. The implementation of the OHS management system itself needs to be planned, as do the inputs and processes required of the system. Thus, the development of the OHS policy, the establishment of organisational arrangements to implement this policy and the processes for measuring, monitoring and controlling OHS performance must all be planned.

At each stage, clear objectives, measurable targets and performance indicators should be set. Objectives are overall goals for OHS performance that state what is intended to be accomplished. Targets, on the other hand, define the performance level to be attained in a specified time frame, and should be clear and quantifiable. Measurable performance indicators are the means by which organisations determine whether objectives have been met. Objectives can be specified for the whole organisation, or be project-specific. Performance indicators can measure outputs (for example injuries) or inputs (for example number of people trained).

For example, if the objective was to eliminate manual handling injuries occurring during steelfixing, the target might be zero injuries to steelfixers during the year, and the performance indicator might be framed as the percentage of reported manual handling injuries involving steelfixers. If the objective was to integrate OHS into purchasing standards, the target might be to include OHS specifications in all purchase orders issued by the end of the year. The performance indicator may in this case be the percentage of purchase orders issued containing OHS criteria.

Multi-level OHS planning

Careful OHS planning is a key component of effective OHS management. Planning occurs at two levels within construction firms. Corporate or strategic planning occurs at the head office. At this strategic level, the company's senior management determine how to meet corporate goals, establish strategic objectives and plan for their achievement. These plans are usually developed by senior executive managers and have long planning horizons. Typically, strategic plans span three to five years. As part of strategic OHS planning, senior management will consider company objectives with regard to OHS outcomes, procedures and standards, training programmes, auditing and assessment, resourcing and costing and organisational culture. Specific goals or targets will be set for each of these areas, and a timeline with milestone achievements will be formulated. Responsibilities for the achievement of strategic objectives will be stated, and resources to be deployed for the achievement of these objectives will be identified. Criteria against which success is to be gauged will be specified.

Strategic plans must be communicated to those who make day-to-day management decisions impacting upon OHS – the firm's project managers – because they must ensure that strategic objectives are reflected in their operational OHS plans. At the project level, managers must plan how to construct a facility and this construction planning requires careful consideration of the OHS aspects of the project. At this level, project managers should consult with those who will undertake the construction work, including subcontractors and undertake a systematic analysis of OHS risks involved. The tools and techniques described in Chapter 5 may be used to ensure OHS risks are identified and assessed systematically.

Although the early identification and assessment of OHS risks is important, it is not always possible to anticipate OHS risks at the commencement of a construction project, especially where design work is not complete when construction commences, as is the case with fast-track or turnkey projects. The use of numerous timings and planning horizons is useful to overcome such uncertainty. The degree of detail contained in a plan should vary inversely with the planning horizon. As the planning horizon expands, the list of activities should become smaller, and the specification of each activity more focused on ideas than precise facts and numbers. Furthermore, upper management should prepare long-term plans with low levels of detail that are infrequently updated, while lower management levels should prepare detailed, short-term plans more frequently (Laufer *et al.* 1994).

For example, a construction company might develop a strategic OHS plan, spanning a time frame of three to five years. This plan would state company objectives on broad terms and be formulated by senior managers and OHS specialists. Large projects might then formulate project-level OHS plans (Figure 4.4). These would identify risks inherent in the major elements of the project and the general methods to be used for controlling these risks. The lifespan of the project-level OHS plan is the duration of the project and the plan would be formulated by project managers in consultation with OHS specialists. For each activity or construction operation, a work method statement would then be developed. This would assess risks inherent in the activity and specify in detail the risk control measures to be adopted. The lifespan of work method statements is the duration of the activity or operation covered by them and work method statements would typically be developed by site engineers under the supervision of the project manager and in consultation with OHS specialists. Finally, the lowest level of OHS planning is a job safety analysis (JSA). This is conducted with the work crew immediately before a new task or operation commences. The foreman usually conducts the JSA, which is very detailed and identifies specific OHS responsibilities of individual crew members during the operation.

Job safety analysis

Job safety analyses, or JSAs, are useful in capturing a safe system of work and defining the procedural requirements for a particular task or operation.

Job safety analysis involves a work team, with their supervisor, analysing the operation step by step to determine the hazards involved in a given

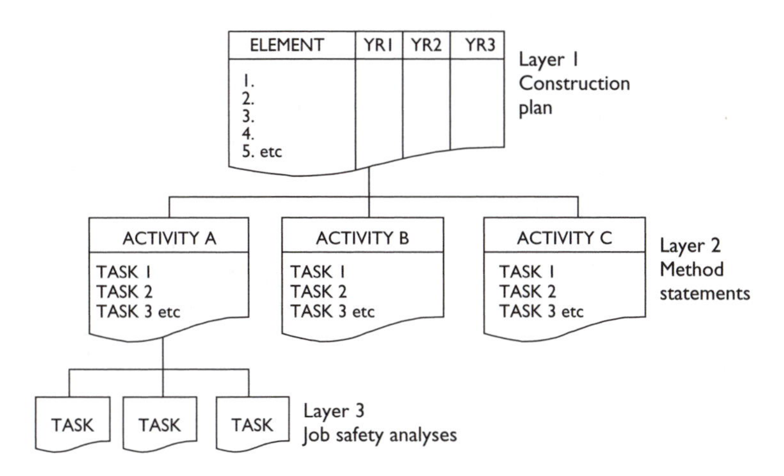

Figure 4.4 Multiple levels of OHS planning.

work procedure. The advantages of JSA are that it draws on the practical experience of the team and can provide creative, practical solutions. Team members are likely to have a strong sense of ownership of the system of work developed.

When preparing a JSA the basic steps of the job are listed in the order in which they occur. For each step all of the hazards that can occur during the job step are listed. Then, for each step of the operation, the group identifies hazards, drawing on the experience of workers who perform the task. Finally, the group considers ways in which the hazards can be controlled to minimise the risk of injury or damage at each step and the outcomes are documented in the form of a safe work procedure.

By carefully studying and recording each step of a job, identifying existing or potential job hazards (both safety and health) and determining the best way to reduce or eliminate these hazards, every component part of the operation is considered.

Resourcing

It is essential that appropriate human, physical and financial resources be allocated to implement OHS plans. In the case of SMEs resources may be scarce. However, there are many sources of external assistance that can be brought in to overcome resourcing constraints. For example, technology or resources could be shared with larger client organisations. The concept of partnering in the construction supply chain lends itself to the sharing of OHS management expertise between contractors and subcontractors, for example. Industry and employer organisations also provide support and guidance in OHS management to members, and government enforcement agencies provide extensive guidance material on how to implement an OHS management system, much of which is now being targeted towards small businesses. For example, in 2003, the Victorian WorkCover Authority distributed a CD to all small businesses in the Australian state of Victoria, explaining OHS management responsibilities. There are also OHS seminars and public courses conducted by consultants and other organisations, and increasingly, workers' compensation insurance agents are extending the services they provide to include OHS advice.

OHS management activities may also be readily integrated into existing processes, reducing the resources required to implement them. For example, OHS information systems can be integrated with existing information systems, OHS documents can be controlled using established processes developed to support quality management systems, and OHS training can be integrated into existing training programmes. The opportunities to use established processes to support the implementation of OHS management plans should always be explored.

Job safety analysis pro forma

JOB SAFETY ANALYSIS	JOB TITLE (and number if applicable)__________ PAGE________ OF_______ JSA NO._____	DATE:	NEW REVISED
	TITLE OF PERSON WHO DOES JOB:	SUPERVISOR:	ANALYSIS BY:
COMPANY/ORGANISATION:	PLANT/LOCATION:	DEPARTMENT:	REVIEWED BY:
REQUIRED AND/OR RECOMMENDED PERSONAL PROTECTIVE EQUIPMENT:			APPROVED BY:
Work Activity (Job):			
Work Team:			

Sequence of Basic Job Steps	**Hazards or Potential Incidents**	**Recommendations to Eliminate or Reduce Risks**

	YES	**NO**		**YES**	**NO**		**YES**	**NO**
Hard Hats?	☐	☐	Work Vests?	☐	☐	Barricades?	☐	☐
Safety Shoes?	☐	☐	Safety Harness?	☐	☐	Fire Extinguisher?	☐	☐
Safety Glasses?	☐	☐	Face Shield?	☐	☐	Lock-Out/Tag-Out?	☐	☐
Cotton Gloves?	☐	☐	Goggles?	☐	☐	Work Permit Required?	☐	☐
Hearing Protection?	☐	☐	Respirator?	☐	☐	MSDS?	☐	☐

Measuring and monitoring performance

Measuring and monitoring OHS performance is a key activity in making sure that the organisation is achieving its OHS policy, objectives and targets. Without measuring performance there is no way of knowing whether OHS is being managed satisfactorily, nor is it possible to hold managers accountable for OHS. Managers should be given the responsibility for measuring OHS performance and monitoring the achievement of OHS objectives in the areas in which they manage. Thus, project managers must ensure that project OHS performance is measured and that key performance indicators specified in project OHS plans are met.

The measurement of OHS performance involves a number of functions. These include:

- providing an indication of how the organisation is performing in OHS;
- identifying problem areas in which improvements are needed;
- providing the ability to track performance over time and evaluate the effectiveness of OHS improvement strategies; and
- providing data that can be used in benchmarking or comparative performance assessments, for example between projects.

As described above, the identification of suitable performance indicators is an important part of the OHS planning process.

There are two different types of performance indicators. Some measures focus on outcomes which are often things that have already gone wrong, for example injury incidence rates and workers' compensation claims. These are sometimes called 'lagging indicators' because they measure events in the past.

Other measures focus on inputs or processes. By inputs and processes, we mean things that are done to prevent occupational injuries and illnesses. For example, inputs might include policy, plans and procedures and processes might include consultative processes, training and auditing. These measures are sometimes called positive performance indicators or 'leading indicators' because they identify things that are done to prevent negative OHS consequences. Measuring inputs and processes is concerned with checking that OHS is being managed systematically and that methods for controlling OHS risk remain in place.

There is no single measure of OHS performance and a combination of leading and lagging indicators is usually recommended.

Organisational injury or incident rates are the most widely used OHS measures. These rates can be compared with industry and national figures to determine the organisation's relative safety performance. Several statistics are normally computed. For example, a common statistic is the accident frequency rate, which is computed per million hours of work, as follows:

$$\frac{\text{Number of accidents} \times 1,000,000}{\text{Number of work hours in the period}}$$

Another common statistic is the accident severity rate, which is computed as follows:

$$\frac{\text{Number of workdays lost} \times 1,000,000}{\text{Number of work hours in the period}}$$

Outcome indicators are widely used because they are easy to collect and easily understood. The links between injury or incident rates and OHS performance are clear and irrefutable and rates can be used for benchmarking or compared over time to identify trends.

However, incident frequency and severity rates should be used with caution, as they are based upon statistically rare events and are known to be subject to random variation, particularly when the number of work hours is low. These rates do not take into consideration the potential seriousness of an incident. For example, a falling brick is a hazard with the potential to cause a very serious injury, but whether it does so or not is largely a matter of chance, depending on whether someone is directly below the brick when it falls. If no one is struck, the incident is unlikely to be recorded, despite the potential for serious injury. As this example shows, incident rates can also be deceptive in that a low incident rate does not necessarily mean that OHS risk is being controlled effectively.

Also, incident rates do not measure occupational illnesses or disease with long latency periods very effectively, and therefore only reflect half of the OHS equation. Outcome measures may also be vulnerable to under-reporting, particularly when indicators include the incidence of near-miss or first-aid-only incidents, which workers often consider too minor to report.

For these reasons a combination of leading and lagging measures of OHS performance should be used. Advantages of using leading OHS indicators include:

- providing the ability to measure the effectiveness of OHS management;
- providing immediate feedback on OHS management activities; and
- permitting improvements to be made to OHS management, before injuries or illnesses occur.

The National Occupational Health and Safety Commission (NOHSC) (1999), undertook a series of case studies in the Australian commercial, civil, heavy engineering and domestic construction sectors, and identified leading OHS indicators that are in use, or which participants think should be in use in the industry. NOHSC reports that participants in all sectors of the construction industry recognised the need for input or process OHS measures, and were ready to implement such measures, with the exception of the domestic construction sector. Participants in the domestic construction sector had a lower uptake of input or process OHS measures, and required simple purpose-specific OHS information. NOHSC concludes that further assistance is needed before leading OHS indicators will be adopted in the domestic construction sector.

The leading OHS indicators identified as a result of the NOHSC case studies are classified under five headings. These are:

1 Planning and design
2 Management processes
3 Risk management
4 Psycho-social working environment and
5 Monitoring.

It is noteworthy that leading indicators for planning and design measures were identified by construction industry participants, and it highlights the need to incorporate OHS into pre-construction activities. The leading performance indicators in planning and design were the following:

- the extent to which the design of the structure enables safe construction rated on a scale of one (safety neglected) to six (safety effectively built in);
- the extent to which site set up contributes to safe construction rated on a scale from one (safety neglected) to six (safety effectively built in);
- the extent to which planning and scheduling contributes to safe construction, rated on a scale of one (safety neglected) to six (safety effectively built in);
- the percentage of design changes required as a result of OHS problems calculated over a specified time frame; and
- the percentage of incidents where poor design was a root cause, calculated over a specified time frame.

It is recommended that performance indicators be carefully selected to reflect organisational and/or project-level OHS objectives and targets.

Workplace inspections

One form of OHS monitoring is systematic inspection of the workplace, plant and equipment. Inspection is a vital tool in recognising existing and potential errors and hazards in the workplace.

Occupational health and safety inspections can be carried out in a variety of ways by various people. There are planned inspections that can be carried out by a supervisor or manager in conjunction with an employee health and safety representative or other relevant employees. These are undertaken at regular intervals and are deliberate, thorough and systematic by design. An established procedure is followed using a checklist to examine the area being inspected in detail. Routine inspections can cover the workplace as a whole or focus on a particular aspect of OHS, such as plant and equipment or office ergonomics. Inspections can also be performed by plant operators or vehicle drivers before they start up the plant or vehicle.

In addition to routine inspections, intermittent inspections can also be made at irregular intervals when the workplace has been modified or changed, or when incident statistics indicate a specific risk exists.

More informal hazard-spotting activities by all employees are also characteristic of organisations with strong safety cultures. This enables employees to bring OHS hazards to the attention of the relevant people, who can then initiate appropriate corrective action.

To ensure a systematic and thorough approach, an inspection checklist should be developed. The checklist must be customised to the needs of the particular workplace. Some things that may be included in a general workplace OHS inspection checklist include:

- the work environment – dust, gases, fumes, sprays, lighting, noise, ventilation;
- buildings and site sheds – floors, stairs, roofs, walls, etc.;
- rubbish and waste bins;
- electrical equipment – switches, generators, cables, distribution points, connectors, grounding, connections, circuit breakers; residual current devices (RCDs);
- fire protection equipment – extinguishers, hoses, alarm systems, access to equipment;
- hand tools – wrenches, screwdrivers, saws, power tools, explosive power tools;
- hazardous materials – flammable, explosive, acidic, caustic, toxic substances;
- lifting equipment – cranes, hoists, conveyors, lifting gear, chains, etc.;
- personal protective equipment – hard hats, safety boots, goggles, respirators, ear protection, etc.;
- pressurised equipment – piping, hoses, couplings, valves;
- access equipment – ladders, trestles, scaffolds, platforms, catwalks, staging;
- powered equipment – compressor equipment, mobile plant, etc.;
- storage facilities – racks, shelves, cabinets, closets, yards, floors;

- walkways and roadways – gangways, ramps, vehicle access routes, etc.;
- machinery guarding – guards, railings, etc.;
- safety devices – emergency switches, cutoffs, warning systems, limit switches, mirrors, sirens, warning signs;
- controls – start-up switches, steering mechanisms, speed controls, manipulating controls;
- lifting devices – handles, eye-bolts, lifting lugs, hooks, chains, ropes, slings; and
- hygiene and first-aid facilities – water supply, toilet facilities, washrooms, lunchroom, first-aid supplies.

Inspections should be recorded, and each hazard found during an inspection should be located and described. Accurate descriptions are important; for example, items of plant and machinery should be identified by their correct names, and unique registration numbers and locations must be accurately described. Specific descriptions of hazards should be given; for example, instead of noting 'trip hazard', the report should give precise details, such as 'electricity cable spanning gangway'.

It is also important to determine which hazards present the most serious consequences and are most likely to occur so that corrective actions can be prioritised. The information and recommendations made after an inspection also provide the basis for establishing priorities and implementing corrective actions. Some of the general categories into which recommendations might fall are setting up safety work processes, redesigning a tool or fixture, re-organising the workplace, for example relocating stockpiles to reduce manual handling requirements, changing workers' work patterns, providing OHS training or providing personal protective equipment. Emphasis should always be on changing the work environment to make it safer wherever possible, rather than changing workers' behaviour. Recommendations may also call for improvements in the system of maintenance or site housekeeping. For example, maintaining a tidy site free from rubbish and debris may not be any individual worker's responsibility, but preventing the accumulation of debris is an important part of an OHS risk control effort. As such, it is ultimately the site manager's responsibility to ensure that this is done. Recommendations arising from workplace inspections should be sent to the relevant personnel for approval and, where possible, a definite time frame within which the corrective action is to be carried out should be set for each recommendation.

Auditing and reviewing

All aspects of the OHS management process should be subject to regular systematic assessment and review. Reviewing OHS performance is the key to learning from experience and continuous improvement. Performance review is the final stage in the OHS management process. There is a very important feedback loop between this stage and all other aspects of OHS management. The review is concerned with determining the adequacy of performance of the OHS management system, and ensuring that improvements are made where possible. Organisational reviews of OHS performance should be conducted at regular intervals, perhaps every year. However, in the construction industry, post-project OHS performance reviews are also helpful because they enable lessons to be transferred from one project to the next.

The OHS performance review should be based upon data collected when measuring and monitoring OHS performance, using both leading and lagging indicators. The results of OHS audits are also useful in the review because they represent an independent and systematic evaluation of the entire OHS management system. Questions to be posed in the review include the following:

- Did the organisation or project reach its targets and objectives? If not, why not? Should the targets or objectives be changed?
- Is the OHS policy still relevant to the organisation's operations?
- Are OHS roles and responsibilities clear and appropriate?
- Are resources being deployed to OHS appropriately?
- Are OHS procedures clear and adequate? Are new procedures needed? Should some procedures be revised?
- Is OHS being monitored effectively – for example, by audits? What do the audit results indicate?
- What effects have any changes in the organisation's materials, products or services had on OHS?
- Have any changes in OHS laws necessitated any changes to the organisation's OHS policy or procedures?
- Since the last OHS review, have any stakeholder concerns been expressed? If so, what are these?
- Is there a better way to manage the organisation's OHS performance? How can the management of OHS be improved?

Auditing is defined as 'the structured process of collecting independent information on the efficiency, effectiveness and reliability of the total safety management system and drawing up plans for corrective action' (HSE 1991, p. 66).

All aspects of the OHS management process should be subject to regular systematic evaluation through OHS audits that reveal how well an OHS management system is functioning. Audits present valuable opportunities for managers to identify weaknesses in the OHS management system, and to develop methods to improve the effectiveness of the system.

In an audit, the entire OHS management system, including the OHS policy, performance standards, planning processes, organisational arrangements and measurement and monitoring activities, is evaluated.

The HSE (1991) identify two approaches to auditing. These are:

1 the 'vertical slice' approach and
2 the 'horizontal slice' approach.

In the vertical slice approach, the OHS management process is evaluated with regard to one specific OHS issue; for example, an audit of the policy, planning and implementation and monitoring of emergency arrangements might be conducted. In the horizontal slice approach, one OHS management stage is evaluated as it applies to all aspects of the organisation's business. For example, an audit of the planning process might be conducted, as it applies to personal protective equipment, OHS training, hazardous materials management, etc. The HSE recommend that both approaches be combined to give a full picture of the adequacy of an organisation's OHS management system. This does not necessarily have to be undertaken in one single audit, but could be performed over time. For example, construction projects or sites to be audited could be randomly selected each year in a rolling programme of audits.

Audits are often undertaken by third-party auditors who may accredit the organisations they audit based on the organisation's compliance with a pre-determined set of system requirements. Several 'off-the-shelf' OHS management systems audit methodologies are available, some of which are specifically designed for use in the construction industry, for example, the Construction CHASE audit developed by HASTAM in the UK. Given the unusual nature of some aspects of OHS in construction, for example the need to manage subcontractors' OHS and the integration of OHS into design decision-making, a construction-specific audit system is likely to be most appropriate.

Incident management

Thus far in this chapter, we have described processes for the prevention of occupational injuries and illnesses. While it is true that 'prevention is better than cure', managers must also prepare for incidents that do occur.

A work-related incident is defined as any unplanned event that occurs as a result of work (or of any activity undertaken at a work place), and that results in, or has the potential to cause, injury, ill health or other loss. This could include:

- personal damage
- property damage
- environmental damage and
- potential damage (near miss).

Sound incident management can minimise the harm, to both individuals and the organisation, resulting from OHS incidents.

First priority

The first priority at an incident scene is to ensure people are safe. For example, hazards should be minimised or removed and an area may need to be cordoned off, barricades erected or the site evacuated. This also helps to ensure that evidence is not destroyed before an investigation. Prompt first-aid treatment can reduce the severity of injuries and improve rehabilitation outcomes. Counselling may also be necessary for injured workers, their family members or witnesses to workplace incidents.

Incident reporting

The reporting of an incident, no matter how minor, is an essential first step in future incident prevention. Over time, incident data can be analysed to identify patterns in incident occurrence and evaluate the effectiveness of prevention strategies.

There are two types of incident reporting, internal and external. All incidents should be reported internally. This includes 'lost time' injuries, 'no lost time' injuries and near misses. This internal notification enables the investigation process to begin. All incidents should be investigated. Because of the fact that near misses will occur far more frequently than injuries, accurate reporting and thorough investigation of minor injuries and near misses is very important, and this information can often be used to prevent more serious injuries from occurring. There are usually legal requirements governing the external reporting of certain types of incidents, for example those causing an absence from work of more than one shift, or which are classified as dangerous occurrences, for example when a crane tips over.

Organisations need to provide a system for incident reporting, both internal and external, and for the storage of incident data – which is also usually a legal requirement. In addition to this, some incidents may need to be reported to other authorities; for example incidents that result in a workers' compensation claim usually need to be reported to the organisation's insurance provider, and damage to overhead power lines or underground services needs to be reported to utility companies.

Incident investigation

Once incidents are reported, the investigation process can commence. The purpose of investigation is to analyse what occurred and identify appropriate corrective actions.

Investigation includes searching for objective facts about the incident, including statements, opinions, physical evidence and related information. Investigation also involves evaluating the available evidence to reduce the possibility of a recurrence. The best time to investigate an incident is as soon as possible after it is reported. Facts and details are clearer and physical evidence is undisturbed. The investigation should therefore begin as soon as the needs of the injured employee have been met.

All internal investigations into workplace incidents should follow these guiding principles. There are three main methods for conducting effective incident investigations. These are:

1 *Observation.* Observing the scene, taking photographs and recording damage are important components of an investigation. Making use of eyewitness observations is also helpful but it is important to ascertain whether they actually saw the incident.
2 *Preparing a description of the incident.* A detailed description of the sequence of events leading up to the incident will be needed. This can be based on evidence such as photographs, statements, plans of the workplace, etc.
3 *Analysing the information gathered.* The incident details should be analysed to determine what were the essential factors in the incident. The key question to ask is 'Would the incident have happened if this factor was not present?'

Physical evidence may be subject to rapid change or obliteration, and it is important that it is recorded immediately. Photographs or sketches can be useful to record features of the incident scene. Some physical evidence, for example items of broken equipment, may need to be removed for further analysis by appropriate experts. Depending on the nature of the incident, the following physical features of the incident site may be checked:

- positions of injured workers
- equipment
- materials
- safety devices
- positions of appropriate guards
- positions of machinery controls
- damage to equipment
- housekeeping
- weather conditions
- lighting levels and
- noise levels.

In gathering data during an incident investigation, it is very important not to use words like 'blame', 'cause', 'careless', 'fault', 'wrong' or 'bad'. Emotive language generates an atmosphere of defensiveness, and may even lead to the cover-up of valuable information. The main focus of the investigation must be to identify contributing factors and methods for their control.

It is important to recognise that no one thing can be solely be blamed for an incident, which is usually caused by a complex interaction of people, environment and equipment factors. Also, the underlying causes of the incident could have occurred quite some time before the incident and may be physically remote from the incident scene, for example in the case when a design decision contributes to an incident on a construction project.

It is important that both immediate and underlying causes are identified. The extent to which incident investigations identify underlying causes was investigated in a contract research study undertaken on behalf of the UK's HSE (Henderson *et al.* 2001). Henderson *et al.* (2001) identified four approaches to incident investigation currently in use. These are:

1 Approach 1 – in which there is a complete lack of documented structure or support for incident investigation.
2 Approach 2.1 – in which there is minimum formal support but where the focus is on identifying the immediate cause of the incident.
3 Approach 2.2 – in which there is a more structured approach and the focus is on identifying the immediate and underlying causes.
4 Approach 3 – in which the causal analysis is supported by the use of more sophisticated tools and methods of analysis.

Henderson *et al.* (2001) report that, within most companies, incident investigations focus on identifying only immediate causes and often focus on the individual concerned. Also, in a large percentage of companies, including some of the bigger organisations participating in the study, incident investigations adopt Approach 1, in which there is no systematic approach or structure. In these cases, the quality of the investigation will largely be driven by the person leading it, and there is unlikely to be any consistency of approach. These findings suggest that, in many instances, incident investigations may not be contributing as much as they should to the incident prevention effort.

A systematic approach to incident investigation should be developed based upon an understanding of incident causation models, such as those described in Chapter 1. Also, the investigation process and outcomes should be documented using a standard format that includes recommended preventive actions, assigns responsibilities and establishes a time frame for their implementation. An example incident investigation report format is provided below.

Example Incident Investigation Report Format

1. Incident was reported on (dd/mm/yy) ____________

2. Incident Report Form reference no. ____________

3. What was the outcome of the incident (i.e. injury, illness, damage, near miss)?__________

4. Provide names of witnesses to the incident

Name	Position

5. Investigation undertaken by:

Name	Position	Signature	Date (dd/mm/yy)

6. Provide a description of the incident . Include what the people involved were doing at the time of the incident, specifying any work process, chemical item of equipment involved. Attach a plan view of the incident location if necessary.

7. Describe the *sequence* of events leading up to the incident. Include approximate timings of events in this sequence.

Date/time	Event

8. Describe the level of supervision in place at the time of the incident.

9. Describe the training provided to people involved in the incident.

10. Describe the safety equipment being used by people involved at the time of the incident.

11. Describe the emergency procedures followed immediately after the incident.

12. Describe the *immediate* causes of the incident. These can be actions (deviations from accepted work practices) or conditions (characteristics of the work environment). Note that all incidents are caused by management system failures and employee carelessnessis not a valid cause.

13. Describe the *underlying* causes of the incident. Detail all management system failures that led to the unsafe action or conditions cited in 12.

14. Identify appropriate corrective action to be taken to prevent a reoccurrence of the incident. Allocate responsibilities for implementation and the proposed completion dates. Those responsible must indicate when actions are complete.

Action	Responsibility (name)	Planned completiondate	Completed on

15. Record the supervisor's comments on the incident

Signature: Date (dd/mm/yy):

16. Record the occupational health and safety representative's comments on the incident.

Signature: Date (dd/mm/yy):

The investigation report should define a plan to prevent a recurrence of the incident. Therefore it is important to identify who is responsible for completing recommended actions and ensure that this information is communicated to all concerned.

Emergency planning

Emergency planning is an important part of incident management to ensure that, should an incident occur, people respond in an appropriate manner. If the OHS management system is effective, emergencies should be rare. However, they do happen and the infrequency with which they happen means that people will not necessarily know how to respond appropriately. It is important that possible emergency situations be identified, and that emergency management plans be put in place to prepare those at the work-site for such situations.

Emergency management plans are a set of written instructions that describe how people at a workplace should respond in an emergency situation. Emergency management plans should be clearly communicated to all workers on the site, and it is important that emergency procedures are communicated to new workers and visitors to the site during an initial site induction process. Emergency procedures should also be displayed on notice boards at prominent locations on site. Workers should be trained in emergency responses and regular emergency drills should be conducted to ensure that appropriate responses are 'over-learned'.

Emergencies that could arise include:

- fire or explosions
- chemical spills or dangerous gas emissions
- structural collapses or failures
- excavation cave-ins
- medical emergencies and
- violence or bomb threats.

Emergency management plans should be developed after an assessment of the hazards and the risks on a site has been undertaken. For example, during the construction of a jetty, where there is a risk of workers falling into the water and drowning, the emergency management plan should include a water rescue procedure and the provision of a rescue boat. It is also important to consider external hazards when developing emergency management plans. Thus, if the site is adjacent to a facility in which large quantities of flammable materials are stored, the possible impact of a fire at the site should be considered in the site emergency management plan.

An emergency management plan should identify a person with suitable skills who will be responsible for co-ordinating the implementation of the emergency management plan in the event of an emergency. It should also specify evacuation procedures and a safe meeting point, and identify the person responsible for ensuring that all workers and visitors to the site are accounted for in the event of an emergency evacuation. Other important responsibilities, for example shutting down power where appropriate, should also be clearly allocated to individuals in the emergency management plan.

The emergency management plan may also provide for the installation and regular testing of a suitable warning or alarm system. Site plans should be available clearly showing the location of emergency exits, fire protection equipment and meeting points. Routine worksite inspections should ensure that these exits remain clear and that fire-fighting equipment is accessible. This is particularly important in construction projects in which the work environment constantly changes and workers often stockpile materials in convenient locations on site without necessarily considering access to emergency exits or equipment.

The emergency management plan should also include emergency contact details and procedures for contacting both people on site and outside emergency services. In some construction projects, the risks are higher than others, and it may be necessary to invite the emergency services onto the site and involve them in the emergency planning process. In such cases, the emergency services should be provided with the site layout, and the location of storage areas of combustible or toxic substances.

A first-aid risk assessment should be performed to identify potential causes of injury and illness in the workplace and assess the risk of these injuries and illnesses occurring. On the basis of this risk assessment, the need for first-aid training, first-aid kits and first-aid rooms should be determined. First-aid facilities should be provided to suit the workplace hazards and proximity of medical facilities. Remote work sites need to have first-aid facilities that are appropriate to stabilise injuries until medical help can be accessed. This may also include transportation arrangements to the nearest medical facility. In construction, it is important to ensure that suitably trained first aiders are available on all shifts, and that their contact details are clearly communicated to all on site and displayed at prominent locations.

Crisis management

A recent study by Loosemore and Teo (2002) revealed that several of the largest construction companies in Australia were ill-prepared for organisational crises. The view that 'it can't happen here' prevailed. Campbell (1999) notes that we live in a world in which crisis is not discussed. Yet, an unwillingness to accept the possibility of crisis, and a belief that crisis management planning is an admission of poor management, leave organisations vulnerable to crisis (Mitroff and Pearson 1993). A crisis is distinct from an emergency in that it is 'an adverse incident or series of events that has the potential to seriously damage an organization's employees, operations, business and reputation' (Campbell 1999, p. 11). A crisis can destroy the reputation or credibility of an organisation, damage its financial performance and threaten its very existence. Fink (1986) suggests that an organisational crisis is happening when the situation is:

- escalating in intensity;
- falling under close media or government scrutiny;
- interfering with normal operation of the business;
- jeopardising the positive public image enjoyed by the company or its management; and
- damaging the company's bottom line (either directly or indirectly).

Thus, crisis management must deal with broader corporate issues than emergency management, and should not be guided by an emergency management mindset. While emergency management focuses on incident response procedures, crisis management must establish protocols and procedures for communicating important information to government agents, the media, the community and other external stakeholders as well as planning for business recovery. As Campbell (1999) suggests, a small fire in a plant might be classified as an emergency, requiring an incident response to extinguish it. However, if the fire spreads and reaches a shed containing flammable liquids, causing an explosion that kills or injures employees and results in considerable damage to the plant, then the emergency has escalated into a crisis.

Crises are low-probability but high-risk events and can have many sources. Reid (2000) suggests that crises can result from natural disasters, such as floods, earthquakes, extreme snow/ice, lightning strikes, hurricanes, typhoons or tsunamis. The potential for some of these natural sources of crisis to impact upon construction activities should not be dismissed as negligible, particularly when working in areas of the world prone to earthquakes, tropical storms or extreme weather conditions. Crises can also emanate from operational issues, including:

- equipment failure
- vehicle accidents
- explosions
- structural collapses and
- design errors.

In a survey of 149 US-based general construction contractors, heavy/highway contractors, subcontractors, manufacturers and suppliers of construction products, engineers and construction clients, the number one type of crisis experienced three years prior to the study was cited as 'on-the-job accident'. This source was top ranked in both 1988 and 1996. On-the-job fatalities were ranked fifth and fourth in 1996 and 1988 respectively (Reid 2000). While exposure to hazardous substances and occupational disease did not appear in the top ten rankings, occupational health impacts should not be ignored as possible sources of organisational crisis. This point is reinforced by the media attention devoted to asbestos-related diseases and the high-publicity court cases arising from occupational exposure to asbestos.

Crisis management process

Campbell (1999) observes that the crisis management process must be flexible in order to cope with a broad range of types of crisis. He identifies the following five stages in this flexible process:

1 identification/discovery
2 preparation and planning
3 response/control
4 recovery and
5 learning.

The *identification/discovery* stage involves identifying the imminent onset of a crisis. Berger (2001) suggests that an important source of warning signs of impending crises is not being observed. This source is the intuition of workers themselves that something is 'just not quite right'. He terms workers' feelings about danger 'the critical eyes and ears on the ground', and suggests that these feelings are the earliest signals, or 'fuse-breakers', that organisations should encourage workers to report. While not all impending crises may be detected by such means, the reluctance of workers to express their unease – because of fear of being branded a 'trouble-maker', fear of job loss or due to earlier failure of management to respond – must be overcome if problems are to be detected early.

Preparation and planning involve identifying threats, formulating crisis plans and fostering good relationships with external stakeholders and the media, so that when a crisis occurs, the organisation and its members are prepared and know how to act appropriately.

The *response/control* stage intends to stop the crisis from escalating and limit the physical damage to people and property caused by the crisis. This step also involves taking control of information concerning the crisis (see also, handling the media, below).

The *recovery* stage is concerned with ensuring a fast and orderly recovery from the crisis. In the event of worker fatalities or injuries, it will involve looking after the welfare and counselling of family members and other affected employees.

Lastly, in the *learning* stage, an evaluation of the crisis management process is undertaken to ensure that lessons are learned and crisis management plans and procedures are improved as a result of the crisis experience. As the Longford disaster case study (p. 175) reveals, serious mistakes were made in the identification, planning, response and recovery stages in the crisis management process. These mistakes have had serious implications for the organisation concerned.

Crisis management planning

It is critical to plan what to do in the event of a crisis because crisis situations require clear decision-making under intense pressure, often in life-and-death situations. Crises place managers who are otherwise highly competent decision-makers under immense stress, and it cannot be assumed that even the best of managers will respond appropriately in crisis conditions without a plan to follow. Despite this, construction organisations do not place great importance on crisis management planning. Reid (2000) reports that, in 1988, only 31 per cent of the construction organisations she surveyed had crisis management plans, compared to 41 per cent in 1996. Loosemore and Teo (2002) reveal that, in the Australian construction industry, 21 per cent of managers believe that it is difficult to plan for crises, and there was a widespread belief that crisis planning was an activity carried out solely in high risk industries such as the petrochemical industry. None of the Australian organisations involved in the study by Loosemore and Teo (2002) reported having an established crisis management team, and crisis management plans were informal and only senior managers received crisis response training. This complacency leaves the construction industry vulnerable and ill-equipped to deal with any organisational crises that arise.

Crisis management plans should be kept simple so that they can be easily comprehended and put into action in the event of a crisis. Owing to the sensitive nature of some of the information they contain, they should be controlled documents, distributed to a selected group of people. They should have tables of contents and clearly numbered pages for ease of use. Key components of the crisis management plan include:

1 The crisis plan should document immediate actions required of the crisis team, and should include a first-hour response list specifying the immediate actions to be undertaken by the senior person on the site, the crisis team leader, the OHS manager, the media spokesperson and the human resources manager. Having this clearly documented allows key people to respond quickly and appropriately, relieving the immediate pressure.

2 The crisis management plan should contain a list of key steps or procedures to be followed by the members of the crisis team during the crisis. Checklists are particularly useful to ensure that steps are not missed out. Procedures for the notification of families of accident victims are especially important due to the sensitivity of this communication. This notification should occur as quickly as possible after the event. Reid (2000) suggests that in the case of fatalities, face-to-face notification should be made unless it is physically impossible to do this. She recommends the use of honesty and simple language to ensure that the message does not cause confusion. Families of injured workers should be transported to the hospital as soon as possible, in company-provided transport and care must be taken not to comment on the severity of injuries.

3 The crisis management plan should contain an emergency contacts list, because a large number of people will need to be informed and in a crisis there is little time to look up contact details. Day, night and mobile phone numbers should be included because crises often happen outside normal working hours. For this reason Reid (2000) recommends that two copies of the crisis management plan be given to the crisis team, one to be kept at home. The crisis management plan must be updated frequently to ensure that team members and their contact details remain current.

Handling the media

Failure to talk to the media in a crisis situation is dangerous because it can be taken as a sign of guilt, a perception that may be difficult to dispel later on. Taking control of information that is transmitted to the media is the recommended course of action. Reid (2000) recommends that in the immediate aftermath of a serious accident, a temporary spokesperson deliver a 'buy-time' statement. This statement should acknowledge the incident and provide information that has been verified. It is designed to buy time during which the company can gather more information and until the corporate spokesperson arrives. Only information that has been verified should be presented, and points should be made succinctly to avoid misinterpretation. It is also important to update employees as to the latest verified information because journalists are likely to ask questions to employees and it is important that a consistent story emerges. In the early statements to the media, questions need not be answered. It is best to provide verified information as to what happened, when, where, who was involved and the status of the site and investigation but not to invite questions. Once more time elapses, however, the spokesperson will have to answer media questions. Likely questions should be anticipated and it is important to determine what message the company wishes to convey. A determination to identify and address the causes of the incident and a concern for the victims and their families are important messages to convey. Note the victim-blaming approach adopted by oil giant Esso in the Longford disaster was not a well-received message. Providing the media with information before it is requested via news releases can also relieve pressure during an organisational crisis because it conveys the message that the organisation has nothing to hide. Unfortunately, most construction companies do not develop good relationships with journalists (Moodely and Preece 1996) and are often depicted in the media as being unscrupulous and corrupt – a situation likely to pose a problem for them in the event of a crisis.

Post-crisis review

A final stage in the crisis management process is to learn from the crisis experience. Post-crisis evaluation is particularly valuable because it is impossible to generate the same communication issues and stresses in a simulated training exercise (Campbell 1999). It is important to pose questions such as:

- Was the plan followed? Was it easy to follow or could it be made more user-friendly?
- Did team members respond in accordance with procedures?
- Was communication adequate?
- Was co-operation with emergency services and government authorities handled appropriately?
- Were there any unintended consequences?
- Did the spokesperson convey accurate and timely information to the media? Was this reported accurately?
- Was information conveyed to all company employees as appropriate?
- Were sufficient resources deployed to handle the crisis?
- Were family members of any victims informed at the earliest possible opportunity? Was this information conveyed in the appropriate manner?

Once this evaluation is complete, the lessons must be incorporated into an updated crisis management plan to ensure that any future crises are handled more effectively.

The following case study is provided as an example of a case in which both emergency and crisis management were woefully inadequate. While the emergency and ensuing crisis did not occur in the construction industry, the case study is valuable because it demonstrates the importance of identifying potential incidents, developing appropriate response procedures, training personnel in the implementation of these procedures and in communicating information to external parties in an accurate and sensitive manner. The case study also highlights the consequences of failing to manage a crisis appropriately.

Case study 4.1: Esso's Longford disaster

On Friday, 25 September 1998, an explosion and fire occurred at Esso's Longford gas and processing facilities. As a result of the explosion and subsequent fire, two men were killed, eight others were injured and many Melbourne homes and businesses were left without gas for a fortnight. The explosion occurred when flow of hot lean oil at around 200 °C was introduced into a pressurised reboiler vessel. The vessel was chilled to a temperature of around –48 °C as a result of a loss of process heating caused by a shutdown earlier in the day. The introduction of the hot lean oil into the cold plant caused a weld to sustain a brittle fracture, and the heat exchanger to fail catastrophically, releasing a volume of explosive vapour. This was ignited, causing a large flash fire. All those in the immediate area who were killed or injured were involved in the repair of the heat exchanger and were caught in the initial explosive release. The fire continued for two days, being fed by successive pipe failures. In early November the state government of Victoria appointed a Royal Commission to investigate the cause of the fire and to recommend measures to avoid a recurrence.

In its evidence to the Royal Commission, Esso claimed the accident was caused by workers, in particular targeting a control room operator.

The Royal Commission received evidence from 66 witnesses, including plant operators and supervisors employed by Esso, Esso managers, emergency response personnel and technical experts. The report concluded: 'The ultimate cause of the accident... was the failure of Esso to equip its employees with appropriate knowledge to deal with the events which occurred. Not only did Esso fail to impart that knowledge to its employees, but it failed to make the necessary information available in the form of appropriate operating procedures.' The Commissioner's report also commented upon the reduction of supervision at Longford, including the transfer of engineers to Melbourne, which had reduced the amount and quality of supervision at the plant.

The report also said that Esso's failure to report a cold temperature incident on 28 August 1998, 'deprived Esso of an opportunity to alert employees to the effects of a loss of lean oil flow and to instruct them of the proper procedures to be adopted in the event of such a loss'. It was such a loss of flow that caused the dramatic reduction in temperature of the heat exchanger that failed upon re-introduction of hot lean oil (*The Australian*, 29 June 1999). Furthermore, the management of Esso should have been aware of the rare catastrophic brittle

Case study 4.1 (Continued)

failure of pressure vessels, because their parent company Exxon published articles on such failures in 1974 and 1983. As a result of research by Exxon, the parent company had included the requirement that special attention be given to the possibility of brittle fracture failure in its hazard identification guidelines (Nichol 2001).

The control room operator was vindicated by this ruling, and commented to the media 'It is a shame that Esso chose to do what they did [shift the blame] and I think that's evident now...Society knows the type of company they are, it knows how they treat their workers and I think they've been condemned because of it' (*The Guardian*, 4 July 2001).

Esso's attempt to shift the blame for the accident onto the dead workers and their co-workers almost certainly backfired on them causing widespread public indignation, fuelled by media exposure of this attempt. Indeed, the head of the Royal Commission, Sir Daryl Dawson, commented, 'The hurt caused by [Esso's] stance that the accident was due to worker error was considerable. It's not difficult to imagine the torment caused by the suggestion that some workers were responsible for their fellow workers' deaths or injuries.'

It was reported that not long before the accident the company had relocated a number of its skilled supervisors to other plants and had done little to provide extra training or support for remaining staff (*The Guardian*, 15 November 2000). Furthermore, the media reported that, despite a string of disturbing incidents prior to the accident, the company had not provided sufficient planning, supervision or training for the remaining staff to deal with emergency situations (*The Guardian*, 15 November 2000).

In July 2001, Esso was fined a record $2.75 million dollars in the Supreme Court after being found guilty of eleven charges under Victoria's Occupational Health and Safety Act. Mr Justice Cummins described the explosion and fire as 'no mere accident' and said the responsibility for the tragedy rested solely with Esso.

In 2002, Victoria's State Coroner handed down findings of an inquest into the deaths of the two workers killed at Longford. He commented 'Clearly Esso is solely responsible for the disaster and tragedy that is known as Longford...Esso failed to conduct a periodic risk assessment which could have prevented the incident' (*The Age*, 15 November 2002).

Esso responded by expressing sorrow for the men's families and saying that, as a result of the Royal Commission's recommendations, the company had spent more than $500 million at the Longford plant. Esso also faced potential costs of billions of dollars in Australia's largest ever class action, brought by the more than 10,000 of those adversely affected by the disaster. Despite Esso appealing against this class action, the Australian Federal Court ruled that the class action should proceed. In addition, eighteen Esso workers and their families mounted a class action against Esso in the Supreme Court, and the Insurance Council of Australia sued Esso on behalf of 120 large businesses that had lost production or had shut down during the crisis.

This case study clearly demonstrates the consequences of failure to heed warning signs, inadequate crisis management, planning and training. It also reveals how clumsy handling of the media can severely tarnish a company's reputation.

Management influence on OHS

It is well established that management actions affect workers' OHS attitudes and behaviour. Research conducted in the 1970s revealed that management involvement and support for OHS activities was a characteristic of workplaces with low accident rates. For example, Simonds and Shafari-Sahrai (1977) examined data about the management activities of companies with high and low accident rates. The comparisons of matched pairs of companies revealed that in companies with lower accident rates, top management was more involved in OHS. In two similar studies, top management commitment to OHS was confirmed to be a common feature of companies with good OHS performance (Cohen 1977; Smith *et al.* 1978). Management commitment is reflected in management's knowledge of OHS issues, belief that high OHS standards are possible and demonstrated efforts to ensure these standards are achieved.

Cohen (1977) also reports that in companies with good safety records, managers frequently communicate with workers about safety issues, and use both formal and informal methods of communication. Managers can communicate what is important explicitly by developing policies, setting objectives, establishing procedures and rewarding behaviours. Cooper and Phillips (1994) suggest that perceptions of managers' OHS attitudes and behaviours can have a direct effect on workers' OHS behaviour as well as an indirect effect through indicators of managements' commitment to safety. Such indicators include the status of the OHS officers, the provision and importance placed on OHS training and the effect of safe behaviour on rewards and promotion.

Schein (1992) suggests that the way that senior managers conduct themselves will be a particularly important determinant of organisational culture; in particular, perceptions of senior managers' OHS attitudes and behaviours is likely to shape OHS behaviour.

Schein (1992) described organisational culture as a pattern of beliefs and assumptions that are shared by organisational members that operate unconsciously and define an organisation's sense of self. Thus, organisational cultures act upon organisational members to establish common values and a shared understanding of what behaviours are acceptable and what behaviours are unacceptable. Kletz refers to this as the 'common law' of an organisation (Kletz 1993). The concept of safety culture describes an organisation's norms, beliefs, roles, attitudes and practices concerning OHS (Turner 1991). Safety culture is a subset of organisational culture. A positive safety culture seeks to establish the norm that everybody is aware of risks in the workplace, feels responsible for their own safety and health as well as the safety and health of others. In a positive safety culture, everybody is continually looking out for hazards and raises any OHS concerns with supervisors and management.

Kletz (1985) suggests that culture is a much more influential determinant of workers' behaviour than many writers on OHS management systems acknowledge. He suggests it is insufficient to implement a paper system, in which formal policy statements and plans establish company objectives because it is essential to win the 'hearts and minds' of workers to elicit a common commitment to the implementation of these policies and plans.

Strong organisational cultures are those in which behaviour is consistent with espoused values. Thus, it is essential that managers 'walk the talk' with regard to OHS. Managers may send conflicting messages about what is important. For example, in the event of a worksite accident, managers may call up the project manager and ask when work at the site can start, rather than asking whether any workers were hurt. This sends the message that adhering to the construction schedule is more important than safety.

One of the authors once observed an extremely untidy worksite, at which 50 per cent of workers were not wearing hard hats, and at which damaged electrical cables were lying unprotected across wet ground, which was being traversed by mobile plant and site vehicles. At this site, scaffold platforms were not fitted with guard rails or toe-boards, and workers were not provided with fall arrest equipment. The site was adorned with posters and banners displaying the message 'Safety first', yet, when the site manager ventured onto the site to speak with the foreman, it was invariably to discuss construction progress. Although this is an extreme example, it serves to demonstrate the problems that may arise when there is confusion over organisational goals. As Turner (1991) notes, false rhetoric is likely to be seen for what it is.

Changing an organisation's culture is not easy and cannot be achieved overnight because traditional customs and practices are usually strongly entrenched and militate against new ways of thinking (Kletz 1985). O'Toole

(2002) describes an attempt by a construction product manufacturing firm to change its corporate safety culture by training managers to demonstrate leadership and commitment to OHS. The managers were trained to set a good example with regard to OHS, to provide employees with positive feedback on OHS issues and to encourage employee participation with regard to OHS. Managers were accountable for performing these activities, and had to report what they were doing in each of these areas to senior management. OHS leadership was linked to managers' bonuses and was incorporated into the managers' performance appraisals, the outcomes of which influenced annual remuneration. Accident rates began to decline following the introduction of these management initiatives, and O'Toole suggests that this is because there is a connection between management's approach to OHS and employees' beliefs about how important OHS is in the company. When managers 'walk the talk' with regard to OHS it sends a clear message to workers that OHS is an important company objective.

Vredenburgh (2002) reports that proactive safety measures, such as the selection and training of workers, are more strongly related to OHS performance than reactive measures, such as the analysis of near-miss incident data and enforcement of OHS rules by supervisors. The selection of workers is an ongoing activity and Vredenburgh's results suggest that the consideration of workers' OHS attitudes and performance in selection may be beneficial. The use of behaviour-based interviews in worker selection is recommended. This approach involves the role of a trained interviewer, who asks prospective employees to describe the types of accidents and near-misses they have experienced in the past and provide examples of when they have called OHS issues to the attention of co-workers. However, it is important that such processes be carefully administered in a non-discriminatory way.

Successful OHS management is clearly about more than simply implementing an OHS management system. The issue of organisational culture is also of critical importance. We return to this issue in Chapters 7 and 10, but it is important to recognise that 'paper' OHS management systems will not ensure that all members of the workforce, including professionals, managers and supervisors, will consistently perform their jobs with OHS in mind.

Benchmarking OHS

Benchmarking can be a useful tool for bringing about improvements in OHS. Benchmarking is the process of identifying, understanding and adapting outstanding practices from other organisations to help your organisation improve its performance. Through benchmarking, organisations can compare their OHS performance and/or management processes with top performing companies.

At its very simplest level, benchmarking can involve comparing simple outcome OHS measures, for example lost time, injury frequency rates or compensation claims, with other companies. This would indicate how much better other companies are at managing OHS and serve as a guide in establishing realistic goals for improving incident rates or reducing losses. While outcome-based benchmarking can be useful, it is usually more beneficial to perform a more detailed benchmarking exercise in which management and work practices are also examined and benchmarked. This type of benchmarking reveals not only how much better top performing organisations are at managing OHS but also how these organisations actually go about managing OHS. Benchmarking is a means of transferring learning between organisations and is a method adopted by organisations that learn effectively. We return to the need to embrace organisational learning with regard to OHS in Chapter 10.

In construction, organisations should look outside the industry to identify suitable benchmarking partners. A great deal can be learned from the OHS management systems in different types of organisations and different industries. The OHS management process is common throughout industry, and the construction industry has much to learn from the management systems utilised in other industries.

Benchmarking can also be a useful tool for stimulating the growth of a positive safety culture. Reichers and Schneider (1990) suggest that in order to bring about cultural change, an organisation must collect data about its own culture and compare this data with the cultures in other organisations. O'Toole (2002) used this approach to identify weaknesses in the OHS culture of one organisation. A perception survey was conducted to gauge employees' views of the organisation's OHS management practices. These results were then compared with normative scores from other industries. Areas in which employees perceived OHS performance to be above average and areas in which performance was below par were identified. From this analysis, cultural impediments to improved OHS performance in the organisation were identified and addressed. Given that organisational culture is an oft-cited reason why OHS initiatives are ineffective in construction organisations, this approach might help construction organisations to overcome some of these barriers through learning from the experience of organisations in other industries.

Learning from quality management systems

The principles of quality management have been widely adopted by many organisations, and quality principles have become part of the corporate cultures of many construction organisations. These are largely client-driven, as more and more construction clients demand compliance with quality standards, such as ISO 9000/9001. Generally speaking, OHS management has not been integrated into the business processes of construction organisations to the degree that quality management has been and is sometimes treated as an 'add-on' or afterthought. This suggests that client interest in OHS may be the best means to ensure that OHS is integrated into the business processes of construction organisations. However, the similarities between the principles of 'total quality' and the requirements of an OHS management system suggest there may be considerable scope for transferring lessons learnt in implementing quality management processes to assist in the implementation of systematic management of OHS.

Total Quality Management systems are based on four principles:

1. *Satisfaction of the client.* Quality is defined in terms of conformance to customer requirements.
2. *Systems focus.* It is understood that problems are caused more by the system than by individuals and improvements are made through designing and implementing well-planned processes.
3. *Zero defects.* The aim is to eliminate defects and reduce uncertainty through statistical methods of control.
4. *Measurement of quality.* The costs of non-conformance are monitored to understand the financial impact of sub-standard processes (Cox and Tait 1993).

These ideas can also be applied to the management of OHS (Manzella 1997). Herrero *et al.* (2002) suggest that employees should be regarded as customers whose OHS expectations (that is, zero incidents) must be met through the systematic management of OHS. Thus, elements of a company's quality programme could be extended to include OHS. For many construction organisations, launching an OHS management system on the back of a pre-existing quality management system is likely to be the easiest route to the implementation of a systematic approach to managing OHS. This system could then be subject to conformity assessment either by the organisation itself or a third-party auditor (Redinger and Levine 1998).

Conclusions

In this chapter we have identified the component parts of an OHS management system. The systematic management of OHS requires that organisations have a clear purpose or policy with regard to OHS, and that OHS is incorporated in business planning at all levels. Ensuring that the entire workforce, including professional, managerial and supervisory levels, has the requisite OHS knowledge, skills and abilities requires that OHS training be part of this systematic approach. Training needs should be carefully analysed, and all training should be evaluated to ensure that it is having the desired effect – changing behaviour in the workplace. OHS performance should also be carefully measured and monitored to ensure OHS objectives, targets and key performance indicators are being met. Measurement of OHS performance should include both outcome (lagging) and input (leading) OHS indicators. Periodically, senior management should review OHS management performance and independent audits of the OHS management system should be conducted to ensure that the system elements are appropriate, are being implemented and are having a positive effect. The management of incidents is also an important part of OHS management system and all incidents should be reported and investigated in an attempt to identify corrective action and prevent similar incidents from occurring in the future. Emergency and crisis management plans are also important for ensuring that incident responses are appropriate and emergencies do not escalate unnecessarily into crises. In implementing an OHS management system organisations can learn from other companies, possibly outside the construction industry, by initiating benchmarking exercises. Also, there may be opportunities to transfer learning from experiences in implementing quality management systems, which bear some similarity to OHS management systems. Importantly though, it must be remembered that while OHS management systems can assist organisations in managing OHS more professionally and consistently, these systems are not likely to be effective unless the 'hearts and minds' of employees at all levels are also won. Attention to the organisational culture and the OHS attitudes of employees at all levels therefore remain of critical importance.

Discussion and review questions

1. What are the components of an OHS management system?
2. What factors determine the extent to which training translates to behaviour change in the workplace?
3. Why are outcome measures of OHS performance of limited use? What alternatives are there to outcome measures?

Chapter 5

Managing OHS risk

Introduction

The term *risk* is often used to describe financial and business chance-taking. Within this broad meaning, the functions of risk management include the identification, analysis and control of risks, which have the potential to threaten the assets or well-being of an enterprise. While the primary objective of general management is to maximise profit, risk management focuses on minimising losses arising from unwanted and unforeseen loss-making events. Such events can result in outcomes such as property damage, liability claims, bodily injury or other consequential losses. The degree to which a company is able to manage these risks will vary. For example, staff recruitment, training and succession planning can control the risk of loss of key employees in the future. Other risks are potentially manageable, such as risks associated with personal injury, business interruptions and property damage. Careful planning and monitoring processes, such as those described in Chapter 4, can help to avoid such events. Some risks, such as natural disasters and adverse economic conditions are unmanageable by individual companies, although their *consequences* can be managed through sound crisis management processes.

Risk management is a technique increasingly used by public and private sector organisations to minimise losses through enhancing the reliability and improving the safety of their business processes and products. Risk management involves the identification and evaluation of business risks followed by sound and rational decision-making based upon this analysis. Safety relates to 'the freedom from risks that are harmful to a person or group of persons, either local to the hazard, nationally or even worldwide' (Warner 1993, p. 6). Consequently, systematic management of workplace risk is an essential feature of an OHS management programme and risk management is a key requirement of OHS legislation in Europe, Australia and other parts of the world. Thus, in accordance with good business practice, and to meet legal requirements, companies of all sizes should ensure that workplace risks are identified, evaluated and controlled.

Although the management of OHS risk is mandatory, firms are generally free to choose the methods through which they will assess workplace risks. This choice poses a problem for many small businesses because they are unfamiliar with risk management concepts (Jensen *et al.* 2001). In this chapter, we examine different approaches to OHS risk management, and present some tools and techniques for assessing risk and selecting appropriate risk control strategies. These tools and techniques vary in their level of sophistication. Some could be implemented relatively easily by small firms, in construction projects of low technical complexity. Other methods are better suited to larger, more technically complex projects. Most importantly, a technique suited to the needs of the situation should be chosen, and risk assessment undertaken in a defensible way.

Despite the recommended use of various tools and techniques in risk assessment, decisions as to the 'acceptability' of a risk are often problematic and involve a complex interplay of technical, social, economic, political and legal issues. In this chapter, we discuss limitations inherent in technical approaches to risk management. In particular, we describe influences on people's perceptions of risk and the extent to which they are willing to tolerate a risk. We also consider a sociological approach to risk. We consider the possibility that those who bear the consequences of a risk collectively have different interests to risk management decision-makers. Furthermore, laypersons' judgements about risk cannot be dismissed. Even the most complex risk quantification processes involve an element of subjectivity. Decisions about acceptability of a risk are essentially value judgements and, in the interests of equity, there is a need to reconcile the views of laypersons and so-called experts in the management of risk. This requires effective two-way communication between risk management decision-makers and those exposed to risks. In this chapter, we also explore different perspectives on risk and risk communication.

The management of risk

Risk management within industrial organisations performs three main functions. These are:

1. to consider the impact of potential risky events on the performance of the organisation;
2. to identify strategies for controlling risks and their impacts; and
3. to relate these strategies to the decision framework used in the organisation (Ridley and Channing 1999, p. 6).

The risk management process, described later in this chapter, involves the systematic identification, evaluation and control of risk. This necessarily involves an estimation of the potential loss associated with identified risks in terms of the likelihood and severity of adverse outcomes. These estimates are then used to guide risk control decisions. Broadly speaking, four strategies for risk control are available. These are:

1. *Risk avoidance*, where an activity is deemed too risky and a conscious decision is made to avoid this risk altogether.
2. *Risk retention*, where financial losses are retained by organisations. Risks may be retained with decision-makers' full knowledge, for example in the case of organisations that opt to be 'self-insurers' with regards to workers' compensation provisions. Risks may also be retained without knowledge, usually because risk identification/evaluation activities have been inadequate.
3. *Risk transfer*, where risk is transferred to another party, for example through insurance policies or contractual clauses. The extent to which OHS risks can be transferred is typically limited because OHS legislation states that responsibilities cannot be 'contracted out'. Thus, in construction, principal contractors are usually unable to transfer legal OHS liabilities to subcontractors, even if other business risks are so transferred. See Chapter 2 for detailed consideration of legal responsibilities for the OHS of contractors and contingent workers.
4. *Risk reduction*, where actions are taken to reduce the risk through the implementation of a programme designed to minimise losses arising as a result of accidents in the organisation.

In industrial activities, it is often not possible to avoid OHS risks altogether. In many cases, known risks are unacceptable and cannot justifiably (or legally) be retained. OHS management practices therefore usually require that risk be reduced to within tolerable limits. Questions as to who decides what level of risk is tolerable, and on what basis, are problematic, and we will return to these issues later in this chapter.

Loss control theory

The control of losses incurred through the mismanagement of OHS has been recognised as an important function of business management (Miller and Cox 1997). Loss control has been defined as 'a management system designed to reduce or eliminate all aspects of accidental loss that may lead to wastage of the organisation's assets [including] manpower, materials, machinery, manufactured goods and money' (Ridley and Channing 1999, p. 9). Loss control is principally an economic approach to risk management. It involves the identification and assessment of the organisation's risk exposures, the selection of risk reduction measures based on economic feasibility and cost-benefit analysis and the implementation of a loss control programme, within economic constraints (Bird and Loftus 1976).

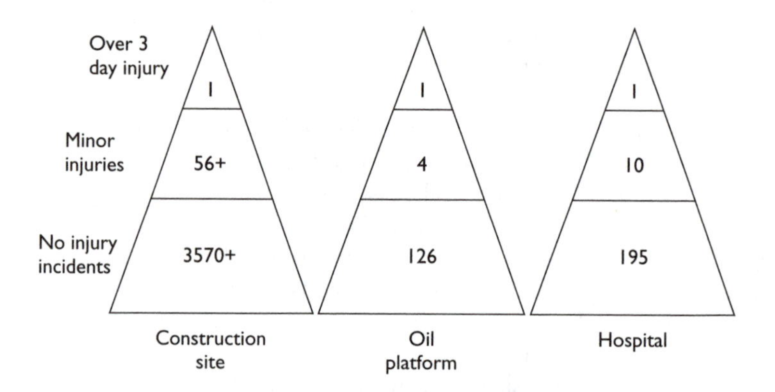

Figure 5.1 Accident triangles (adapted from HSE 1993, p. 13).

Loss control programmes aim to prevent loss-making events, such as accidental injury, occupational illness, property damage, fire, security breaches, environmental damage and product safety issues. Loss control programmes also have a damage control element, seeking to identify potential problems using 'near-miss' incident recalls to improve loss control strategies before actual losses are incurred.

Loss control principles hold that for every event resulting in a serious occupational injury there are many more incidents resulting in minor injury requiring first-aid treatment. Minor injuries are similarly outnumbered by incidents resulting in damage to property, and these in turn are exceeded by incidents in which no visible injury or damage resulted, but in which the potential for loss was present. Accident triangles have been estimated to determine the ratios between these different types of occurrence (see Figure 5.1). Different ratios have been estimated for different industries. A recent study revealed that in the UK construction industry, for every injury resulting in an absence of more than three days, there were more than 56 first-aid-only injuries, and more than 3,570 non-injury accidents (HSE 1997).

The implication of these accident triangles is that OHS performance indices that focus on serious or reportable injuries fail to show the true potential for loss in an organisation's business activities. It is argued that loss prevention strategies must take into account the alarmingly frequent occurrence of incidents which have the potential to cause serious injury but which do not do so, largely due to chance. Loss control theorists argue that non-injury incidents often incur losses through such things as downtime and other production process disruptions. Thus, failing to account for all such OHS incidents does not accurately reflect the true cost of OHS incidents to an organisation.

Loss-control theory has prompted a growing interest in quantifying the financial losses arising from workplace OHS incidents (Brody *et al.* 1990; Veltri 1990). In the UK, Davies and Teasedale (1994) undertook a detailed analysis of these costs, based upon interviews conducted as part of the 1990 Labour Force Survey. In their analysis, Davies and Teasedale identified costs to victims and their families, to employers and to society as a whole. These costs are listed in Table 5.1.

These costs may be significant, and some may be particularly salient in construction. For example, there are heavy penalties for time-overruns on construction projects. Construction clients and the community are increasingly

Table 5.1 Costs of OHS incidents to employees, employers and society (adapted from Davies and Teasedale 1994)

Costs to employees	*Costs to employers*	*Costs to society*
Short and long-term loss of earnings. Cost of hospital attendance, medical treatment and other expenses. Cost of pain, grief and suffering.	Compensation for injuries or illness and associated costs. Usually covered by insurance. Loss of output due to absence of staff, disruption of work, damage to equipment or structures, impaired working ability or overtime to make up lost work time. Financial penalties for failing to meet contractual deadlines. Costs of hiring/training replacement workers. Costs of accident investigation and administrative costs. Costs of medical treatment provided by the employer. Costs of clearing up and repair. Fines and legal costs. Costs of administration of sick pay. Loss of goodwill and reputation with workforce, unions, community and clients.	Cost of loss of current resources, such as labour services, materials and capital. This includes the cost of medical treatment, as there is an opportunity cost associated with their use. Losses resulting from infrequent major events, such as fires, explosions, etc. The temporary or permanent loss of the labour services of victims. The cost of pain, grief and suffering to victims and their families.

concerned about contractors' OHS performance, and many large clients are building OHS provisions into pre-qualification criteria or linking OHS performance to tendering opportunities. Workers' unions also devote considerable time and attention to OHS issues, and poor OHS performance could result in costly industrial problems. At an industry level, the construction industries in many industrialised countries are facing chronic skills shortages and thus the loss of skilled labour services, through injury or illness, is likely to impact upon the competitiveness of these industries. Given the industry's obsession with cost-efficiency, it seems surprising that little effort is devoted to understanding these costs in construction. One reason for this may be that many of the costs of OHS incidents are 'hidden' and difficult to quantify. Consequently, construction firms may not be aware of their magnitude.

An OHS incident cost audit was developed by the UK's HSE (HSE 1997). This comprises a pro forma for the collection of comprehensive OHS incident costs. This pro forma measures a wide range of costs, including indirect costs, such as time spent processing insurance claims, liaising with head office support staff and so on. This pro forma is a useful tool for the quantification of the costs of OHS incidents to organisations.

Employers are able to 'transfer' some of the costs arising from workplace accidents to insurers. For example, costs such as public liability, damage to property and damage to vehicles are covered by insurance. Also, statutory employees' compensation schemes require that employers take out employees' compensation insurance policies (see also Chapter 2). However, the increasing tendency for workers' compensation insurance premiums to be linked to past claims performance, in terms of both the number and duration of claims, means that this financial risk is not completely transferred. Moreover, as the accident 'triangles' presented in Figure 5.1 indicate, the vast majority of workplace incidents do not result in 'reportable injuries' leading to employees' compensation claims. Thus, the ability of businesses to transfer the financial losses arising from OHS incidents to insurers is limited to a small proportion of the total number of loss-making incidents.

Loss control theorists also argue that most of the costs of OHS incidents actually cannot be insured against. For example, losses arising from damage to the company's reputation, damage to relationships with the company's clients, diminished employee morale, industrial relations problems, administrative costs associated with the replacement of injured workers and time lost during the incident investigation, overtime and legal costs or fines must all be borne by the company. In the UK, the ratio of insured to uninsured costs was found to vary from industry to industry but ranged from 1:8 in a transport company to 1:36 in a creamery. The construction industry ratio was found to be 1:11 (HSE 1997). Loss control theory holds that OHS risks must be reduced if losses are to be kept in check.

The costs of OHS

Andreoni (1986) identified the following four categories of safety-related expenditure:

1 routine expenditure incurred before occupational injuries happen;
2 expenditure following the occurrence of an occupational injury;
3 expenditure associated with transferring the financial consequence of an occupational injury to an insurer; and
4 exceptional expenditure on prevention.

Routine expenditure incurred before occupational injuries happen can be expressed as the costs of prevention. Expenditure associated with transferring the financial consequence of an occupational injury to an insurer can be expressed as the cost of insurance premiums.

Some of these costs are fixed and some are variable. For example, total expenditure on OHS infrastructure, such as committees, representatives, administrative costs of record-keeping and so on, will be determined by statutory requirements, and are likely to be fixed. Similarly, insurance costs will vary from country to country depending on methods used for their calculations. Some workers' compensation insurance costs are fixed; however, a portion of insurance costs is variable because some costs are experience-rated, that is, linked to an organisation's recent claims history. Some expenditure on prevention will also be variable depending on the nature and severity of incidents, identified training needs and so on. Preventive OHS costs are likely to be particularly variable in the constantly changing construction environment in which new hazards can emerge on an almost daily basis. Expenditure on occupational injuries is also variable, depending upon injury experience. This includes the costs of activities such as medical treatment, first-aid provision, transport and legal costs and the costs of wages paid during workers' absence or overtime to make up for lost production. As the outcomes of safety incidents are largely a matter of chance, expenses arising from damage linked with occupational injury can vary considerably. Finally, some activities will require exceptional expenditure on prevention. Again, this is likely to be the case when one-off construction projects pose unique, project-specific risks.

Andreoni (1986) suggests that the sum of all of these costs represents an organisation's total safety expenditure, and cumulative safety expenditure is an important part of organisational costs. Using these cost categories, organisations can undertake more meaningful cost–benefit analysis and examine whether expenditure devoted to incident prevention is commensurate with the expenditure arising from material damage or occupational injuries.

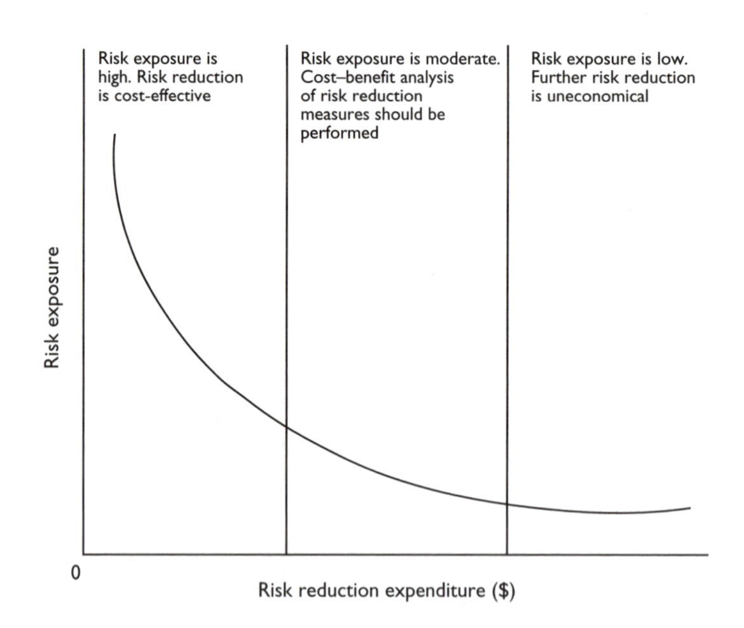

Figure 5.2 Costs of risk reduction (adapted from AS/NZS 4360: 1997, p. 19).

An economic approach to OHS risk management

Theoretically, there is an inverse relationship between expenditure aimed at prevention of OHS incidents and the risk of accidental losses. This relationship is depicted in Figure 5.2. Figure 5.2 also shows that the law of diminishing marginal returns applies to prevention or risk reduction expenditure (costs). Thus, when risks are high, a small increase in prevention expenditure can result in a considerable reduction in the risk level. Under these conditions, few would disagree that risk reduction measures should be implemented. However, as expenditure increases, and the risk level falls, each unit of risk reduction expenditure yields a progressively smaller reduction in the overall risk level. In this situation some judgement as to the acceptability of the risk is required and some risk reduction measures may be deemed to be uneconomical.

Hopkins (1995) suggests that in much of the Western world, a philosophy of deregulation has gained prominence since the 1980s. This way of thinking has been termed 'economic rationalism', and is based on the notion that if markets are left to operate freely with minimal government interference, optimal outcomes will be achieved. Hopkins says that, in Australia, economic rationalism has informed many policies of deregulation of workplace relations and OHS. The UK HSE seems to embrace an economic rationalist

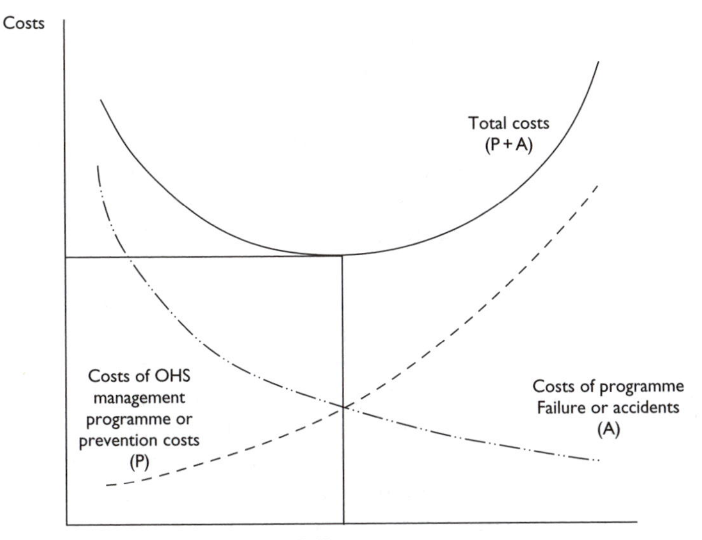

Figure 5.3 Economic analysis of OHS management expenditure.

perspective in suggesting that it is possible to identify a level of OHS risk that represents the optimum economic level of prevention and incident costs. This risk level is the point at which the cost benefits from improving OHS are just equal to the additional costs incurred.

In Figure 5.3, the accident (incident) level curve (A) is a measure of OHS risk relating to the direct and indirect costs of safety incidents. The preventive expenditure curve (P) represents money spent on OHS before the event. The cost of any level of OHS risk is sum of A and P at that point. This will have a minimum value at a particular level of OHS risk.

The HSE do not go so far as to suggest that once this economic optimum point is reached, no further risk reduction measures should be implemented, and it is not clear whether the corresponding level of risk is deemed tolerable. However, adopting a purely market approach to OHS is dangerous. The assumptions that the 'market knows best' and that economic efficiency should be the ultimate criterion against which all policy should be measured are dubious – not least because moral and ethical considerations must guide value judgements concerning tolerable levels of risk exposure.

One difficulty in using market forces to guide risk management decision-making is that many of the costs of occupational injuries and ill health are 'hidden'. This means that they are indirect, difficult to quantify and usually go unreported in company financial records and accounting systems. This makes it very difficult for managers to consider these costs and respond to them in a 'rational' manner. Indeed, it is likely that, because many of these costs are hidden, they are underestimated. Without knowing the magnitude of losses arising as a result of OHS incidents, managers may behave 'rationally' in failing to reduce risk to a level reflecting the true economic optimum point (Hopkins 1995).

A second problem with a reliance on market forces to guide risk management decision-making is the inequitable way in which risks and losses are currently borne. Risk management decision-makers are usually middle managers and professionals who bear significantly less risk of bodily harm than construction workers. In addition, many of the costs of OHS incidents are externalised. For example, the Industry Commission of the Commonwealth Government of Australia reported that employers bear around 30 per cent of the average cost per OHS incident, workers incur 30 per cent and the community bears the remaining 40 per cent (Industry Commission 1995). Similar disparities were reported in the United Kingdom (Davies and Teasedale 1994). In these circumstances, the economic optimum risk level for employers behaving rationally may not be considered to be tolerable by other affected parties – particularly the victims of OHS incidents and their families.

Finally, organisational accounting systems operate to nullify economic incentives to reduce OHS risk, because post hoc incident costs, such as legal fees, human resource administration, workers' compensation and rehabilitation costs, are often dealt with at corporate level and distributed evenly between the different profit centres or business units. The managers of individual business units, who make risk-management decisions, will therefore have little incentive to reduce these costs. The decentralised construction industry is even more likely to be subject to this 'organisationally structured immunity' (Hopkins 1995, p. 44) than other industries.

For these reasons, economic rationalism is likely to be limited as a means of promoting sound OHS risk management decision-making, and markets are not an acceptable means of determining tolerable levels of risk.

The risk management process

The risk management process depicted in Figure 5.4 is an iterative process, involving a continuous cycle of risk identification, assessment and treatment. The assessment of risk informs risk treatment (or control) decisions, the implementation of which is monitored and reviewed to ensure that risk is controlled and remains within tolerable limits.

Risk management should be undertaken by a cross-disciplinary team, and be supported by free and open communication and consultation. We discuss the importance of effective risk communication in greater detail later in this chapter.

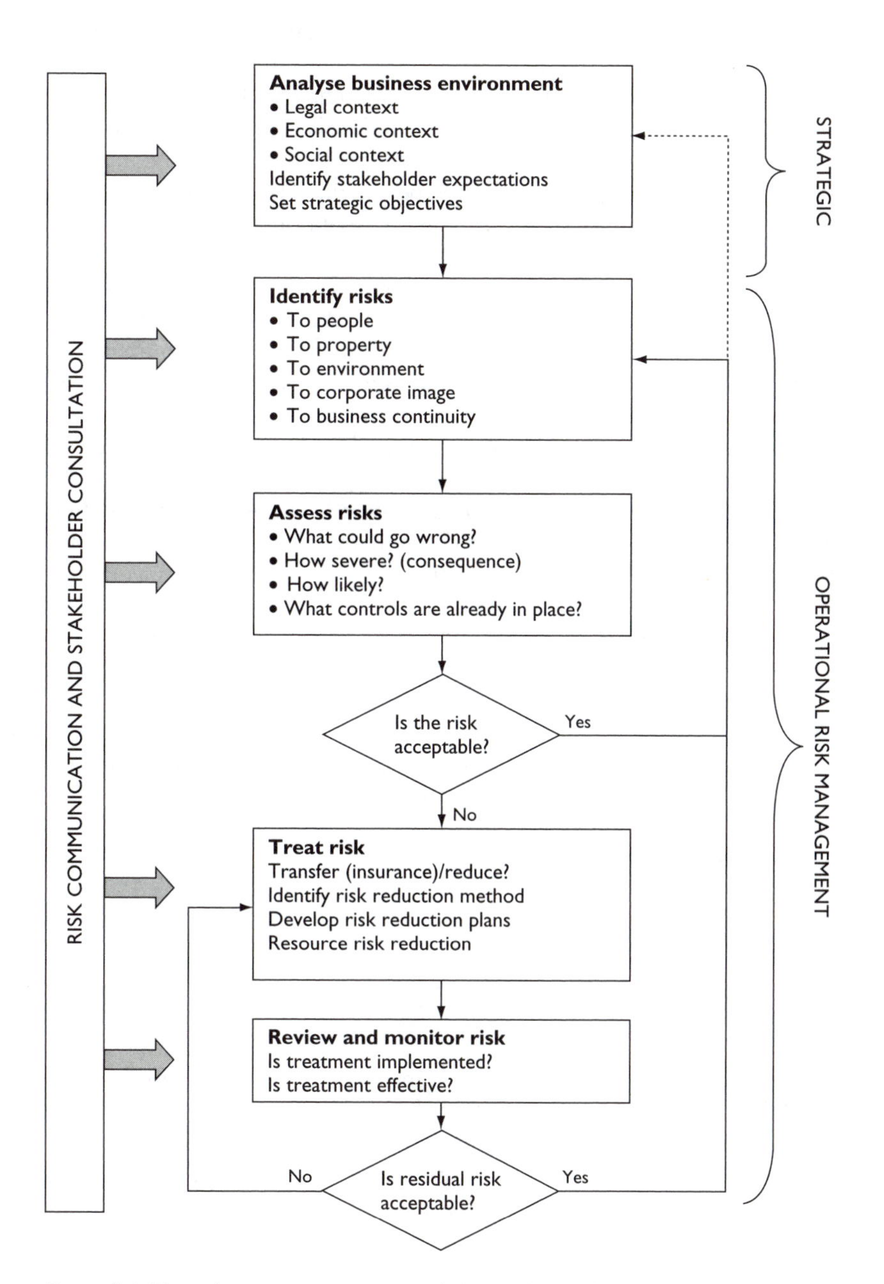

Figure 5.4 The risk management process (adapted from AS/NZS 4360: 1997, p. 11).

Risk management is not a linear process that is undertaken once. The cyclical nature of risk management is particularly important in the constantly changing construction environment in which new or emergent risks must often be assessed and controlled. It is critical that risk be assessed at every stage in the life of a construction project, and that the input of key stakeholders and project participants be sought. Involving designers in OHS risk assessment exercises can provide opportunities to 'design out' features of a building that pose a threat to the health or safety of operatives during the building's construction phase. Similarly, the involvement of subcontractors in the risk assessment and decision-making can facilitate the development of practical risk reduction solutions, and promote a clear understanding of workplace risks and safe and healthy work methods and procedures. Worker participation in risk decision-making is also a critical aspect of two-way risk communication that is important in developing workers' trust in OHS risk decision-making.

The risk management process is described in more detail below, but first we define some of the basic concepts in the field risk management. The term *hazard* usually refers to something that is capable of causing harm or loss to society. Risk is a more diffuse concept dealing with the consequences or exposure to the possibility of harm or loss. As such, it is often taken to be a function of the probability that harm or loss will occur, the frequency of exposure to the hazard and the magnitude of the consequences of the harm or loss should it occur. Defined in this way, hazards and hazard management are subsets of risk and risk management. This approach is consistent with the definitions contained in the Australian and New Zealand standard on risk management. Some key definitions are presented in Box 5.1.

Box 5.1 Definitions

Consequence – The outcome of an event expressed qualitatively or quantitatively, being loss, injury, disadvantage or gain. There may be a range of possible outcomes associated with an event.
Frequency – A measure of the rate of occurrence of an event expressed as the number of occurrences of an event in a given time. See also Likelihood and Probability.
Hazard – A source of potential harm or a situation with a potential to cause loss.
Likelihood – Used as a qualitative description of probability or frequency.
Loss – Any negative consequence, financial or otherwise.

Probability – The likelihood of a specific event or outcome, measured by the ratio of specific events or outcomes to the total number of possible events or outcomes. Probability is expressed as a number between 0 and 1, with 0 indicating an impossible event or outcome and 1 indicating an event or outcome is certain.
Risk – The chance of something happening that will have an impact upon objectives. It is measured in terms of consequences and likelihood.
Risk analysis – A systematic use of available information to determine how often specified events may occur and the magnitude of their consequences.
Risk assessment – The overall process of risk analysis and risk evaluation.
Risk control – That part of risk management that involves the implementation of policies, standards, procedures and physical changes to eliminate or minimise adverse risks.
Risk evaluation – The process used to determine risk management priorities by comparing the level of risk against predetermined standards, target risk levels or other criteria.
Risk identification – The process of determining what can happen, why and how.
Risk management – The culture, processes and structures that are directed towards the management of risk.
Risk management process – The systematic application of management policies, procedures and practices to the tasks of establishing the context, identifying, analysing, evaluating, treating, monitoring and communicating risk.
Risk reduction – A selective application of appropriate techniques and management principles to reduce either the likelihood of an occurrence or its consequences, or both.
Risk retention – Intentionally or unintentionally retaining the responsibility for loss, or financial burden of loss, within the organisation.
Risk transfer – Shifting the responsibility or burden for loss to another party through legislation, contract, insurance or other means. Risk transfer can also refer to shifting a physical risk or part thereof elsewhere.
Risk treatment – Selection and implementation of appropriate options for dealing with risk.

(Source: AS/NZS 4360: *Risk Management*: 1997)

Figure 5.4 shows the risk management process. This process is cyclical, involving constant monitoring and review to ensure that residual risk – the part of risk that is remaining after risk controls have been implemented – is acceptable. The process also ensures that risk management is strategic and responsive to the external environment by requiring that the context in which business is undertaken is analysed. This enables the organisation to respond in an appropriate way to legal obligations and aspects of the economic and social environment. We describe each of the stages in the risk management process below.

Stage 1: Analyse the business environment

Risk management is context-specific. Thus, the first step in the risk management process is to analyse the context within which risk is to be managed. There are several aspects to this context, including the organisation's position within the broader social, political, legal and economic environments. Thus, the organisation's strategic context with regard to OHS should be identified at an executive level in order that more detailed risk management decisions are in line with the organisation's strategic OHS objectives. The goals and objectives of the organisation will define the organisational context within which risk management decisions are made. In Chapter 4, we discussed how genuine and committed safety leadership from senior management is a requisite for excellent OHS performance (Lee 1997). Thus, it is important that senior managers communicate their commitment to OHS by setting out the company's high expectations with regard to OHS in a formal policy statement. This policy statement then guides strategic objectives set by the company's executive management, and sets organisational criteria for middle managers who must make operational decisions about OHS risk.

The scope and boundaries of decision-making, resources available, and the costs and benefits of risk control measures in a particular circumstance need to be balanced against organisational OHS objectives. In the construction industry, operational risk management decisions are made by managers at the project level, and must be made with an understanding of the risk management context of the particular project. However, decentralised decision-making can also be dangerous as short-term project goals may be more salient in the minds of project managers than the organisation's values and strategic objectives with regard to OHS. It is therefore essential that project managers be regularly reminded of corporate OHS policies so that their risk management decisions remain consistent with strategic OHS goals.

Strategic risk management objectives provide a basis against which to decide upon the acceptability of a risk, which in turn provides guidance as to the degree of effort that must be directed towards controlling it. Such decisions may be based upon financial, technical and operational issues. However, in the case of OHS, risk management decisions should also be guided by legal, ethical and humanitarian considerations. The source of risk evaluation criteria can be internal to an organisation, for example the organisation's own values, policies or goals. Increasingly however, the external environment is shaping organisations' risk evaluation criteria. The pressure

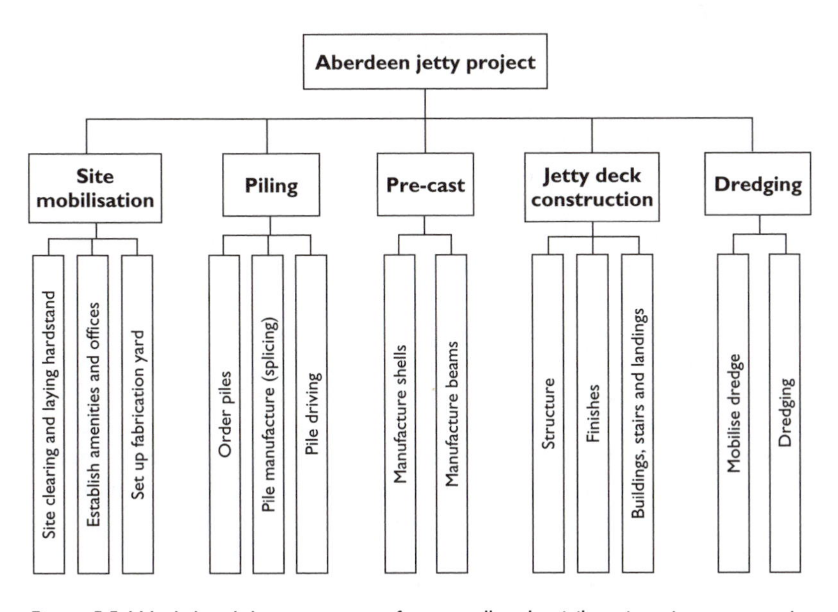

Figure 5.5 Work breakdown structure for a small-scale civil engineering construction project.

on businesses to be 'good corporate citizens' is having a growing influence on organisations' risk evaluation criteria with regard to their conduct in OHS, as well as other areas of business performance (Vredenburgh 2002).

In order to identify and analyse risks in a logical way, activities and projects need to be broken down into their component parts. This decomposition should be undertaken in a logical way, and care taken so as not to overlook any significant risks, particularly at the interface between activities. In construction projects, basing the risk management decision-making structure on elements in the project work breakdown structure is highly recommended because the work breakdown structure is closely related to the management of the project. This way, risk management planning and reporting can be fully integrated with other aspects of project planning and reporting. For example, Figure 5.5 shows the work breakdown structure for a small-scale civil engineering construction project. In this project, comprehensive risk evaluations should be undertaken for each of the work elements, such as pile splicing, pile driving and dredging.

Stage 2: Identify risks

In this step of the risk management process, OHS hazards are identified. Hazards can be identified by using checklists or by examining past injury/incident records. Hazards can also be identified through relying on judgements made by people with experience of tasks or through the use of systems engineering techniques. It is important to ask the people who perform the work on a daily basis about the hazards associated with their tasks because these people are likely to know more about the job than scientists or experts who are removed from the work itself. This is particularly true in the case of work undertaken by specialist tradesmen in construction, whose work may involve task-specific hazards that are not readily apparent to others. The human element should also be included in hazard identification, so cultural, organisational, group level and individual hazards should also be identified (Crossland *et al.* 1993). For example, hazards arising from group dynamic processes, the interface between trades, or individual workers' attributes and attitudes should be considered.

Questions asked at this stage are 'What could happen?' and 'How or where could OHS problems arise?' When answering these questions, the elements of the project work breakdown structure should be systematically considered to avoid missing any hazards. At this stage it is very important that all hazards are identified because incompleteness of information seriously jeopardises the accuracy of the entire risk assessment exercise.

Tait and Cox (1998) identify three approaches to hazard identification:

1 intuitive
2 inductive and
3 deductive.

Brainstorming

One method of intuitive hazard identification is brainstorming. If brainstorming is to be used, it is important that participants be carefully selected and that ideas be allowed to flow freely in an open atmosphere. Team members could include site management and supervisors, technical specialists, OHS advisors, subcontractors and representatives of equipment or material suppliers. The involvement of the people who 'do the work' will also be beneficial. Ideas should be recorded and further investigated by the hazard analysis team.

Zonal analysis

Inductive approaches to hazard identification identify what could go wrong. They include fault tree analysis and job safety analysis. One tool that may be of particular value in the construction industry is *zonal analysis*. This is a method used to identify hazards in the aerospace industry. An aircraft is divided into zones within which all items of equipment are identified. Interactions within and between the functional aspects of the zones, for example electrical faults or mechanical problems, are then carefully examined to identify potential failures (Tait and Cox 1998). In the construction industry, in which many work processes occur simultaneously in a relatively small physical area, there is the possibility that an activity occurring in one location could be hazardous to work in adjacent areas. For example, welding on an upper level of a building may cause hot particles to fall onto a lower level causing injury or fire if people or materials are unprotected. Zonal analysis of the construction site by geographical area could identify hazards presented at this interface. Alternatively, the zonal analysis could be undertaken by construction trade or activity to achieve similar results.

Accident databases

Deductive methods for identifying hazards include the use of accident databases. These can either be company databases or industry databases. The nature of hazards can be deduced from understanding the circumstances of previous incidents. The HKHA accident information database described in Chapter 9 is particularly powerful, because the uniformity of design of HKHA buildings ensures that hazards inherent in the design and construction methods used to construct these structures can be identified.

Stage 3: Assess risks

Following the identification of hazards, risk assessment is conducted to separate risks that are minor and tolerable from those that are major and must be controlled. The likelihood that a risk will result in harm and the severity of the consequences of risks posed by hazards identified in Stage 2 are evaluated in turn. Different risk assessment methods may be used depending on the type of risk being considered and the availability of data about this risk. Three types of risk assessment are described below.

Qualitative risk assessment

Qualitative analysis is the simplest and least costly method of risk assessment. It can be used where the level of risk does not warrant the cost involved in applying a more detailed analysis. It can also be used as an initial screening method, to identify risks that require more thorough analysis, or where numerical data are so inaccurate as to render quantitative analysis of risk meaningless. Risk matrices are a commonly used method of qualitative risk analysis. An example risk matrix is shown in Figure 5.6. Risks are rated according to the likelihood or probability of their occurrence and their likely consequences. Probability and consequence are rated using verbal descriptors and cross-referenced to establish the position of a risk in the matrix. These positions indicate the magnitude of the risk, which can then be used to guide the selection of suitable risk control methods and to establish priorities for the implementation of these controls. The greater the magnitude of the risk, the more effort should be expended in its control, and the more urgently risk control actions should be implemented.

→Severity→ ↓Probability↓	Catastrophic (4)	Critical (3)	Marginal (2)	Negligible (1)
Frequent (A)	High (4A)	High (3A)	High (2A)	Medium (1A)
Probable (B)	High (4B)	High(3B)	Medium (2B)	Low (1B)
Occasional (C)	High (4C)	High (3C)	Medium (2C)	Low (1C)
Remote (D)	High (4D)	Medium (3D)	Low (2D)	Low (1D)
Improbable (E)	Medium (4E)	Low (3E)	Low (2E)	Low (1E)

Figure 5.6 Risk assessment matrix.

Semi-quantitative risk assessment

Risk assessment can also be semi-quantitative. In this type of analysis, qualitative risk descriptors are assigned numbers or ranked to produce a more detailed prioritisation of risks. An example of a semi-quantitative risk estimation method is presented by Bamber (1996). Bamber's method involves assigning numerical values to represent the Maximum Possible Loss (MPL) that could arise as a result of the risk (consequence), the frequency with which the risk is identified during inspections and the probability with which the risk will result in the harm. Risk magnitude is then calculated using the following formula:

$$\text{Risk} = \text{Frequency} \times (\text{Maximum Possible Loss} + \text{Probability})$$

Maximum Possible Loss is rated on a 50-point scale with outcomes ranked in descending order of importance. For example, Bamber suggests points may be allocated as follows:

Multiple fatality	50
Single fatality	45
Total disablement	40
Loss of eye	35
Arm/leg amputation	30
Hand/foot amputation	25
Loss of hearing	20
Broken/fractured limb	15
Deep laceration	10
Bruising	5
Scratching	1

Probability is also rated on a 50-point scale as follows:

Imminent	50
Hourly	35
Daily	25
Once per week	15
Once per month	10
Once per year	5
Once per five or more years	1

Thus, where an unsafe action or condition, for example failing to wear eye protection when cutting or grinding metal, is identified once during an inspection, the worst possible outcome was considered to be the loss of an eye and the probability of the occurrence is rated as once per day, the risk rating is calculated as follows:

$$\begin{aligned} \text{Risk} &= F \times (MPL + P) \\ &= 1 \times (35 + 25) \\ &= 60 \end{aligned}$$

The magnitude of the risk is then compared with a previously agreed risk reduction implementation guide to determine the degree of urgency with which the risk should be controlled.

Semi-quantitative methods can give the false impression of being scientific and objective. However, the assigned numbers do not reflect realistic values of probability or consequence of risk, and therefore the rankings produced usually do not reflect the relative importance of risks. They may therefore lead to inconsistent decisions about the immediacy with which risks should be controlled and the deployment of resources in implementing these controls. It is also difficult to rank the relative severity of injuries/illnesses, because such rankings fail to account for proven qualitative determinants of people's risk perceptions, such as the dread associated with 'fates worse than death'.

Quantitative risk analysis (QRA)

The most resource-intensive approach to risk assessment is quantitative risk analysis (QRA). QRA differs from semi-quantitative analysis in that it uses numerical values to express both the consequences and likelihood of a given risk. Data sources for quantification include:

- past records;
- relevant experience;
- industry practice;

- relevant published data, including reliability databases;
- test marketing and research;
- experiments and prototypes;
- economic, engineering or other models; and
- specialist and expert judgements (AS/NZS 1998).

Consequences can be expressed in terms of monetary, technical or human criteria while likelihood is usually expressed as a probability or combination of probability and exposure. In safety risk analyses, consequences are usually expressed in human terms. For example, mortality rates are often expressed for occupations or activities as the number of deaths per annum from an activity divided by the total population at risk. Some risks can be expressed in terms of different measures of activity rather than a unit of time, for example the number of injuries per total number of man-hours worked or per building constructed. Fatal Accident Rates (FARs) have been used to quantify occupational risks. The FAR is defined as the number of deaths per 100 million man-hours of exposure. Alternatively, this is expressed as the number of deaths per 1000 people involved in an activity during a working lifetime of 10^5 hours. The relative ordering of risks is known to be sensitive to the choice of risk index, and therefore risk indices should be carefully selected (Crossland *et al.* 1993).

The estimation of risk probabilities and consequences invariably involves making certain assumptions. For example, in assessing the risk associated with the use of a formwork system, assumptions as to the reliability of component parts and the correct use and storage of the equipment to date may be made. Assumptions may also be made about the reliability of workers erecting this formwork system. In the case of risks in which failure could be catastrophic, the sensitivity of QRA results to the assumptions made should be determined.

Some common pitfalls in risk assessment have been identified by Gadd *et al.* (2003). They report that in some instances risk assessment is used to justify decisions that have already been made instead of being used to perform a systematic estimation of the magnitude of the risk and a comparison of alternative risk control options. The use of generic risk assessments is another cause for concern. Owing to a lack of competence in performing risk assessments, many organisations use generic risk assessments for similar operations or similar sites. While generic risk assessments can be a useful starting point, it is essential that the site-specific circumstances are taken into consideration. This is particularly true of construction work in which no two sites are identical and site hazards, such as ground conditions or the location of underground services or overhead power lines vary from site to site. Another common pitfall in quantitative risk assessments is the use of generic data in analysing the risk of failure of large items of plant or equipment, for example cranes. Such equipment has many component parts and there are many different protective systems that can prevent the failure of these items of plant. Generic failure data for similar items of plant will not be valid unless the particular item of plant has the same component parts and protective systems as those on which the generic risk data were estimated. Consequently the use of generic risk data for large items of plant can be misleading. When performing risk assessments, it is very important that good practice be followed and that those performing the risk assessment are competent to do so.

Stage 4: Evaluate risks

Once the level of a risk has been estimated, these levels must be compared with previously established risk criteria to create a prioritised list of risks to be controlled. The evaluation of risks should be undertaken with an understanding of organisational OHS goals. In some circumstances, risks are borne by parties external to the organisation undertaking the risk evaluation. For example, in construction, subcontractors or members of the general public may be exposed to safety and health risks posed by construction operations. In the interests of equity, the tolerability of risks to all of the risk-bearing parties should be considered in risk control decision-making. Certain risks may be deemed to be sufficiently low to be tolerated at this stage. However, these risks should be carefully monitored to ensure that the risk analysis was accurate. This is particularly true in construction, in which unforeseen risks can emerge once an activity has commenced. For example, asbestos may be discovered in a building once demolition or renovation work has started, or unexpected geological conditions may render the ground less stable than expected in an original risk assessment.

Stage 5: Control risks

Where risks are deemed to be unacceptably high, an organisation must decide how to manage these risks posed. In the case of the most serious risks, organisations might decide not to proceed with an activity. However, in the case of most OHS risks, organisations take steps to treat or control the risks evaluated in Stage 4.

In OHS risk management there is an established ‘hierarchy’ of risk control measures. This hierarchy is depicted in Figure 5.7. The hierarchy of controls is based on the principle that control measures that target hazards at source and act on the work environment are more effective than controls that aim to change the behaviour of exposed workers (Matthews 1993). Thus ‘technological’ control measures – such as the elimination of hazards, the substitution of hazardous materials or processes, or engineering controls – are preferable to ‘individual’ controls – such as the introduction of safe work practices or the use of personal protective equipment. This is because

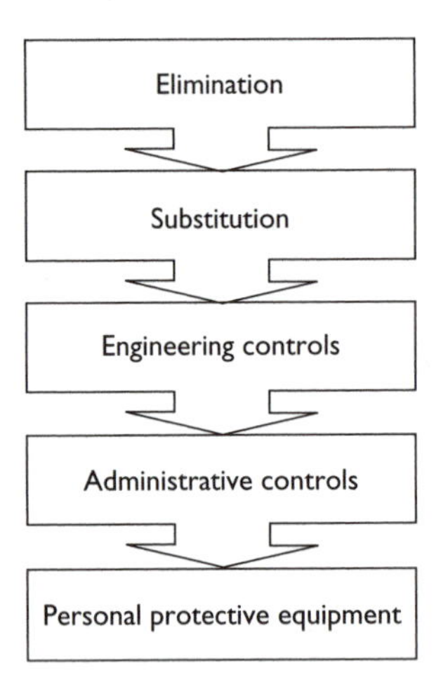

Figure 5.7 Hierarchy of OHS risk controls.

controls that rely on human behaviour are less reliable, since human beings are fallible and prone to error. Wherever possible, managers should focus on the identification and implementation of 'technological' risk controls.

The hierarchy of risk controls is contained in OHS codes of practice and standards and is implied in the legal interpretations of which risk control measures are 'reasonably practicable'. Thus, in selecting methods to reduce the risk of occupational injuries or ill health, organisations must understand the control methods available and select the uppermost control measure that is reasonably practicable to implement. It is insufficient to rely solely on lower order controls for OHS risk. It is unlikely that setting site safety rules, developing OHS procedures and issuing personal protective equipment to workers will be sufficient to demonstrate compliance with the general duties requirements of OHS legislation.

Table 5.2 presents an application of the risk control hierarchy to the risk of falling from height in construction.

Assessing risk control options

Selecting appropriate risk controls involves an assessment of the costs and benefits of implementing different risk control options. The extent to which risk control options provide additional benefit (in terms of reduced risk) is weighed against the costs of implementing them. OHS law and standards refer to the concept of 'reasonable practicability', which is similar to the principle 'as low as reasonably practicable' or ALARP. These principles are narrower than 'physically possible'. They recognise that safety comes at a cost, and they imply that technically feasible measures to reduce risk need only be implemented if the benefit to be gained by implementing them is commensurate with the costs involved. AS/NZS 4360 adopts the ALARP principle, but the question 'what is an acceptable level of safety?' is a difficult one to answer. Horlick-Jones (1996, p. 174), cites the judge in a mine safety case in 1949 commenting on the concept of reasonable practicability. The judge stated:

Table 5.2 Example risk control hierarchy for falling from height

Risk control category	*Control measures*
Eliminate the hazard	Structures should be constructed at ground level and lifted into position by crane (e.g. prefabrication of roofs or sections of roofs).
Substitute the hazard	Non-fragile roofing materials should be selected. Fragile roofing material (and skylights) should be strengthened by increasing their thickness or changing their composition.
Isolate the process	Permanent walkways, platforms and travelling gantries should be provided across fragile roofs. Permanent edge protection (like guard rails or parapet walls) should be installed on flat roofs. Fixed rails should be provided on maintenance walkways. Stairways and floors should be erected early in the construction process so that safe access to heights is provided.
Engineering controls	Railings and/or screens guarding openings in roofs should be installed before roofing work commences. Temporary edge protection should be provided for high roofs. Guard rails and toe-boards should be installed on all open sides and ends of platforms. Fixed covers, catch platforms and safety nets should be provided. Safety mesh should be installed under skylights.
Safe working procedures	Only scaffolding that conforms to standards should be used. Employers should provide equipment appropriate to the risk like elevated work platforms, scaffolds, ladders of the right strength and height, and ensure that inappropriate or faulty equipment is not used. Access equipment should be recorded in a register, marked clearly for identification, inspected regularly and maintained as necessary. Access and fall protection equipment such as scaffolds, safety nets, mesh, etc. should be erected and installed by trained and competent workers. Working in high wind or rainy conditions should be avoided. Employers should ensure regular inspections and maintenance of scaffolding and other access equipment, like ladders and aerial lifts. Employers should ensure that scheduled and unscheduled safety inspections take place and enforce the use of safe work procedures.

Table 5.2 (Continued)

Risk control category	*Control measures*
	Employees should be adequately supervised. New employees should be particularly closely supervised. Employees should be provided with information about the risks involved in their work.
	Employers should develop, implement and enforce a comprehensive falls safety programme and provide training targeting fall hazards. Warning signs should be provided on fragile roofs. Ladders should be placed and anchored correctly.
Personal protective equipment	Employees exposed to a fall hazard, who are not provided with safe means of access, should be provided with appropriate fall arrest equipment such as parachute harnesses, lanyards, static lines, inertia reels or rope grab devices. Fall arrest systems should be appropriately designed by a competent person. Employees should be trained in the correct use and inspection of personal protective equipment (PPE) provided to them. Employees should be provided with suitable footwear (rubber soled), comfortable clothing and eye protection (for example, sunglasses to reduce glare).

> a computation must be made in which the quantum of risk is placed in one scale and the sacrifice involved in the measures necessary for averting the risk (whether in money, time or trouble) is placed in the other, and that, if it should be shown that there is a gross disproportion between them – the risk being insignificant in relation to the sacrifice – the defendants discharge the onus upon them.

The ALARP concept has been criticised for treating employees' health and safety as being secondary to production goals (Willis 1989) and there remain difficult issues concerning social and power relationships between those who bear the risk and those who benefit from optimising production efficiency. AS/NZS 4360 concedes that in some rare cases, the risks may be so severe that risk reduction measures are warranted, even where these are not justified on economic grounds.

The concepts of 'reasonable practicability' and ALARP imply that there is need for professional judgement in decisions concerning what risk controls should be implemented. Decision-makers must be aware of all the available technically feasible risk control measures in making these decisions because ignorance is no defence for failing to implement a cost-effective risk control measure. Furthermore, what is deemed to be cost-effective for a large company is also cost-effective for a small business. Small firms cannot therefore argue that risk control measures cannot be implemented due to a shortage of resources. Employers operating small construction sites are expected to control equivalent OHS risks in the same way as contractors operating large commercial construction sites, though in practice this is rarely the case.

Once risk controls are selected, their implementation must be managed. If the budget does not permit the implementation of all risk controls at the outset, risks may need to be prioritised according to the costs of their implementation and the magnitude of the risks they are intended to control. Obviously, the most serious risks require the more urgent attention. The implementation of risk controls needs to be undertaken in accordance with sound management practice. Thus, an implementation plan should be developed clearly stating the risk control actions to be taken, assigning responsibilities for these actions and establishing a time frame within which the actions must be carried out. Those responsible for the implementation of risk controls should be held accountable, and the implementation process therefore needs to be monitored to ensure that planned risk controls are put in place.

Stage 6: Reviewing and monitoring risks after controls have been implemented

It is essential to monitor the effectiveness of risk management activities. Thus, once risk controls have been implemented, risks should be re-examined to determine whether the residual risk – that remaining after controls have been implemented – is tolerable. The risk management process is a continuous cycle, and the risk situation must be regularly reviewed to account for changing circumstances. As noted earlier, this is likely to be particularly important in the construction project environment. Reviews must also take into account new technologies or materials that may present better ways to control risks than those initially implemented. Thus, it is incumbent upon operational managers to keep abreast of OHS technologies or to ensure that they receive sound professional advice on the risk control possibilities.

Communicating and consulting

It is important that risk information and management decisions be communicated to all internal and external stakeholders. Employees, community groups, subcontractors, regulatory authorities and trade unions are all likely to have an interest in OHS risk management decisions. Information about how and why risk management decisions were made should be communicated. This communication is important in building trust in risk management decisions but, in order for it to be effective, communication must be a two-way process. The opinions of the different stakeholders must also be solicited and considered in risk management decision-making. We discuss the importance of risk communication later in this chapter.

The technical approach to risk

The technical approach to risk is founded upon scientific and technical expertise. Applying this approach, scientists and engineers seek to model real-life events to determine the nature and magnitude of the risk associated with a given activity, facility or system. Accident sources and sequences are systematically identified. In many instances, statistical probabilities are calculated for events or chains of events. As we shall see later in this chapter, the assumption of the technical approach to risk is that all causal pathways to events can be identified, which, owing to the complex interplay of human and technological factors involved, seems unlikely to be the case.

Probability and risk

Probability is a key concept in risk estimation and is widely used in QRA. Probabilistic risk analysis enables the risk of system failure to be quantified based on the identification of failure routes and an assessment of the probability of their occurrence. Consider this simple example.

A contractor has a contract to undertake fit-out work on a lift shaft of a high-rise building under construction. The procedure set out by the main contractor requires the subcontractor to check that at each floor there is a sign notifying other workers that people are working in the lift shaft below. Also, the subcontractor is required to check that toe-boards and fencing are in place so that nothing can fall into the lift shaft. If this did not happen, the following sequence could occur:

1 Subcontractor does not check signage and barriers in the floors above.
2 Toe-boards were removed on floor above, were not installed at all or were installed partially.
3 A heavy steel scaffold tube is accidentally kicked on the floor above and falls down the lift shaft.
4 There is a worker in the lift shaft at the time that the scaffold tube falls.
5 The worker in the lift shaft does not hear or see anything and does not leave the lift shaft.
6 The worker is struck by the scaffold tube.
7 A death or serious injury results.

Assume that the subcontractor consistently fails to check the upper floors because he has become complacent about this risk. The chance of death or serious injury then becomes the product of the chances of stages 3 to 6 happening.

Therefore, if:

- the chance of a toe-board being removed or being only partially installed is 1 in 5;
- the chance of a scaffold tube lying on the floor and being accidentally kicked or pushed into the lift shaft is estimated to be 1 in 10;
- the chance that someone is working in the lift shaft is 1 in 2; and
- as the lift shaft is a fairly small area, the chance of being struck should something fall is 1 in 4,

then the probability of death or serious injury is:

$$1/5 \times 1/10 \times 1/2 \times 1/4 = 1/400 = 0.0025$$

In the analysis of critical engineering systems, such as aircraft or chemical process plants, QRA has typically focused on the probability of failure of a system's components. However, Baron and Pate-Cornell (1999) suggest that human errors, such as operator failure to follow correct procedures, can be part of failure modes. We contend that, in the construction industry, human errors are likely to be a major source of system failure because of the labour-intensive, non-repetitive nature of the work. Thus, human errors should be included in probabilistic risk analysis wherever possible.

Quantitative risk analysis (QRA)

Many techniques have been developed to undertake risk analysis and often these techniques involve QRA. It is argued that quantitative risk estimates are essential to informed risk management decision-making (Cohen 1996). Cohen suggests that not to use QRA would be to 'drive blind' and 'to be moved by organised pressure groups with their own agendas, or by people's uninformed fears' (Cohen 1996, p. 96). Even in the absence of 'hard' probability data and with incomplete information, QRA techniques are still an attempt to improve our understanding of a significant subset of the pathways to failure (Crossland *et al.* 1993). As such, they still have value even in the face of the uncertainty characteristic of a construction project. Some of the most commonly used methods of QRA are described below.

Fault tree analysis

Fault tree analysis is a 'top-down' procedure, starting with an undesirable event. Conditions or factors that can contribute to a specified undesirable outcome, called the 'top event' are deductively identified and organised in a logic 'tree'. The faults in the tree can be component failures or human errors. Proximate causes of the top event are first identified. Then each of the proximate causes is traced back to identify ways in which this cause eventuated. Each of these contributing factors is traced back and so on until the beginning of the chain of events is reached. Some standard symbols are used in fault tree analysis. Rectangular boxes represent events or contributory factors. Domed AND gates and pointed OR gates indicate whether two or more conditions or whether either one or another condition must be satisfied for an event to occur. In the example depicted in Figure 5.8 the 'top event' is the overturning of a mobile crane during a lifting operation.

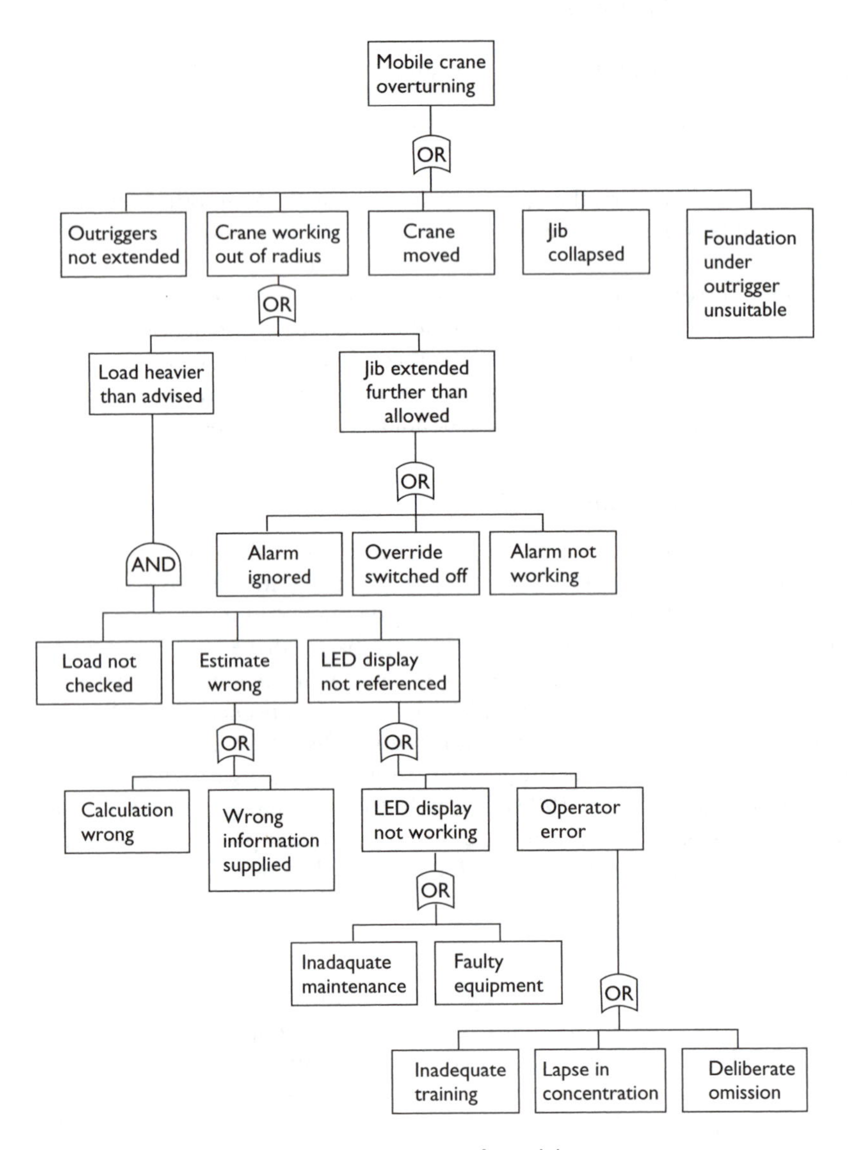

Figure 5.8 Example fault tree for overturning of a mobile crane.

While quantitative fault tree analysis is commonly used to model risk in complex engineering systems, qualitative versions of the same process can also be useful where incomplete probability information is available. Qualitative fault trees can assist with the identification of pathways that could result in OHS incidents, and may also be usefully applied by accident investigators.

Event tree analysis

Event tree analysis uses inductive analysis to identify possible outcomes and sometimes their probabilities, given the occurrence of an initiating event. A simple event tree for the overloading and lifting of a brick skip is presented in Figure 5.9. Probabilities have been assigned to this event tree. The question posed at each stage of the analysis is 'what happens if...?' Like fault tree analysis, event tree analysis may be either quantitative or qualitative.

One possible problem with this technique is that if an important initiating event is overlooked, the risk assessment will not be comprehensive. The technique is useful for calculating the probabilities associated with a sequence of events leading up to a particular outcome.

Failure modes and effects analysis

Failure modes and effects analysis (FMEA) is a 'bottom-up' approach to QRA. An example FMEA for a concrete slab is provided below. In FMEA the consequences of individual component failures are identified by posing the question, 'what happens if...?' In order to be effective, all system components

Initiating event	Scaffold collapse	Skip free fall to ground	Person beneath scaffold	Outcome	Frequency (per year)
Brick skip overloaded and lifted onto scaffold 0.05	Yes 0.8	Yes 0.99	Yes 0.7	Fatality, serious injury	2.7×10^{-2}
			No 0.3	Possible equipment damage and broken scaffold	1.2×10^{-2}
		No 0.01	Yes 0.7	Damage to scaffold only	2.8×10^{-4}
			No 0.3	Damage to scaffold only	1.2×10^{-4}
	No 0.2			No damage	

Figure 5.9 Event tree analysis example.

must be identified. How these components may fail and what effect the failure would have on the overall system must also be systematically considered. This analysis can involve the assignment of probabilities, but is sometimes descriptive. Information can be presented in a table or spreadsheet.

Example FMEA for a concrete slab

The concrete slab has been cast and is supported by formwork. In order to remove the formwork and for the slab to be self-supporting, it is required to apply a 100-t force to the stressing strand. This force will lead to an extension of 50 mm to the strand at the end. When stressing occurs there are three requirements:

1 that the slab has reached a strength of 25 MPa;
2 that there is 80 mm between the strand and the top of the slab (as depicted in Figure 5.10a). If not, the slab will fail and the strand will crack the surface; and
3 that the jacks must be normal to the strand.

If the jack is not normal to the strand (as depicted in Figure 5.10b) then the 100-t force will not achieve the 50-mm extension. If this occurs, the required level of pre-stress will not be achieved. This will cause more deflection. In itself this will not be catastrophic but would reduce serviceability of the slab.

It is a requirement of the job that the stressing occurs three days after the concrete is cast. The concrete mix has been designed so as to achieve 25 MPa in three days. If this does not occur, for example due to bad batching or cold weather, then the strand will slip in the concrete. If the strength is between 20 and 25 MPa then there will be additional deflection reducing the serviceability of the slab. However, if the MPa is less than 20, then the level of slippage will be such that when the formwork is removed, significant cracking will occur and the slab may fail. This is very dangerous as it will not be obvious externally and could be catastrophic. If less than 80-mm cover is achieved (as depicted in Figure 5.10c), then there is a problem in that the resultant uplift force of the strand will cause cracking at the surface. If the cover is less than 65 mm, then the strand will explode through the concrete causing a dangerous situation, meaning the slab will have to be removed and re-cast.

The failure modes effects and probabilities identified in the placing of the concrete slab are presented in Table 5.3. Note failure probability represents the probability that each failure-inducing item will occur, not the probability of slab failure once this condition has occurred.

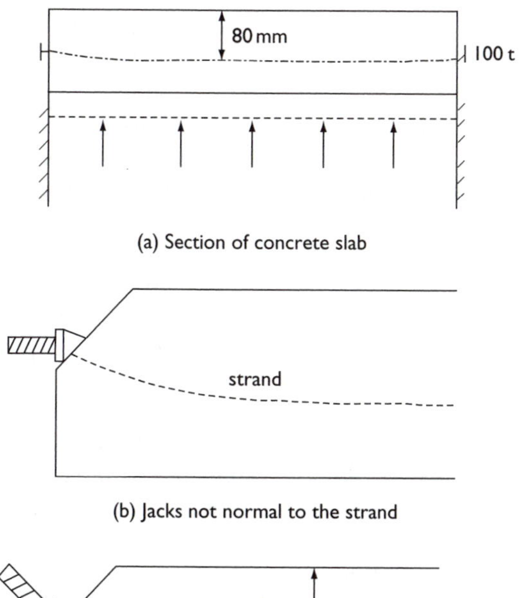

(a) Section of concrete slab

(b) Jacks not normal to the strand

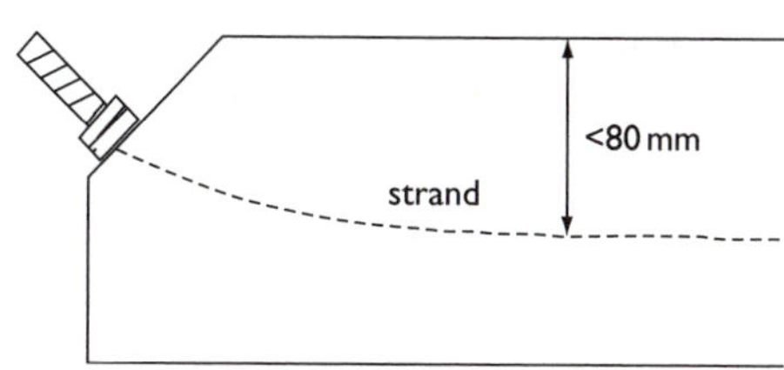

(c) Less than 80 mm cover archieved

Figure 5.10 FMEA for a concrete slab.

Table 5.3 Probability of slab failure

Item	*Failure probability*	*Failure to danger probability*	*Failure to serviceability probability*
Jack not normal to strand	0.2		0.2
Concrete strength 20–25 MPa	0.1		0.1
Concrete strength <20 MPa	0.01	0.01	
Cover 65–80 mm	0.05	0.05	
Cover <65 mm	0.001	0.001	
Total	0.361	0.061	0.3

Hazard and operability studies (HAZOPs)

HAZOP studies are a systematic method for identifying hazards and operability problems in a facility. They are widely used in the chemical and process industries and are particularly helpful in identifying hazards at the design stage of a facility, or for anticipating hazards associated with a change in plant operating procedures. A full description of the process or facility is produced. Every part of the facility or process is then carefully scrutinised to identify areas in which deviations from the intended process or operations could occur. These deviations are then considered to determine the extent to which they present a hazard or could cause operability problems.

At the detailed design stage, documentation, drawings and process specifications are examined. The design intent of each major item of equipment is defined. Guidewords such as *no*, *more*, *less*, *as well as* and *reverse* are applied to process variables such as *temperature*, *speed*, *pressure*, *level*, *flow* and *chemical composition*. The possible causes and consequences of deviations from normal operations for each component are then analysed and the means by which the deviations could be detected or prevented are identified. Thus, the sources and consequences of quantitative increases or decreases and qualitative increases or decreases of process variables are systematically identified. The consequences if intended processes fail to occur, occur only in part or act in reverse, are also examined using the guidewords and process variables. Because of their focus on processes, HAZOP studies may be useful for engineers designing building services, particularly in complex projects, such as hospitals.

HAZOP studies in construction design

In Australia, WorkCover New South Wales has developed a HAZOPs process for evaluating the OHS risk implications in construction design (WorkCover New South Wales 2001). The process, known as Construction Hazard Implication Review (CHAIR), involves a three-stage review of design by multi-disciplinary teams, involving all stakeholders involved in the design, construction and use of the facility. Participants might include project managers, architects, the construction foreperson, safety specialists, the client, engineers and service consultants. The first CHAIR review occurs at the conceptual design stage. At this stage, the design is divided into logical components and, for each component, sources of OHS risk are identified and assessed. The magnitude of the risks and the adequacy of controls are considered. Guidewords and prompts are used to stimulate the discussion. For example, guidewords such as *Load/force* are considered alongside prompts such as *high/excess*, *low/insufficient*, *additional loads* (construction), *dynamics* and *temporary weakness*. Using guidewords and prompts, the design implications of the size, heights/depths, position/location, movement/direction and load/force of design elements are systematically considered. Other features of design that could give rise to a risk to health or safety that are considered include energy, timing, access and egress, and maintenance and repair. The outcome of this review is documented, and design changes are made where necessary.

The second and third CHAIR reviews take place immediately prior to construction, when the detailed design is complete. The second review focuses on OHS issues arising in the construction and demolition phases of the project, while the third review focuses on maintenance and repair of the facility. The first of these detailed design reviews focuses on the construction or demolition sequence, and aims to make sure that as much as can reasonably be considered has been incorporated into the design to reduce OHS risks during the construction and demolition stages. At this stage, there is not much opportunity to make fundamental design changes, but construction methods may be modified. CHAIR guidewords and prompts are again used to facilitate this review. These reflect the hierarchy of risk controls. For example, guidewords such as *avoid* are considered alongside prompts such as *construction/lifting*, *sequence*, *timing/location*, *temporary instability*, *access/egress*, *delays/confined spaces*, *erection/dismantling* and *heat/cold/noise*.

The third CHAIR review considers in detail, the implications of design for maintenance and repair of the facility. Issues such as access and egress, heights, slips, trips and falls, and the weight and manual handling of equipment are all systematically considered at this stage.

The CHAIR process has been trialled by two prominent construction companies on their projects, and both comment favourably on the process, suggesting that it integrates OHS planning into design decision-making effectively (WorkCover New South Wales 2001). This innovative adaptation of HAZOP studies method to construction demonstrates that risk evaluation methods that have not been widely used in construction can have genuine value if they are thoughtfully implemented.

Human reliability assessment (HRA)

The importance of the human element is highlighted by the role of errors and omissions in catastrophic systems failures. Human error is also involved in the majority of occupational accidents, highlighting the dangers of ignoring the human element in risk assessments. Human reliability assessment focuses on the interface between technological aspects of the system and the humans who operate and maintain it. In particular, it seeks to classify and understand common types of human error, consider human reliability within the context of technological systems and quantify human reliability. The aim is to ensure that human reliability is maximised through appropriate organisational and systems design.

Human error can be classified into:

- errors of omission, in which a required action is not carried out;
- errors of commission, which include a failure to perform a required action correctly, for example at the wrong time, in the wrong sequence or with too much or too little force; and
- extraneous or unrequired actions (AS/NZS 3931: 1998).

Human reliability assessment is a hybrid discipline, which spans psychology, human factors and engineering. Several techniques have been developed for assessing human reliability. In its most basic, HRA involves the analysis of tasks, the identification of possible errors and sometimes the quantification of error probabilities. Task analysis is a method by which jobs and tasks are analysed to identify likely error situations and/or quantify the probability of human error. Human error probability (HEP) (Tait and Cox 1998) may be calculated using the following equation:

$$\text{HEP} = \frac{\text{Number of human errors}}{\text{Total number of opportunities for human error}}$$

The identification of the possible consequences and causes of human error should then be examined and suggestions made as to how to measure or reduce human error probability, improve opportunities for recovery from human error or reduce the consequences of human error in the system. We consider human error in greater detail in Chapter 7.

Objective versus subjective risk

Technical models of risk and methods for its estimation assume that an objective or statistical measure of risk can be derived. However, Blockley (1992) suggests that developments in modern physics, quantum mechanics and new theories of deterministic 'chaos' demonstrate limits to what we can know. The assumption of objective risk, which underpins the technical approach to risk, is questioned by theorists who adopt a 'social science' approach to risk. Blockley, himself an engineer, suggests that a tension is emerging within engineering practice between the dominant scientific way of thinking and the need to address the less predictable human and organisational components of complex systems. This tension relates to conflicting epistemological perspectives. Scientists' and engineers' slavish pursuit of technical rationality leads them to adopt a closed-world model in which all component parts are known and relationships understood. This view does not reflect the fact that when man-made disasters happen, they are usually the result of a complex interaction between people, their social arrangements and technological hardware over time (Turner 1978). These interdependencies do not lend themselves readily to mathematical modelling.

The assumption that all pathways to failure in complex socio-technical systems can be clearly identified is unlikely to be appropriate. Green *et al.* (1991) draw a distinction between 'what you know you don't know' and 'what you don't know you don't know'. The latter source of error poses particular difficulties for QRA because it cannot easily be mathematically factored into risk calculations. In complex dynamic systems, some pathways to failure remain difficult to foresee. Accurate assessment of the probability of an event is likely to be particularly elusive in the organisationally complex and constantly changing environment of a construction project.

Furthermore, we contend that risk assessment can never be totally objective. Pidgeon (1996) dismisses the notion that science is 'ethically agnostic'. He argues that science is essentially a human and social activity that is not value neutral (see also Latour 1987). Scientists and engineers work within institutional contexts, including large companies, private sector consultancies and government agencies. Even university-based scientists are increasingly looking to corporate sponsorship. It is inconceivable to think that these affiliations and dependencies do not influence scientists' values and research agendas.

Pidgeon *et al.* (1993) suggest several areas in which subjective judgement plays a role in risk estimation. These are:

- *The selection of a risk index.* The outcomes of 'objective' risk assessment are likely to differ depending on which operational measure of risk is used.
- *The assessment of the disutility resulting from the consequences of risk.* Assessing the acceptability of social and individual costs associated with injury or illness involves making value judgements about the value of life and other contentious matters.
- *The way the risk problem is framed.* For example, in considering the risks posed by construction activity, the question as to whether only on-site construction activities should be considered arises. If so, greater pre-fabrication may be a recommended risk reduction measure. However, pre-fabrication may simply transfer the risk to another location, resulting in no overall reduction in OHS risk.

Pidgeon *et al.* (1993) also argue that the validity of assessments of the probability of undesirable consequences is reliant upon the pedigree of technical information used in risk calculations. They suggest that frequency estimates of mechanical equipment reliability, which are based on rigorous testing, are likely to be of a high level of accuracy. An acknowledged expert's judgement of failure of the same component might be of medium accuracy, but the opinion of a recent engineering graduate as to the likelihood of failure in the same component may be much less accurate. Pidgeon *et al.* (1993) suggest that the overall validity of any combined assessment should be taken to be equivalent to the pedigree of its weakest input.

Slovic *et al.* (1980) suggest that experts' risk judgements are subject to error and overconfidence, in a similar way that laypersons' judgements are commonly found to be wanting. For example, a study by Hynes and VanMarcke (1976) revealed a worrying lack of consensus among 'expert' risk assessors. They asked seven geotechnical engineers of international repute to predict the height of an embankment that would cause a clay foundation to fail and to specify bounds around this estimate that were wide enough to have a 50 per cent chance of enclosing the correct failure height. None of the bounds identified by any of the engineers enclosed the true failure height. Similarly, Slovic *et al.* (1980) suggest that the collapse of the Teton Dam in 1976 was attributable to the overconfidence of engineers who believed that problems identified during construction had been rectified. Thus, Slovic *et al.* (1980) conclude that technical experts are almost as prone to subjectivity and misjudgement in their assessment of risks as laypersons.

Pidgeon (1996) rejects the narrow participation model of risk decision-making. In this model, risk assessment is undertaken by an elite group of scientists and policy-makers and the public has no direct role in this decision-making process. The public can, of course, influence risk decisions through normal democratic political processes. However, there are strong arguments for adopting a broader model of participation in risk decision-making because risk decisions are inherently value judgements about the fairness and distribution of harm and consequences in society. To exclude those who will be affected by these decisions from the decision-making process cannot be justified. In an organisational context, it is equally important that risk decisions affecting workers should not be made by technical 'experts' and imposed upon a disenfranchised workforce.

Social and cultural factors are reported to determine perceived risk, and researchers report that different social groups, for example employers and employees, think differently about risk. Pidgeon (1996) writes of 'plural rationalities' with regard to risk and suggests that the technocrats' emphasis on objective risk is no longer applicable because there exist competing and equally legitimate viewpoints concerning risk. Unlike the technical approach to risk, which assumes an objective truth known only to scientists and experts, social scientists' approach to risk is characterised by descriptive analyses rather than formal statistical models. Social scientists do not treat risk as something that can be represented by a single index; instead, they seek to understand the complex interplay of psychological, social, organisational and technological factors involved in accidents. Two social science approaches to risk are described below. These are the individual (psychological) approach and the group (sociological) approach.

Psychological approaches to risk

The Decision Research Group in Oregon, comprising Paul Slovic, Baruch Fischhoff and Sarah Lichtenstein, pioneered early empirical psychological studies of risk perceptions. This work drew upon the traditions of cognitive psychology, involving the study of human memory, sense perception, thought and reasoning, and the study of human decision-making. The Group's research followed the publication of a seminal paper by Starr (1969). Starr considered 'accepted' accident levels as an indicator of society's preferred trade-offs between the risk and benefits of a hazard and argued that risk acceptance in the United States was influenced by the extent to which the risk was undertaken voluntarily (for example, smoking) or involuntarily (for example, nuclear generation of energy).

The findings of Slovic *et al.* (1980) demonstrated that people's understanding of a wide variety of risks was influenced by a number of qualitative factors, which were not easily modelled using the mathematical methods of risk assessment. These qualitative factors included Starr's notion of voluntariness of exposure, but also included the extent to which individuals believed a risk to be within their personal control, and their familiarity with the source of the risk.

The study of perception is concerned with the ways individuals develop an understanding of their environment through assimilating current inputs with stored experiences or knowledge structures (Pidgeon *et al.* 1993). However, the study of risk perception cannot regard the human perceiver as an isolated being because people are exposed to many sources of information about hazards and their benefits. Thus, many influences shape people's understandings of risk. These sources include the scientific community, the mass media, family and friends, employee groups and other trusted sources.

An early study by Slovic and his colleagues (Slovic *et al.* 1978) sought to comprehend people's understandings of risk by asking them to judge the annual frequency of deaths arising from 40 hazards, against the reference point of annual motor vehicle accident deaths, which was made known to participants. The results were then plotted against the best available public health estimates, and differences between judged frequency and statistical estimates were examined. Two types of systematic difference were observed. First, respondents overestimated the number of deaths from infrequent causes, such as tornados and botulism, but underestimated deaths due to frequent causes, such as cancer and diabetes. Second, the risks for which fatalities were judged higher than they actually were usually involved vivid or imaginable causes of death. Lichtenstein *et al.* (1978) suggest that this may be due to the operation of the 'availability' heuristic whereby events are judged more likely if they can be easily recalled or conjured up in mental images. This has implications for the presentation of risk information. For example, it is possible that people's perceptions of risk are distorted by the reporting of dramatic events by the media. More insidious but less dramatic health problems, such as occupational cancer, are likely to be much less prominent in people's understandings of OHS risks.

Further psychometric studies supported the importance of qualitative characteristics of risk in people's perceptions of the magnitude of these risks. The complexity of people's qualitative understandings of risk is considerable. For example, Green and Brown (1978) investigated the influence of the timing of the effect of exposure to a risk, that is, whether it is associated with immediate or delayed consequences on understandings of risk. They report on the effect of low morbidity but high disability consequences, such as permanent paralysis or brain damage, on people's attitudes towards risk. Societal risks, such as release of nuclear radiation, are also distinguished from individual risks, such as suffering a snakebite. Pidgeon *et al.* (1993) argue that these qualitative characteristics of risk have important implications for risk management and communication because they are likely to underlie people's understandings and responses to these risks.

The Decision Research Group in Oregon undertook many studies to identify qualities that characterise risks and the effect of these qualities on people's perception of risk. In particular, this group asked whether the ways people understand risks can be modelled. In the Oregon studies, respondents were typically asked to provide judgements of risk for a range of hazards and rate risk sources or attributes in terms of their similarity or dissimilarity or their perceived characteristics. In these studies, respondents were not provided with an *a priori* definition of risk to ensure their personal understandings were elicited. Factor analysis was applied to the data collected and risk attributes were grouped according to their underlying dimensions. These dimensions could then be plotted to represent the psychological meaning of risk. Slovic *et al.* (1980) identify a systematic pattern in risk ratings. Two primary dimensions emerged from the analysis. These are depicted in Figure 5.11. The first factor was labelled 'dread' risk and was associated with uncontrollability, dread or fear, involuntariness of exposure and inequitable distribution of risk. Hazards that scored highly on the dread dimension included crime, nuclear weapons and nerve gas. The second factor was termed 'unknown risk' and was associated with delayed health effects, the familiarity of the risk, whether risks are known to science or not and judgements of the observability of the risk. Solar electric power, DNA research and satellites rated high on this dimension of risk. The studies revealed that perceptions of risk were closely associated with the position of the risk in this two-dimensional factor space. In particular, the higher a hazard's score on the dread dimension, the more serious people perceived the risk to be, and the more they desired strict regulation to reduce the risk.

A study of construction workers' perceptions of risk indicated that these two dimensions of risk apply. This finding has implications for risk communication and worker training programmes in the construction industry. The results of this study are described in the following case study.

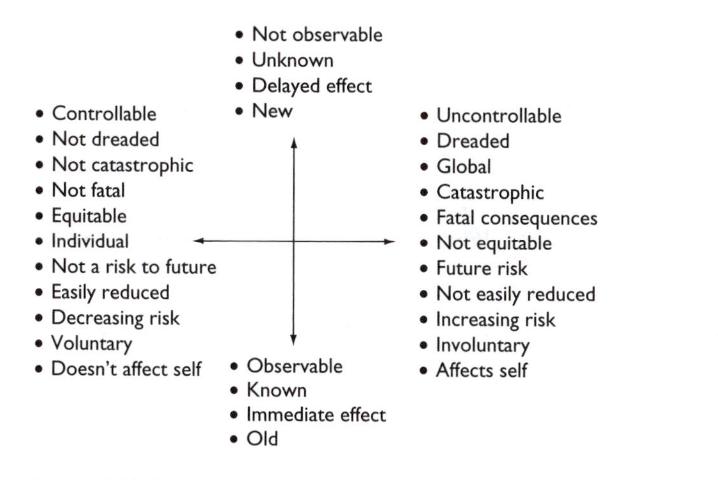

Figure 5.11 Two-dimensional model of risk.

Case study: Risk perception in the Australian construction industry

Tradesmen and workers in small business construction firms in Australia were asked about two different OHS risks. These risks were chosen because one risk was associated with an immediate outcome (falls from height) and the second (occupational skin diseases) with a delayed outcome.

Both risks were selected by trade union and employer representatives as being highly relevant to construction work. The OHS literature identifies the construction industry as a high-risk population for falls from heights (Davies and Tomasin 1990; Helander 1991; Kisner and Fosbroke 1994) and occupational dermatoses, particularly allergic dermatitis (Rosen and Freeman 1992; Geier and Schnuch 1995).

In-depth interviews explored how the participants understood these risks in relation to their workplaces. The participants understood these risks in very different ways.

All participants stressed the importance of falls from height, in terms of both the seriousness of the risk and the probability of its occurrence. The risk was said to be serious because of the immediacy of the effect and the possibility of the fatal consequences of falls. Most participants could easily imagine a fall happening and several recalled a fall they had witnessed or heard about. Falls were also regarded as being difficult, in fact almost impossible, to control given resource constraints and contractual relationships in the construction industry. The risk of falls was also considered to be something that was involuntary in that the risk was accepted as being an 'inevitable part of working in the construction industry'.

Case study (Continued)

In contrast, most participants knew little about the risk of occupational skin disease, and could not imagine it happening. The participants also expressed little trust in sources of information about the risk of occupational skin disease – that is, scientists. They thought that the risk of skin disease was related to individual susceptibility, only affecting those with an allergy. Many expressed the view that they were not at risk of occupational skin disease. The risk was perceived to be voluntary, in that workers could opt to use protective equipment if they were susceptible to the risk of skin disease. Therefore, it was believed that the risk could be easily reduced.

A comparison of the participants' understandings of the two risks indicates that the risks of occupational skin disease and falls from height are positioned at opposite diagonal corners of the two-dimensional model depicted in Figure 5.11. Fall risks were known to workers and are associated with serious or fatal immediate consequences. In contrast to this, the risk of occupational skin disease was associated with a prolonged effect and a lack of knowledge or awareness. The risk of occupational skin disease was understood in terms of qualitative characteristics of low-dread risks.

In accordance with the findings of the Decision Research Group, the participants perceived the risk of falls to be significantly greater than that of occupational skin disease.

These different ways of understanding the two OHS risks have implications for effective risk control strategies. For example, it is likely that an effective risk control strategy for occupational skin disease would have to include a strong educational component. This would be designed to address the lack of knowledge concerning the nature and potential severity of the occupational skin disease risk.

Conversely, the risk of falling from height is well known, and its consequences are dreaded. The response of participants to the risk of falls from height was also consistent with the findings of Slovic *et al.* (1980) who report that the greater the perception of a risk on the 'Dread risk' dimension, the greater the perceived need for external regulation of the risk. Participants in the Australian study suggested that there was a need for improved regulation and enforcement of OHS issues related to the risk of falls from height.

(Source: Lingard and Holmes 2001)

Sociological approaches to risk

Industrial sociologists have suggested that, rather than focusing on the behaviour of individual employees, the social organisation of work should be examined as a determinant of occupational illness and injury. Sociologists argue that risk is socially constructed by organisational structures, payment systems, industrial conflict and other industrial and social factors. The general pattern is that mortality rates for blue-collar, unskilled and semi-skilled workers is higher than that of professional, managerial and clerical workers (Bohle and Quinlan 2000). Sociologists argue that occupational injury and illness are inherent features of the social organisation of work, and that existing models of accident causation fail to address the clash between employer and employee interests, between profit and safety, that is the essence of workplace social relations.

There is also a growing understanding that people's perceptions of risk are not generated solely by individual psychological processes, but are shaped by their membership in social groups whose members share common attitudes, beliefs and behaviours. Thus, membership of a particular social category influences how risks are perceived, and ultimately determines people's beliefs about the adequacy of existing arrangements for risk management.

Empirical evidence exists to support group differences in risk acceptance. For example, Vlek and Stallen (1981) identify occupational differences in risk acceptance. Similarly, Holmes and Gifford (1996) explored the understandings of OHS risk among business owners and their employees in the Australian construction industry. They report that the power hierarchy, based on economic transactions from builders to subcontractors, was related to different understandings of risk. Employers focused on employees as the source of risk and placed responsibility for risk control onto individual workers. Employees were deeply pessimistic about the extent to which they would be protected from workplace risk and perceived their only method of control to be the 'tools of their trade', which were deemed to be inadequate due to employer decisions to hire cheaper, but poorly maintained, equipment and purchase unsafe chemicals. Holmes and Gifford (1996) conclude that social differences in assumptions about OHS risk have a direct impact upon risk management decision-making in the construction industry. These group differences in thinking about risk threaten the effectiveness of consultative risk management activities in construction. Risk communication tools need to be developed to develop a mutual understanding, if not agreement, of the risk factors involved in occupational injury and illness in construction.

Theorists have tried to develop an integrated model of psychological and social approaches to risk perception. One such theory is the social amplification theory. This theory holds that all the information individuals receive about risk is second hand, and that risk information is interpreted according to the beliefs and values of individuals and other social groups which influence the meaning of these risk messages (Kasperson *et al.* 1988). These groups may include the mass media, government agencies, employers' groups or workers' groups. The theory holds that the amplification stations attenuate or intensify risk according to their position in the social structure. This theory is used to explain the individual and social differences in perceptions of risk.

The need for a multi-disciplinary approach to risk

The understanding that psychological and social forces play an important role in shaping beliefs about risk means that a multi-disciplinary perspective to risk management must be adopted. Risk management processes, at workplace, industry or government policy-making levels, should not accept the hegemony of the technical approach because reliance on this approach is likely to lead to inequitable and unsatisfactory outcomes. As discussed earlier, risk assessment is never entirely objective. Even technical experts are members of a social grouping, the scientific community, who have been exposed to a common set of explicit or implicit assumptions, which inform their judgements. These judgements are likely to be just as much the product of social processes as those of employee or community groups. It is essential to recognise that the social construction of risk renders risk management an inherently political process, within which the interests and perspectives of different social groups must be reconciled. Given that risk management has a political aspect, it is important that risk communication is made a critical component of effective OHS risk management.

Risk communication

Risk management decision-makers have both legal and moral responsibility to provide information to people who are exposed to risks. In the workplace, this means that workers have a 'right to know' about OHS risks associated with work processes and materials, and employers have a duty to provide this information. In exercising this duty, employers must establish clear and effective risk communication channels. Appropriate forms of risk communication can generate a mutual understanding of risk and help to resolve conflicts that may arise concerning risk management decisions (Pidgeon *et al.* 1993). However, Otway and Wynne (1989) identify a dilemma inherent in risk communication: it is designed both to warn people of risks and to reassure them as to the tolerability of these risks. They call this the reassurance-arousal paradox. This paradox makes the communication of risk information challenging, particularly when risk messages are to be received by different audiences who are exposed to multiple information channels and competing messages. For example, OHS risk information may be provided to employers and employees by government agents, professional institutions, trade union representatives, OHS advisers, suppliers, consultants, the mass media, trade associations and family and friends.

The technical approach to risk management implies a 'top-down' model of risk communication, in which one-way communication occurs between an expert communicator and a non-expert recipient. This model has been rightly criticised because it devalues the knowledge or viewpoint of risk bearers, and ignores the underlying socio-political determinants of concerns about risk. In contrast to this approach, the US National Research Council (NRC 1989) advocates a two-way risk dialogue or an exchange of information between parties to a risk issue. The consultative processes required by OHS legislation are key to the development of two-way communication regarding OHS risks. Employee OHS representatives and workplace health and safety committees should play an important role in disseminating risk information to employees and acting as employee advocates in the risk decision-making process. Owing to organisational power structures and differing levels of educational attainment, training employees to act as advocates of employees' interests in the OHS risk communication process is also essential. Research suggests that 'empowering' workers to negotiate more effectively for safety improvements in the work environment is beneficial for both workers and organisations (Lippin *et al.* 2000).

The NRC suggests that improving risk communication is as much a matter of improving risk communication processes as it is about improving the quality of risk messages to be communicated. Thus, the NRC recommends involving interest groups early in the risk evaluation process, thereby allowing these parties to identify points of particular salience from their perspective. The NRC also recommends that risk information be presented in plain language so that it can be easily understood. Technical jargon should be avoided, especially where it uses terms such as 'morbidity', which appear familiar to laypersons but which have a special meaning in risk management. The use of plain language is particularly important in construction, where many workers have undergone limited formal education and may be unfamiliar with technical specifications. For example, MSDSs have been widely criticised for their overly technical hazard information, and it is important that their contents be explained to employees exposed to hazardous chemicals. In addition, in some countries, such as Australia, many workers are from non-English-speaking backgrounds, so that risk information may need to be provided in languages other than English. While risk information should be provided in plain language, it should not be oversimplified. Information should be comprehensive and any uncertainties associated with risk exposure made clear. Full information about the nature of the risk, information about alternatives, uncertainties in knowledge about the risks and information about responsibilities for the management of the risk should be provided.

Trust also plays an important role in risk communication. The credibility of communicators is dependent upon the trust that message recipients have in the communicator. Openness is essential to the development of trust because perceptions that decisions are made 'behind closed doors' are likely to fuel suspicion and hostility.

All risk management decisions are inherently political because they involve decisions concerning who should bear the risks, and who should benefit from them. They also involve answering the contentious question 'how safe is safe enough?' It is therefore likely that OHS risk management decisions will remain controversial. However, effective two-way risk communication may mitigate some of this controversy. The NRC concludes that 'Even great improvement in risk communication will not resolve risk management problems and end controversy (although poor risk communication can create them)' (NRC 1989, p. 146). Construction organisations should therefore develop processes for early, open, complete and clear two-way risk communication with employees, subcontractors and other interested parties in the community.

Risk compensation

Despite technological advances that have made the world safer and healthier, accidents still occur, and humans expose themselves to known sources of illness or disease. In the past, road safety researchers have tried to explain why the mandatory use of safety devices, such as seat belts, has done little to reduce national death rates associated with motor vehicle accidents (Wilde 1982). One explanation for this is that drivers whose cars are equipped with safety features may believe that driving cautiously is unnecessary as a result of these other protective measures. In a sense, they 'compensate' for the additional safety measures by behaving in a more risky way. The same logic can be applied to an industrial situation. For example, the engineering solution to the risk of flying particles striking an operative's eye during a grinding operation may be to install a protective shield to prevent such particles from coming into contact with the worker. However, this may lead to a failure to wear eye protection, such as safety glasses or goggles, because the operative may regard the additional benefit of wearing this protective equipment, which is the maximisation of eye protection, to be insufficient to warrant the costs, perhaps in terms of discomfort or restricted vision, of wearing the protective equipment.

Wilde (1982) developed a psychological theory to explain such compensatory risky behaviours, which he termed *risk homeostasis theory* (RHT). Wilde suggested that a person making a decision as to how to behave in any activity would act in a way that would maintain the risk to which they were exposed at a constant level. Thus, 'at any moment in time the instantaneously experienced level of risk is compared with the level of risk the individual wishes to take, and decisions to alter ongoing behaviour will be made whenever these two levels are discrepant' (p. 210).

The operation of RHT has received some empirical support – for example Streff and Geller (1998) found that subjective risk assessments were related to compensatory behaviours; however, they did not assess whether compensatory behaviours returned participants to a 'target' level of subjective risk. Stetzer and Hofmann (1996) undertook an experimental study to test RHT more rigorously. Their results suggest that risk compensation mechanisms occur on a widespread basis. In two separate studies, participants adjusted behavioural intentions in a manner consistent with risk compensation expectations. However, only a small proportion of participants were found to compensate enough to maintain a homeostatic level of risk, as predicted by RHT. Thus it appears that when the environment is made safer, a large proportion of individuals change their behaviour to compensate for this enhanced safety.

In a workplace context, RHT has implications for the implementation of engineering technological controls described earlier in this chapter (see selection of risk controls). While such controls are preferable according to the hierarchy of risk controls, attempts to make the workplace safer could lead employees to behave in a way that undermines the safety benefits of these improvements. Thus, those involved in occupational risk management decision-making should recognise that a purely engineering approach to OHS is not likely to yield the best results. Instead, it is also important to address the psychological factors impacting upon workers' perceptions of OHS risk and behaviour. Psychological determinants of workers' OHS behaviour are considered in detail in Chapters 7 and 8.

Conclusions

OHS legislation and standards are based upon risk management concepts. In the management of OHS, construction firms must therefore follow the process of hazard identification, risk assessment and risk control. There are many available tools and techniques for risk assessment, ranging from qualitative methods to more sophisticated probabilistic quantitative methods. Risk management is an iterative process and OHS risks must be constantly monitored to ensure they are reduced so far as is 'reasonably practicable'. This requires decision-makers to have an understanding of the technologically feasible methods to control OHS risks, and to implement appropriate controls given the magnitude of the risk, the cost of implementation of risk controls and the benefit (risk reduction) achieved. A 'hierarchy of controls' dictates that it is unacceptable to rely on administrative controls or personal protective equipment where technological solutions are cost-effective, although administrative controls and personal protective equipment may supplement technological controls. Decision-makers must also be aware of the psychological and social influences on risk perceptions, as these are likely to impact upon workers' attitudes towards risk, behaviour in the face of danger and acceptance of risk management decision-making. Modern OHS legislation requires that workers have the opportunity to participate in OHS decision-making, which means that two-way risk communication is essential. Construction organisations, irrespective of their size, should implement OHS risk management processes in order to meet their legal obligations.

Discussion and review questions

1. Is it reasonable to assume organisations have cost incentives to reduce OHS risks to tolerable levels?
2. Can experts' risk estimations be assumed to be objective and scientific?
3. Why is it important that cross-disciplinary teams be involved in the risk management process?

Chapter 6

Ergonomics in construction

Vivienne A. Monk

Ergonomics

The science of ergonomics is a comparatively recent phenomenon; however, there is evidence to suggest that ergonomic principles were known to the ancient Greeks (Marmaras *et al.* 1999). The word *ergonomics* is derived from the Greek words *ergon*, meaning work, and *nomos*, meaning physical law. The term was coined by Professor Hywell Murrell following a working group meeting in 1949. This meeting formed a society for 'the study of human beings in their working environment' (Marmaras *et al.* 1999). Ergonomics combines the knowledge of human abilities with those of tool design, equipment and the organisation of work. People from many different backgrounds work in ergonomics, such as physiotherapists, occupational therapists, psychologists, engineers, physiologists, doctors and others.

Occupational ergonomics has to do with the design or modification of the workplace and the organisation of work to match the worker. The aim of occupational ergonomics is to decrease injuries at the workplace and to improve productivity. Ergonomic principles can be used at many stages of work design. They can be used in tool and equipment design, workplace layout and the planning of work processes. The aim of this chapter is to consider the relevance of ergonomics to the construction industry, as well as identify particular problem areas for construction trades, and suggest ways in which the ergonomic aspects of construction tasks can be assessed in order to prevent injury. The chapter also considers the legal framework for regulating ergonomic issues, and identifies the need for greater regulation of ergonomics in design processes.

Construction work

Construction work is physically demanding, requiring frequent manual handling of heavy loads. Within the industry, the organisation of work and the work environment are constantly changing. Construction is quite unlike most other industries in this regard. Much of the work is performed in poor or awkward postures and, consequently, construction workers are among those at highest risk of injury.

This is borne out in injury statistics derived from workers' compensation claims. In Australia, national 1998–1999 workers' compensation statistics indicate that the incidence rate (IR) (incidents/accidents per thousand employees) for the construction industry was 35.3. The comparable all-industries average was 17.4. Frequency Rate (FR) for construction and mining labourers was 28.6, while the FR for building tradespersons was 20.2. Construction and mining labourers and building tradespersons experienced IRs of 54.6 and 41.2 respectively. The frequency rate measures the number of incidents/accidents per million hours worked (AS 1885.1–1990).

Further analysis of the statistics suggests that many of these injuries may be related to ergonomic issues. For example, in 1998–1999, there were 14,112 manual handling injuries reported nationally. These accounted for 33.8 per cent of all workplace injuries. These cost $226.2 million and 126,862 weeks in lost time. The principal agencies of these injuries were: crates, cartons and boxes (2,591), other persons (1,397) and metal objects (663).

Many injuries and illnesses that affect construction workers are musculoskeletal disorders (MSDs). These are primarily labelled as sprains and strains, which can be categorised as *acute* (that is, of recent onset) or *chronic*. Chronic problems are often described in terms of cumulative trauma disorders or repetitive strain injuries.

Legislation and standards pertaining to ergonomic issues, and in particular manual handling, adopt a hazard management approach, in which it is incumbent upon organisations to identify and assess the hazards associated with manual handling tasks and to apply a hierarchy of risk controls in the selection of injury prevention measures. However, research in the construction industry suggests that construction firms, particularly smaller firms, are not well versed in hazard management.

For example, as part of the 'Safework 2000' project, Trethewy *et al.* (2000) identified some key problems in the Australian building and construction industry. They found that most employers had difficulty in formally identifying hazards and in assessing the level of risk in relation to a hazard. The research team appraised safety documentation from approximately fifty different sub-contractors and found vast differences in the interpretation of risk. Similarly, another Australian report titled *Safely Building NSW* (2001) identified the gaps in effective OHS management in the industry. These included poor programming practices, poor design and unrealistic scheduling and problems arising between interacting trades. Arguably, all of these problems can be improved using ergonomic interventions.

There is relatively little work done in the area of quantifying ergonomic exposures in the construction industry. A notable exception is the work of Schneider, Griffin and Chowdhury (1998) who performed a detailed analysis of ergonomic exposures of construction workers based on an analysis of

Table 6.1 Ergonomic exposure variables and types of injuries from Schneider *et al.* (1998)

Ergonomic exposure	*Types of injuries*
Strength (lifting heavy weights)	sprains/strains
Climbing (agility)	strains
Balancing (strength)	falls
Stooping	
Kneeling	lower extremity cumulative trauma disorders (CTDs) of knees, ankles, hips
Bending	
Crawling	
Reaching	shoulder CTDs
Handling	
Fingering	carpal tunnel syndrome, tendonitis

the US Department of Labor Employment and Training Administration Database on job demands. The ergonomic exposure variables and types of injuries identified by Schneider *et al.* (1998) are as in Table 6.1.

Bernard (1997) conducted a comprehensive review of epidemiological evidence for work-related MSDs of the neck, upper extremity and lower back. He included over six hundred studies in his review process. Table 6.2 presents the evidence for a causal relationship between certain physical work factors and MSDs. Many of these risk factors are evident in the work of construction trades.

Table 6.2 Evidence for a causal relationship between physical work factors and MSDs (adapted from Bernard 1997)

Body part *Risk factor*	*Strong evidence*	*Evidence*
Neck and neck/shoulder		
Repetition	—	+
Force	—	+
Posture	+	—
Shoulder		
Posture	—	+
Force	—	—
Repetition	—	+
Elbow		
Repetition	—	—
Force	—	+
Posture	—	—
Combination	+	—

Table 6.2 (Continued)

Body part *Risk factor*	*Strong evidence*	*Evidence*
Hand/wrist carpal tunnel syndrome		
Repetition	—	+
Force	—	+
Posture	—	—
Vibration	—	+
Combination	+	—
Tendonitis		
Repetition	—	+
Force	—	+
Posture	—	+
Combination	+	—
Hand-arm vibration syndrome		
Vibration	+	—
Back Lifting/forceful movement		
Awkward posture	+	—
Heavy physical work	—	+
Whole body vibration	—	+
Static work posture	+	—

Note: 'Evidence' indicates that some convincing epidemiological evidence shows a causal relationship for intense or long duration exposure to the specific risk factor(s) and MSD, when the epidemological criteria of causality are used. 'Strong evidence' indicates that a causal relationship is shown to be very likely between intense or long duration exposure to the specific risk factor(s) and MSD, when the epidemiological criteria of causality are used. Epidemiological studies, such as this one, provide indirect evidence of a relationship between health or disease and other factors. This process is referred to as a causal inference.

From these studies, Bernard concludes that there is a substantial body of credible epidemiological research that provides evidence of an association between MSDs and certain work-related factors when there is a high level of exposure to more than one physical factor; for example, repetition, frequency, duration, intensity and posture. The results suggest that, while exposure to a single risk factor alone may cause injury, injuries are more likely where multiple risk factors occur.

In construction work, many of these physical risk factors are present, making ergonomics particularly important. In construction, applying ergonomic principles to achieve a good match between workers and their jobs should reduce the incidence and severity of occupational injuries. MSDs, and particularly back problems, are common and costly in the construction industry in which the common risk factors include repetition, force, weight, posture and vibration.

Legislation relevant to ergonomic issues

With the growing understanding of the prevalence of musculoskeletal injuries, and the importance of ergonomic issues, most developed countries have implemented legislation and standards applicable to issues such as manual handling and repetitive strain injury. For the most part, the legislation and standards adopt a hazard management approach, requiring, for example, that the risks presented by manual handling tasks be identified and assessed and appropriate control strategies selected. In the UK, the *Manual Handling Operations Regulations* (1992) require that manual lifting tasks be assessed and, where possible, avoided. If these tasks cannot be avoided, employers are obliged to reduce the risks presented so far as is reasonably practicable.

Manual handling

Manual handling is a particularly important area for ergonomic intervention because of its association with a very large number of work-related injuries and compensation claims. For example, in the Australian state of Victoria, strains and sprains affecting shoulder, neck, arm, hand or back, account for 55 per cent of all workers' compensation claims; 62 per cent of all workers' compensation costs and 70 per cent of long-term workers' compensation claims (WorkCover Authority 1999). While it is difficult to identify a direct one-to-one relationship with these outcomes and manual handling activity, sprains and strains are probably the best proxy of manual handling related injuries (Bottomly and Associates 2003).

Traditionally, limits were placed upon the weight of loads to be handled. However, this simplistic approach has fallen out of favour because it is now understood that weight is only one factor contributing to injuries in manual handling. Loads on anatomical structures are variable over time and the capabilities of structures are difficult to estimate without the use of complex biomechanical models. It is now believed to be almost impossible to specify a threshold limit below which injury is unlikely and above which injury is probable. This is why manual handling codes and standards now emphasise a hazard management approach and avoid specifying threshold weight limits. Despite this, some remnants of threshold limits are retained. For example, in the UK, the guidance material produced for the *Manual Handling Regulations* (1992) suggests a threshold value for lifting of 25 kg in ideal situations, for men. This value is then reduced as a function of the load distance, lifting height and lift frequency. Lower limits are stipulated for women.

Perhaps the most blatant use of threshold limits is in the use of the NIOSH lifting equation. The NIOSH equation is a method developed by the National Institute of Occupational Safety and Health in the USA to assess the risks of manual handling associated with lifting and lowering tasks. It is used to answer two questions:

1 What is considered a safe weight to lift in a particular circumstance?
2 What can be done to make this lifting safe?

The equation generates a recommended weight limit for a particular lifting task being assessed. This limit represents the weight that nearly all healthy workers performing this task could handle without increasing their risk of lower back pain.

In order to use the NIOSH equation, the following information about the task is considered:

- the weight of the object being lifted;
- how the lifting is done;
- how long the lifting is done for;
- the height of the hands at the start and end of the lift;
- how far the hands are away from the body at the start and end of the lift;
- how good a grip the employee can get on the object; and
- the degree of twisting of the body (WorkSafe Victoria 1998).

The NIOSH equation is recommended for use in some Australian jurisdictions, including Victoria. However, it has been criticised. The equation is based upon a threshold value of 3.4 kN for spinal compression, which, it is argued, is not justifiable because the compressive tolerance of lumbar failure varies considerably with factors, such as age and sex. Thus, the NIOSH equation may not adequately protect female or older workers (Burgess-Limerick 2003).

In Australia, legislation dealing with manual handling is based on a *National Code of Practice for Manual Handling* and a *National Standard for Manual Handling* which were published by the National Occupational Health and Safety Commission (NOHSC) in 1990. The NOHSC provides a forum for the Commonwealth, state and territory governments, employer organisations and trade unions to develop national approaches to OHS matters. In the area of OHS legislation, the National Commission has the power to declare national OHS standards and codes of practice, which are developed as the basis for nationally consistent OHS Regulations and codes of practice, but these are not legally enforceable unless state and territory governments adopt them as regulations or codes of practice under their principles.

The objective of the *National Standard for Manual Handling* (NOHSC: 1001, 1990) was to prevent or reduce the frequency of manual handling injuries at work by a three-stage process of identifying, assessing and controlling risks associated with manual handling. The *National Code of*

Practice for Manual Handling (NOHSC: 2005, 1990), explains how to comply with the standard. A checklist included in this code suggests the following threshold limits:

Is the weight of the object:

- more than 4.5 kg and handled from a seated position?
- more than 16 kg and handled from a seated position?
- more than 55 kg?

In addition, NOHSC also developed core-training elements for the *National Standard for Manual Handling*, which aim to set the standard for the development and delivery of quality training on manual handling legislation, and advise on the training required to support the successful implementation of a workplace manual handling strategy.

In April 2001, NOHSC approved a Continuous Improvement Program designed to improve national standards and related materials and, consistent with this Program, the *National Standard for Manual Handling* and the code of practice are being reviewed.

Occupational Overuse Syndrome

In 1994, NOHSC produced the *National Code of Practice for the Prevention of Occupational Overuse Syndrome* (NOHSC: 2013, 1994). The purpose of which is to provide practical guidance for the prevention of risks, and the identification, assessment and control of risks, arising from tasks undertaken in the workplace which involve either or both of the following:

- repetitive or forceful movement;
- maintenance of constrained or awkward postures.

This code of practice is consistent with, and complementary to, the *National Code of Practice for Manual Handling*. Both of these codes adopt the same approach, which can be summarised in three stages: identification, assessment and control.

Risk identification is undertaken through:

- an analysis of workplace injury and incident records;
- consultation with employees; and
- direct observation of the worker, the task and the workplace.

Risk assessment includes an assessment of:

- the workplace and workstation layout;
- the working posture;

- the duration and frequency of activity;
- the force applied;
- the work organisation;
- the level of skills and experience of workers; and
- individual factors.

In *Risk control* the emphasis is placed upon:

- job design and redesign;
- modification of the workplace layout;
- modification of the object or equipment;
- maintenance; and
- the provision of task-specific training.

Thus, the standards and codes broadly follow the risk management process described in Chapter 5 and focus on controlling risk by changing the work tasks and the work environment to better match individual workers' characteristics.

Hazard-based standards and codes, such as the manual handling standards and codes, have been criticised because they do not provide detailed guidance about how to control risks in a particular industrial context. Thus, it is left to people in the workplace to interpret the codes in the design of appropriate risk control strategies for manual handling. In an industry such as construction, guidance that is more practical may be needed.

Ergonomic risks in construction

Job-related activities and behaviours that increase the risk of MSDs in the construction industry are similar to those identified in other industries (Schneider *et al.* 1995). These include:

- the frequency with which a task is performed (or *repetition*);
- the amount of physical force that is used;
- the need for lifting or moving heavy loads;
- the amount of prolonged static muscular tension;
- working posture and position;
- vibration from tools or machinery; and
- the need for working overhead or at extreme ranges of movement.

Each of these is discussed in turn below.

Frequency with which a task is performed (or repetition)

Repetition can be defined as the number of movements that occurs in a given amount of time to complete a task. Highly repetitive tasks have been defined as those that take 30 seconds or less to complete, or a concentration of more than 50 per cent of the time performing the same task.

When muscles are engaged in highly repetitive activity, they do not have sufficient time to recover properly, and muscle fatigue occurs. The speed at which this fatigue occurs depends on how often the muscles move, the force required and how long the activity is maintained without a break. When any other ergonomic risk factor is added to repetition, such as exerting force, weights or awkward posture, a longer recovery time is required. When muscles fatigue, the risk of pain and injury is greatly increased. Tendons and sheaths are especially vulnerable to repetitive movement.

Many jobs within construction involve repetitive movements. One obvious example of the need for repetitive movement in construction is in the task of bricklaying. Bricklayers are constantly moving to pick up bricks and mortar and to lay the bricks. This frequency can be worsened by the weights, forces and postures involved.

The amount of physical force that is used

Movement is caused by muscles contracting and relaxing. If muscle strength is inadequate to provide the strength required for the activity, over-exertion occurs. If sufficient recovery time is not allowed, additional joints and muscles will be used, resulting in additional energy expenditure and onset of fatigue. The amount of fatigue experienced is related to the amount of force applied and the duration of the force. The greater the relative force, the greater the risk of tissue damage. The term 'relative' is used to describe the amount of force, as the closer a muscle is to its strength capacity, the greater the risk of tissue damage.

Common causes of high relative force in the construction industry are the manual handling of heavy loads, which is common to all occupational groups, the sizes and shapes of loads, and the location of objects. Objects can often be made easier to work with – by changing their location, for example, so that they do not have to be carried so far, or by changing the orientation of the load.

Lifting or moving heavy loads (manual handling)

In the NOHSC National Code, *manual handling* is defined broadly as any activity requiring the use of force to lift, lower, push, pull, carry or otherwise move, hold or restrain any animate or inanimate object.

Risks from manual handling include those associated with the working environment, the task, the load and the individual's capacity. The spine is particularly vulnerable to manual handling injuries. Other vulnerable tissues include discs, muscles and ligaments.

In the construction industry, the working environment often presents elements which increase the risk of injury, such as slippery, uneven surfaces, inadequate lighting, confined spaces, extremes of temperature, noise, proximity of other tradespeople and workers carrying out different tasks.

Prolonged static muscular tension

When muscles or joints are maintained in fixed or static postures, fatigue develops. Static loading is the continuous tensing of muscles, which results in decreased circulation causing localised pain and muscle fatigue. This can lead to an increased risk of developing a cumulative trauma disorder. The duration of exposure pertains to both the length of time a static posture is held and the length of the exposure during the working shift. The longer the worker is exposed to the risk factors during the working day, the greater the risk of injury.

Concreters in particular are exposed to many risks associated with prolonged static muscular tension. For example, workers who are holding a heavy hose full of concrete may maintain static postures for long durations and have long exposures during the length of their shifts.

Working posture and position

Awkward or uncomfortable postures occur when a part of the body is positioned outside its neutral range. The risk of pain and injury is increased:

- where the work height varies greatly from the optimum level;
- where there are frequent actions that require extremes of reach, bending or twisting (for example, when working outside a neutral, or 'low risk' position);
- when maintaining a single posture for long periods, for example standing or sitting;
- when maintaining fixed body postures without support, for example when sitting without back support;
- when the head, neck and/or trunk are inclined forwards, for example in a stooped position;
- when arms are used in a raised position, causing static load on the shoulders;
- when the spine is twisted, which increases loading on the spine; and
- when using poorly designed tools.

The key to improving work postures lies in understanding why they are being adopted. By observing workers performing their tasks, this can become apparent. Often working materials are not placed near to the working area, necessitating unnecessary manual handling of loads, either in a horizontal or vertical direction. Sometimes tools or products might be poorly designed or not appropriate for the job or the worker. The emphasis on changing the work environment is an important feature of the hazard-based approach of manual handling codes and standards and legislation.

Vibration from tools or machinery

Vibrations from tools/machinery displace the body from its resting position. The oscillations can cause changes in the position of limbs and internal organs (Grandjean 2000). Grandjean describes the following factors, which are important in understanding the effects of vibration:

- where vibrations enter the body;
- the direction of oscillations;
- the frequency of oscillations;
- the acceleration of oscillations;
- the duration of effect;
- the resonance; and
- the 'damping' or reduction of amplitude by body tissues.

Operating powered hand tools involves high levels of vibration in the hands, wrists and arms. High level exposure to vibration from tools can lead to hand–arm vibration syndrome (HAVS), recognised as disturbances to blood circulation, and to neurological and locomotor functions of the hand and arm. This is due to a cramp-like condition of the blood vessels known as Reynaud's disease. Reynaud's phenomenon is a classical condition attributed to vasospasm triggered by exposure to external cold or emotional stress. It is a manifestation of vasomotor instability, which may be initiated by hand-tool vibration. Pneumatic drills, jackhammers, power saws and other tools used in the building and construction industry can cause vibration problems.

Working overhead or at extreme ranges of movement

Tasks involving work with the hand or hands at or above shoulder level result in a greatly increased risk of acquiring rotator cuff tendinopathy. Sporrong *et al.* (1999), in a study of ceiling fitters, found that 60 per cent of fitters estimated they worked with their hands above shoulder level more than three-quarters of their working time. They concluded that workers undertaking ceiling fitting were at a high risk of suffering shoulder pain, due mainly to rotator cuff tendinopathy.

There are many construction trades in which operatives frequently work with their arms overhead or at extreme ranges of movement. These include painters, plasterers, ceiling fixers, electricians, plumbers and air-conditioning fitters.

Monk (1998) found overhead working to be a particular problem for workers engaged in painting ceilings. Preparation for painting also produced a high number of awkward postures with the 'one arm above shoulder' position occurring 54.9 per cent of the time in this activity. Painters also handle a roller filled with paint while working overhead, adding to torque on their spine. In the same study, plasterers were reported to be in the 'two arms above shoulder level' while sheeting 18.3 per cent of the time. While they are doing this, the risk factor is exacerbated because they are also supporting the weight of the plasterboard.

Summary of ergonomic risks in construction

Workers are at increased risk of injury by any of the following:

- repetitive bending;
- bending their backs while lifting a weight;
- twisting;
- twisting while lifting a weight;
- working above shoulder height;
- working above shoulder height while lifting a weight;
- working overhead;
- working in prolonged static postures;
- absorbing excessive vibration; and
- working in poor conditions.

The above risk factors are central to many jobs in the construction industry. Many of these risk factors are very common. For example, in an analysis of work-related risk factors among unionised construction workers, Cook and Zimmerman (1992) identified the following as the most troublesome elements of construction jobs:

- maintaining awkward postures for prolonged times (69 per cent of respondents);
- working in adverse environmental conditions (76 per cent of respondents);
- having to work 'very hard' (44 per cent of respondents); and
- having to reach overhead (43 per cent of respondents).

Workplace planning and organisation

The flow of materials, co-ordination of tasks and scheduling of work can all affect the ergonomic risk factors on a construction site. Work should be designed and organised so that employees are able to regulate their tasks to meet work demands. Meeting tight deadlines and peak demands will increase time pressures, reduce control over workflow, and may contribute to the risk of pain and injury. Factors, such as the industrial relations climate, wage-/salary-negotiating system, influence of workers on decision-making, stress and the level of workers' skill utilisation can all affect productivity, well-being and health in an organisation.

Psychosocial factors

The construction industry culture could also influence workers to overexert themselves when feeling pressure from supervisors or peers to work harder. Production pressures may lead to unreasonable increases in workload, and corners may be cut when it comes to evaluating the safety of work. Often ergonomic hazards are not perceived in the same way as any other hazard, but as an inherent part of the job. Also, there may not be someone on site who understands ergonomic risk factors and who can work to correct these situations in a proactive fashion.

Environmental factors

Extremes in temperature, excessive noise or inadequate lighting can all affect workers' ability to perform their tasks. Excessive noise can affect concentration and communication with others. It can also increase stress to the body causing static muscle loading (Kjellberg 1990). Glare can cause squinting, eyestrain and headache, while poor lighting can mean that the worker may have to adopt a poor posture in order to see. These factors all require additional physiological resources over and above the task demands.

Organisational issues also present ergonomic hazards in construction. Within the building and construction industry, the organisation of work and the work environment are constantly changing and workers have to deal with uncertainty of employment and a variety of ergonomic hazards at the workplace.

These aspects of the organisational and physical work environment make managing ergonomic issues more challenging in construction than in other industry sectors, such as manufacturing, in which workstations and organisational issues are more predictable and controlled.

Specific ergonomic risks for construction trade groups

The need for an effective ergonomic risk assessment tool has been highlighted. In a report to the NOHSC, Burgess-Limerick (2003) suggests that many of the checklists currently used are inadequate because they consist of a list of yes/no questions about manual handling tasks. While these checklists may be helpful in identifying ergonomic risks, they are of limited utility in assessing the relative contribution of different risk factors to enable the prioritisation of risk controls. Burgess-Limerick (2003) argues that there is a need to develop a manual handling risk assessment tool that:

- is applicable to the complete range of manual handling tasks;
- provides an integrated assessment of biomechanical risk factors;
- provides an independent assessment of injury risk to different body regions;
- provides an overall risk assessment which allows prioritisation of tasks and incorporates suggested action thresholds, but does not imply a misleading level of precision;
- facilitates the effective targeting of controls by providing an indication of the relative severity of different risk factors within a task; and
- is suitable for use by workplace staff with minimal training and equipment.

One risk assessment tool used for manual handling is the Ovako Working Posture Analysing System (OWAS). The OWAS was developed to analyse a wide range of varying postures and has been used by Mattila (1989) and Kivi and Mattila (1991) to analyse work postures in building construction. On the basis of this analysis, construction jobs and tasks were classified according to the severity and generality of awkward postures involved. Mattila (1985) tested the OWAS method for two years at a construction site and found the method to be useful and reliable.

The OWAS method uses a four-digit code to describe the position of the back, arms and legs and the need to use force at frequent intervals during task performance. It is based on a work-sampling procedure that provides information on the amount of time spent in each posture. Sampling intervals are usually between 30 and 60 seconds. The task is assigned to one of four action categories based on combinations of postures, or the proportion of time spent in each posture. The postures are classified by the four-digit code that gives an indication of their discomfort. The OWAS posture set is depicted in Figure 6.1.

Figure 6.1 The OWAS posture set (adapted from Karhu *et al.* 1977, 1981; reprinted with permission from Elsevier).

The OWAS allows for some differentiation of risk between back, arms and legs, based on posture and is intended to serve as a practical tool for analysis at the worksite. The OWAS not only helps to identify the problem, but also points the way to correctional measures and is therefore a useful tool to aid in the control of manual handling risks. For example, if a task is observed to be undertaken in a high-risk posture, alternative ways of performing the task can be considered. However, despite the potential usefulness of the OWAS, Burgess-Limerick (2003) points out that the OWAS does not measure other known risk factors, for example repetition, duration and vibration.

Monk (1998) used the OWAS to assess 25 tasks within seven occupational groups in the building and construction industry. For each task, the proportions of the workers' time spent in each posture category (1, 2, 3 and 4) are shown in Table 6.3.

These results indicate that bricklaying, screeding concrete and steelfixing floors had the highest levels of poor postures of all the occupational groups. These tasks involved a high proportion of time working in a bent-back posture. Bricklaying, in particular, involves a lot of bending and twisting to pick up bricks and mortar. Screeding concrete involves a bent back with two arms above shoulder level for much of the time, whilst handling the straight-edge. The range of results suggests the need for task specific analysis of ergonomic risk factors and the need to design trade-specific intervention strategies for ergonomic risks within the construction industry.

Job and task analysis

The OWAS is just one of many methods for identifying and assessing ergonomic risk factors in construction industry tasks. Another method is to perform a job and task analysis. The principle behind job and task analysis is that the design of a job should be based on a task analysis of an operator's actions. A task is a unit of activity within the operator's work situation.

Task analysis

Task analysis aims to identify all of the sub-tasks involved in performing a work-system's objective and to analyse the demands these place on the human operator. This analysis can use existing documentation, such as operating manuals, job/task descriptions and so on. Direct observation of operators, conducting interviews and discussing with operators exactly how they carry out their tasks are also extremely helpful ways to undertake ergonomic task analysis. Below is an example task analysis for the occupational group plasterers.

Table 6.3 Percentage of time spent in postures assigned to each category for construction tasks performed (adapted from Monk 1998)

Occupation	*Task*	*Category 1*	*Category 2*	*Category 3*	*Category 4*	*Number*
Bricklayers	Blocklaying	55.6	38.3	5.0	0.6	10
	Brick labourer	66.7	29.3	3.2	0.2	10
	Bricklaying	38.3	54.9	5.1	0.8	10
Concreters	Linesman	68.8	27.7	3.5	0.0	10
	Vibrator operator	71.0	25.9	2.8	0.3	10
	Screeding	50.5	41.2	7.7	0.4	10
	Pulling hose/vibrator	75.9	17.9	5.8	0.4	10
Formworkers	Constructing formwork	55.6	37.3	5.7	0.6	10
Carpenters	Constructing frames	73.7	23.6	1.7	0.3	10
Labourers	Stripping formwork	77.2	19.7	2.3	0.1	10
	Removing supports	70.9	26.7	2.1	0.2	10
Painters	Painting walls	70.5	20.6	7.1	0.8	10
	Preparing paint	63.0	27.2	6.2	3.0	5
	Painting ceilings	87.4	10.9	0.2	0.1	10
Plasterers	Rendering walls	67.0	24.1	6.3	1.6	10
	Sheeting plasterboard	73.2	20.9	4.0	1.0	10
	Rendering patchwork	69.4	24.4	4.0	2.8	5
	Setting plasterboard	80.6	14.9	3.0	0.8	10
Scaffolders	Dismantling scaffold	77.7	18.0	3.0	0.4	10
	Constructing scaffold	74.8	19.7	4.4	0.4	10
	Chaining scaffold	86.6	11.9	0.8	0.1	10
Steelfixers	Steelfixing floors	42.2	44.6	12.8	0.4	10
	Steelfixing prefab	60.7	38.3	0.7	0.1	10
	Steelfixing walls	57.9	34.0	7.1	0.5	10
	Steelfixing columns	63.0	30.0	4.6	2.2	5

Note: 'Number' equals the number of workers observed performing each particular task.

***Case study 6.1:* Plasterers' task analysis**

The job of plastering involves the following tasks:

- patchwork,
- rendering walls,
- sheeting, and
- setting.

Patchwork and rendering walls are known as 'wet' plastering, while sheeting and setting are known as 'dry' plastering.
The task of rendering walls is made up of the following sub-tasks:

- smoothing walls,
- handing hook,
- applying cement,
- handling buckets,
- cleaning equipment,
- handling straight edge, and
- handling scaffold.

Analysis using the OWAS method revealed that the task of rendering walls involves more of Categories 3 and 4 postures than other tasks. Workers also perceived rendering walls to be the most difficult task they perform. They rated the tasks as follows in terms of perceived difficulty. (The maximum difficulty score was ten.)

Task	*Difficulty*
Rendering walls	7.1
Sheeting	6.4
Setting	4.4
Patchwork	3.6

'Bent' and 'bent and twisted' postures were most frequently used in patchwork and rendering walls and 'twisted' postures were also involved during setting. The sub-task sheeting involved 'two arms above shoulder level' for 18.3 per cent of the time.

This example shows how job and task analysis allows ergonomically problematic sub-tasks to be analysed with a view to developing injury risk control solutions.

Techniques for assessing ergonomic risk

Situations need to be identified which pose ergonomic risks to workers at the workplace. Routine workplace assessments can help to determine where these situations exist. Assessments can take the form of:

- analysing workplace injury records;
- consultation with employees; or
- direct observation.

Each of these methods will be discussed in turn.

Analysing workplace injury records

Analysing workplace injury records can be a good first step in identifying problematic jobs/tasks within an organisation. However, the organisation will probably need to be quite large before any patterns become apparent. To help to overcome this in smaller workplaces, it is a good idea to supplement the analysis of injury records with workplace assessments following an injury. The emphasis should be placed on the task the worker was performing at the time the injury was sustained. The types of questions to ask when investigating an injury resulting from an ergonomic exposure are as follows:

- What was being handled?
- What was the physical sequence of events (that is, what actions were being performed)?
- What was the weight of the object?
- How was the object being handled?
- Where was the object (and the worker) situated?
- Why was the worker performing this task?
- Had the worker been trained to perform this task?
- Was there another (better) way this task could have been performed?
- Why was the task not performed this better way?
- Was the worker aware of the risk of injury?

The above information is collected in order to reduce the risk of a similar incident happening in the future; that is, to reduce the risk of injury to other workers. In order to prevent similar incidents, the risks need to be minimised or managed appropriately (see 'Risk Control' later in this chapter).

Consultation with employees

It is important to note that codes of practice on manual handling require that consultation be undertaken with employees and their health and safety representatives regarding manual handling tasks. Consultation with employees should occur:

- when planning for the introduction of new tasks or work processes;
- when reviewing existing tasks;
- during the identification and assessment of risks; and
- when planning and reviewing risk control measures.

The consultation process does not need only to be a formal one. It can also be useful to speak directly to the workers involved in performing manual handling tasks. Using their knowledge and experience of doing the job should increase the likelihood of developing appropriate and acceptable risk controls. This process should be allowed adequate time to be truly effective. Needs of employees who do not speak English as their first language should also be taken into consideration.

Direct observation

Direct observation of building industry tasks can be used in both identification and assessment stages of manual handling risk reduction. Observing workers performing their everyday tasks is one of the easiest ways to identify manual handling risks. Likewise, observation of the workplace can highlight potential areas of risk. Observation can help identify tasks which require further examination, which could include workplace inspection, audits or the use of checklists.

The Victorian *Code of Practice for Manual Handling* includes a risk assessment worksheet which can be used for manual handling risk assessment. This is reproduced below.

MANUAL HANDLING RISK ASSESSMENT WORKSHEET – LONG VERSION

Task: ______________________ Date: ______________________

Management rep: ______________________ Health and safety rep: ______________________

Step 1a – Does the task involve repetitive or sustained postures, movements or forces?

Tick yes if the task requires any of the following actions to be done more than twice a minute or for more than 30 seconds at a time (see section 12)

	Yes	Comments
Bending the back forwards or sideways more than 20 degrees	r	______________
Twisting the back more than 20 degrees	r	______________
Backward bending of the back more than 5 degrees	r	______________
Bending the head forwards or sideways more than 20 degrees	r	______________
Twisting the neck more than 20 degrees	r	______________
Bending the head backwards more than 5 degrees	r	______________
Working with one or both hands above shoulder height	r	______________
Reaching forwards or sideways more than 30 cm from the body	r	______________
Reaching behind the body	r	______________
Squatting, kneeling, crawling, lying, semi-lying or jumping	r	______________
Standing with most of the body's weight on one leg	r	______________

	Yes	Comments
Twisting, turning, grabbing, picking or wringing actions with the fingers, hands or arms	❐	
Working with the fingers close together or wide apart	❐	
Very fast movements	❐	
Excessive bending of the wrist	❐	
Lifting or lowering	❐	
Carrying with one hand or one side of the body	❐	
Exerting force with one hand or one side of the body	❐	
Pushing, pulling or dragging	❐	
Gripping with the fingers pinched together or held wide apart	❐	
Exerting force while in an awkward posture	❐	
Holding, supporting or restraining any object, person, animal or tool	❐	

Step 1b – Does the task involve long duration?

Tick yes if the task is done for more than 2 hours over a whole shift or continually for more than 30 minutes at a time

Yes	Comments
❐	

Step 2 – Does the task involve high force?

Tick yes if the task involves any of the following high force actions

	Yes	Comments
Lifting, lowering or carrying heavy loads	❐	
Applying uneven, fast or jerky forces during lifting, carrying, pushing or pulling	❐	
Applying sudden or unexpected forces (e.g. when handling a person or animal)	❐	
Pushing or pulling objects that are hard to move or to stop (e.g. a trolley)	❐	
Using a finger-grip, a pinch-grip or an open-handed grip to handle a heavy or large load	❐	
Exerting force at the limit of the grip span	❐	
Needing to use two hands to operate a tool designed for one hand	❐	
Throwing or catching	❐	
Hitting or kicking	❐	
Holding, supporting or restraining a person, animal or heavy object	❐	
Jumping while holding a load	❐	
Exerting force with the non-preferred hand	❐	
Two or more people need to be assigned to handle a heavy or bulky load	❐	
Exerting high force while in an awkward posture	❐	

Tick yes if your employees report any of the following about the task

	Yes	Comments
Pain or significant discomfort during or after the task	❐	
The task can only be done for short periods	❐	
Stronger employees are assigned to do the task	❐	
Employees think the task should be done by more than one person, or seek help to do the task	❐	
Employees say the task is physically very strenuous or difficult to do	❐	

(Continued)

(Continued)

Step 3 – Is there a risk?

	Yes	Comments
Does the task involve repetitive or sustained postures, movements or forces, and long duration? (Did you tick yes in step 1a and step 1b?) **If yes, the task is a risk. Risk control is required.**	❐	________________
Does the task involve high force? (Did you tick yes in step 2?) **If yes, the task is a risk. Risk control is required.**	❐	________________

Step 4 – Are environmental factors increasing the risk?

Tick yes if any of the following environmental factors are present in the task

	Yes	Comments
Vibration (hand-arm or whole-body)	❐	________________
High temperatures	❐	________________
Radiant heat	❐	________________
High humidity	❐	________________
Low temperatures	❐	________________
Wearing protective clothing while working in hot conditions	❐	________________
Wearing thick clothing while working in cold conditions (e.g. gloves)	❐	________________
Handling very cold or frozen objects	❐	________________
Employees are working in hot conditions and are not used to it	❐	________________

Sketch the task or attach a photograph, if helpful

(Reproduced with permission of WorkSafe Victoria).

The checklist above is a very simple way for anyone to assess ergonomic risks in a job, and is useful in that no specialist ergonomic knowledge is needed to use the checklist. However, as Burgess-Limerick (2003) points out, while these checklists may be helpful in identifying manual handling risks, they are not sophisticated enough to assess the many risk factors involved in manual handling related injuries. It is important to recognise that, where 'yes' answers are appropriate, a more detailed assessment of manual handling risks is necessary.

Managing ergonomic risk

Hierarchy of risk control options

There are many different types of controls for ergonomic risks. However, manual handling codes and standards emphasise the elimination of risks through job re-design wherever possible. Where ergonomic risks cannot be eliminated, they are to be reduced so far as is practical, again with emphasis on re-designing jobs through the modification of workplace layout, object, equipment or task. Risk reduction can also involve the use of mechanical lifting or handling aids or devices. The Case study 6.2 below shows how different risk control options can be used to eliminate or reduce the risk of injuries during manual handling.

Case study 6.2: Cement bag handling

A shop assistant at a builders' supply shop was making up customer orders. He was lifting a bag of cement from a pallet and carrying it between five and ten metres to a customer's vehicle when he sustained a partial rupture of a disc in his lower back. He was off work for more than six months as a result of the injury. An investigation of the incident revealed the following factors contributed to the injury.

Workplace layout

The location of storage areas and customer vehicles determines the distance objects must be moved. The need to carry objects over long distances increases muscle fatigue and the risk of injury. In this case, five to ten metres was a considerable distance. Furthermore, bags were stored on pallets at floor level, meaning that the employee had to bend down to pick bags up from lower layers. The aisles of the store area were too narrow to permit the use of a forklift truck, and poor racking design meant that awkward postures while lifting were often adopted.

The object being handled

The cement bag being lifted weighed 40 kg, and was of a compact design. This gave the impression that the bag could easily be handled by one person, despite its lack of handles. The bag could not be carried flat, close to the body and the risk of injury increases the further the load is carried away from the body. If the bag was lifted on its end, its weight would be largely supported by the hands, arms and shoulders and the use of these small muscle groups would increase the risk of injury. Thus, poor lifting techniques were encouraged as a result of poor bag design.

Equipment

No mechanical equipment, such as trolleys, wheelbarrows or pallet lifting devices were provided.

Task design

The injured worker was expected to make up several customer orders at one time which meant that he sometimes had to handle 40 kg bags of cement for a considerable time. Thus, rest and recovery was not always possible.

Case study 6.2 (Continued)

The physical environment

The work environment was dusty and bags were sometimes slippery. Inclement weather meant that loading jobs were often done hurriedly.

Investigators recommended that the following measures be taken to prevent further incidents of this nature.

Elimination To eliminate the need for manual handling of cement bags, customers should be encouraged to buy ready-made concrete and have this supplied by truck directly to the site.

Substitution 40 kg bags of cement could be substituted with larger containers or bulk bags, which necessitate the use of mechanical aids to deliver them to site. Alternatively, the weight or size of cement bags could be reduced to allow only 20 kg bags to be sold.

Engineering controls Engineering solutions could include the use of mechanical aids to transport bags. Adjustable trolleys could allow bags to be pushed and slid into customers' cars rather than lifted and carried.

Job re-design Cement bags could be stored close to waist height on shelves or racks to reduce the need to bend when lifting and pallet lifters could be used to reduce the need to bend when re-organising stock.

Administrative controls Finally, administrative controls, such as manual handling training and adequate supervision are required, particularly when young, inexperienced workers are engaged in manual handling tasks.

(Adapted from: WorkSafe Victoria 2001a,b)

The case study illustrates the need to examine the work environment and job design in addressing manual handling risks. In particular, it illustrates that efforts to eliminate the hazard or minimise the risk of injury through changing the nature of the task or the work environment or providing mechanical lifting devices are the best means to avoid injury.

Administrative and personal protective equipment risk controls for manual handling

Administrative and personal protective equipment risk controls for manual handling are very limited. In the past, back belts have been used by people performing manual handling tasks. These belts are made of an elasticised material, and are generally worn around the lower back. However, the use of these belts is not recommended because research suggests that the support from back belts does not significantly reduce the stress on the spine and the surrounding muscles and ligaments during lifting. Neither do back belts have an effect upon muscle fatigue or the maximum weight that can be lifted. In fact, back belts can give users the impression that they have increased lifting ability and encourage lifting weights that are too heavy. There is also evidence to suggest that wearing back belts during lifting can result in exaggerated or stiff postures and increase blood pressure and breathing rate, presenting additional risk factors. According to WorkSafe Victoria guidance, the only legitimate use of back braces is when injury has occurred and a brace is used to restrict movement during the wearer's recovery period (WorkSafe Victoria 2002).

Job rotation

Job rotation is one administrative risk control sometimes used to reduce the risk of injury in manual handling tasks. However, it is important to recognise that job rotation does not eliminate manual handling risk but simply reduces exposure time to the risk. Job rotation is only suitable to reduce exposure to repetitive or sustained postures and is not effective in manual handling tasks that involve high force. In these tasks, job rotation can actually increase the risk of injury as more workers are exposed to the risk of an immediate or acute injury. Furthermore, job rotation is often used incorrectly because people are rotated through jobs with similar movements or postures. This negates the benefits of job rotation because muscles are unable to recover. Thus, it is important to assess tasks in job rotation to ensure that these tasks are different.

Also, job rotation measures must be designed to take account of workers who may be absent or on sick leave. If one worker on a team is absent, the others may need to spend more time on each task exposing them to greater manual handling risk. WorkSafe Victoria (2002) caution that job rotation is only permissible as a method of risk control for manual handling if other control measures to eliminate or reduce the risk of injury have already been implemented as far as is practicable. This involves modifying the workplace, designing jobs appropriately and providing mechanical aids where possible.

Ergonomic risk control in construction trades

The following series of case studies is taken from work conducted by Monk (1998). They illustrate how ergonomic risks can be controlled in three common construction trades – steelfixing, concreting and bricklaying.

Case study 6.3: Steelfixing

Steelfixing involves the tying of steel reinforcement bars to make a 'mat'. Perpendicular intersections are tied by steelfixers using tie-wire. Ergonomic hazards in steelfixing include:

Equipment: Lifting of reinforcement (reo) bars presents an ergonomic hazard due to their weight and bulkiness. Rods can be approximately 9-m long and weigh from 0.2 kg per 0.3 m for the smallest rods, up to 6.2 kg per 0.3 m for larger rods. The most commonly used rods weigh about 0.45 kg per 0.3 m and thus weigh around 13.6 kg each.

Manual handling: The piles of bars may not be close to where the work has to be carried out and access to them may often be impaired by other work going on in the area. Workers often carry several rods at a time. Several studies have highlighted the serious ergonomic risks faced by steelfixers.

Posture: Tying steel bars involves repetitive twisting of the wrist, often in a bent posture. Steelfixers often report back problems, sciatica and lumbago, as well as aches, fatigue and stiffness from an ordinary working day (Wickstrom *et al.* 1978).

Control solutions: The Glim Scan G – 91 Tying Tool, a steel tying tool which is also known as the Najomat, allows the worker to stand upright while tying steel which is placed on the formwork deck. The worker places the device over the point where two reo bars cross at 90 degrees to each other and pushes down on the handles which are located at arm's length. The worker remains in an upright position while he pushes down with his arms and no bending is required. The worker can then move on to the next position without having to bend down to the spot where the tying is done. This device also reduces stress on the wrists from repeated twisting of wires. The use of such a device removes the need to continually bend down to the ground level to perform the steel tying task. This results in a reduction of the awkward postures usually generated by this task (Figure 6.2).

Research has found that the Najomats allow more erect postures with much less disc stress and less stretching of the muscles, fascia and ligaments, as well as less tension on spinal nerve roots.

Modification of workplace layout: Control solutions are often aimed at redesigning the work environment. A storage device for reo bars can result in shorter carrying distance and storing of bars at waist height, reducing the need for bending. Monk (1998) divided steel-fixers into three groups to represent the conditions in which the task

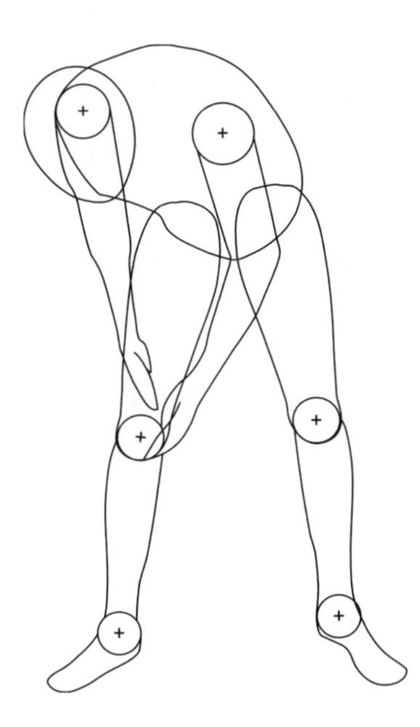

Figure 6.2 Example position 2141.

was performed. Some performed floor work with steel placed close to them, some were steelfixing some distance away from the location of the steel and some were required to handle the steel across difficult terrain. The OWAS was used to analyse workers' postures. The high occurrence of category three postures was due to a high incidence of the 'bent back, arms both below shoulder level', loading on both limbs bent and handling weights below 10 kg, which occurred in almost all work phases, but was especially present while tying steel and handling steel.

Handling steel that was placed some distance away from the working area produced the greatest occurrence of awkward postures. When the steel was placed close to the working area, the occurrence of awkward postures was seven; when the steel was placed far away from the working area, the occurrence of awkward postures was 48 (Table 6.4).

Case study 6.3 (Continued)

Table 6.4 Tying and handling steel under different conditions

Condition	*Work phase*	*Posture*	*Occurrence*	*OWAS category*
Steel is placed close to steelfixing area	Handling steel	2141	5	3
		2142	1	3
		2232	1	3
	Tying steel	2141	3	3
		2261	1	3
		4161	2	4
Steel is placed far away from steelfixing area	Handling steel	2123	2	3
		2141	42	3
		2142	2	3
		2172	1	3
		4172	1	3
	Tying steel	2141	19	3
Handling steel over an uneven terrain	Handling steel	2141	6	3
		2151	1	3
	Tying steel	2141	29	3
		2151	7	3
		4141	1	4
		4161	1	4

Case study 6.4: Concreting

In the concreting task, concrete is pumped onto the deck from a large-diameter hose, spread out and pushed into place using hand rakes and then smoothed with a straight edge, while the workers are standing in 15–20 cm of wet concrete. Finishing is often done by hand, while a hand-held vibrator is used to distribute the concrete and prevent honeycombing. The surface is mechanically finished again after it has partly dried. Ergonomic hazards in concreting include:

Equipment: Holding the concrete hose involves static postures combined with a heavy weight.

Manual handling: Pouring and finishing concrete requires significant lower back, leg and upper-body strength. Raking the surface is difficult due to the nature of the cement.

Posture: Use of the straight edge involves working in a bent-over posture along with twisting of the spine. Trowelling of edges is also performed in a bent-back posture. The hand-held vibrator is bulky and requires static postures and results in some hand–arm vibration.

Control solutions: Fitting a handle to the straight edge allows the job to be performed in a standing position.

Use of mechanical aids or devices

Mechanical aids or devices are controls that make changes to tools or machinery, which alter the work process in a positive way.

The screeding sub-task, within concreting, involves a significant amount of high risk postures. Using the OWAS, Monk (1998) observed a high occurrence of Category 3 and some Category 4 postures during screeding. The first work phase of 'handling the straight edge' displayed the highest incidence of Category 3 rated postures. This phase was also the only work phase in which Category 4 rated postures were observed.

The most common postures to generate a Category 3 rating involved a bent back, loading on both limbs which are bent, and handling a load below 10 kg. Arm positions could be below shoulder level or above. Such a position is depicted in Figure 6.3.

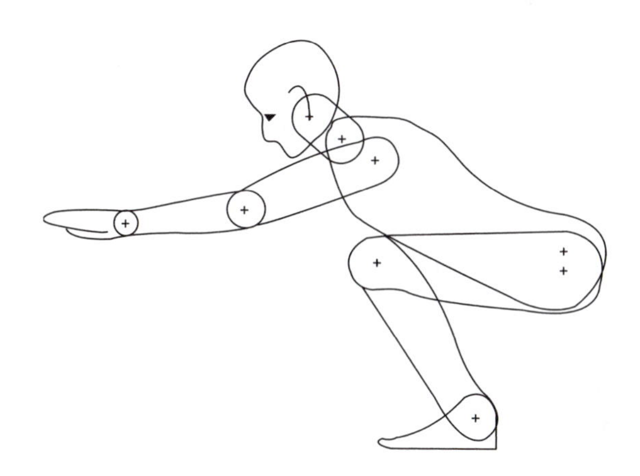

Figure 6.3 Example position observed during screeding.

To perform the screeding task, the worker must bend over and handle the straight edge at floor level. Modification of this task would require the worker to handle the straight edge from an upright position in order to remove the bent-back work, which is the cause of the awkward postures.

A solution to this problem may be found with the use of a mechanical vibrating screeding device. A device such as this, would allow the worker to screed in an upright position. The worker would only be

Case study 6.4 (Continued)

required to guide the device across the concrete. This would remove the need for the worker to handle the equipment at the source of its application, and therefore remove the need to bend and twist the back to physically move and shape the concrete.

This modification, when retested using the OWAS, dramatically reduced the number of higher risk postures during screeding. There was a reduction in the amount of bending required to perform this task, and therefore the associated risks were also reduced.

Figure 6.4 shows the average action category results of the initial screeding task, and the results from the observations of the modified screeding task. While the initial results suggest that the modified screeding method is preferable, there is also a need to determine whether the modified task is acceptable and satisfies the criteria outlined in standards on hand/arm vibration. However, the modification has significantly reduced the bent-back occurrences.

Administrative controls

Administrative risk controls are measures taken to limit the exposure to the ergonomic hazard. This may alter the way a worker performs the job. These could include job rotation, training, changing work methods, exercise programmes, alteration of shift patterns and allowing for regular rest breaks.

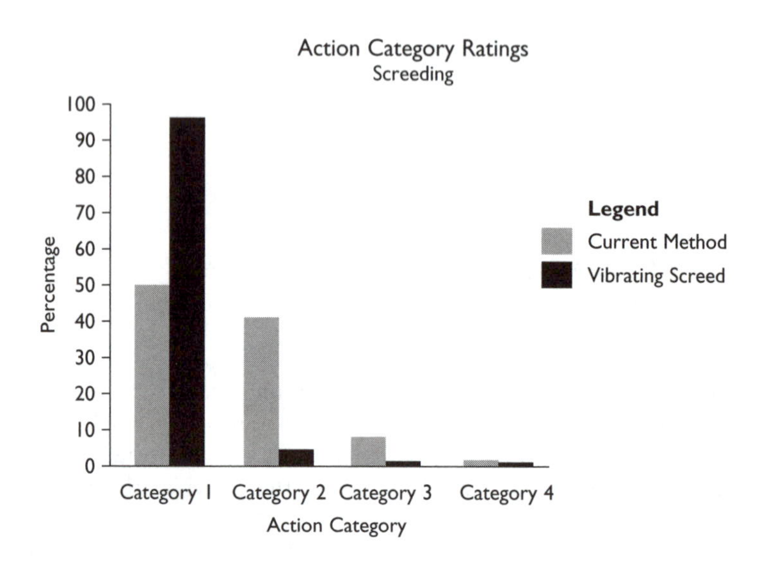

Figure 6.4 Average OWAS action categories for traditional and modified screeding methods.

If the work phases of the various concreting tasks are looked at together, it appears that there is an uneven distribution of work. That is, the linesmen and vibrator operators were found to spend 70–77 per cent of their time handling the concrete hose or vibrator. These are heavy pieces of equipment, the handling of which can involve static postures. In comparison, the task of 'pulling' involves handling these pieces of equipment for only 12–18 per cent of the time, with 65 per cent of the time spent pausing or waiting.

Job rotation could be used as an additional administrative form of risk control in concreting. If concrete workers could be rotated through the heavy positions of linesman/vibrator operator and the awkward postures involved in screeding, with the task of 'pulling', this would decrease the frequency with which static or awkward postures and heavy loads would be borne by individual workers. However, job rotation should only be used as a supplementary risk control measure when efforts to reduce the risk so far as practicable have been made.

Case study 6.5: Bricklaying

Bricklayers can bend up to 1,000 times per day to pick up a brick and mortar to lay. 'Brickies' labourers' frequently load and unload wheelbarrows and can handle a load of more than 12 kg up to 750 times per day. Ergonomic hazards in bricklaying are described below.

Equipment: Bricks/blocks of varying weights and mortar.

Manual handling: Bricklaying involves the repetitive handling of bricks and mortar, often with large vertical reaches. The weight of bricks and blocks varies considerably and this variation can exacerbate ergonomic risks, as does the workplace layout in which the placement of materials is not always ideal.

Posture: Bricklaying involves a considerable amount of bending and twisting.

Control solutions: A simple improvement would be to heighten the stack of bricks and mortar to waist height to prevent bending. Hoist-Console-Scaffolds can be moved with a crane, which can reduce the biomechanical load placed upon bricklayers by 30 per cent. Masonry blocks can be manufactured with handholds to make them easier to lift and lightweight blocks can also be used.

Case study 6.5 (Continued)

Workplace modification

The work phases in bricklaying include:

- handling bricks,
- handling cement,
- cementing bricks,
- levelling, and
- finishing.

Categories 3 and 4 postures observed during the bricklaying task are displayed in Table 6.5. All of the work phases of this task have produced postures falling into Category 3 and all but work phase 5 'finishing' have produced Category 4 postures. These are examples of awkward postures that should be avoided if possible.

The posture depicted in Figure 6.5 has a bent back, both upper limbs below shoulder level, loading on both lower limbs which are bent, and handling a weight less than 10 kg. This posture was observed to occur more often than other Category 3 or 4 postures and was observed in every work phase of bricklaying.

Table 6.5 The most frequently occurring action Categories 3 and 4 occurrences for bricklaying

Work phase	*Posture*	*Occurrence*	*Category*
Handling bricks	2141	2	3
	2151	3	3
	3141	1	3
	3151	2	4
Handling cement	2141	6	3
	2151	9	3
	4141	1	4
Cementing bricks	2141	3	3
	2151	8	3
	3151	2	4
	4241	1	4
	4141	1	4
Levelling	2141	8	3
	2241	1	3
	4151	1	4
Finishing	2141	4	3
	2241	1	3

The first and second work phases involve the handling of materials, either cement or bricks. These phases were observed to involve the greatest incidence of Categories 3 and 4 postures. 'Cementing bricks' involved the highest incidence of Category 4 rated postures. The OWAS analysis indicated that 'twisted' and 'bent and twisted' back positions occur frequently during the cementing of bricks.

Reducing the amount of 'twisted' and 'bent and twisted' postures from the first three work phases, that is, handling brick, handling cement and cementing bricks would help to remove the high risk postures. The first two work phases can be simplified into one area: handling materials. Handling materials and cementing bricks should be targeted for modifications.

A solution to the occurrences of awkward postures while handling materials would be modifications to the workplace layout. If labourers were able to place the bricks and cement in a better position, such as on a platform and in a position where twisting is not required, and the bricklayers were trained to move to the materials instead of twisting from the spot they are in, then the task would require fewer awkward postures.

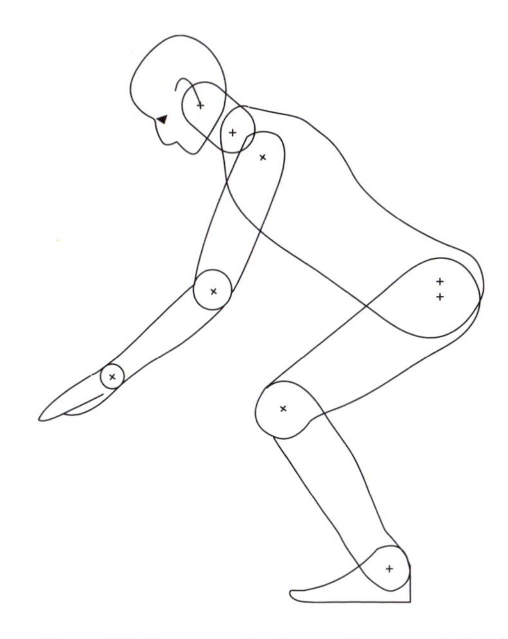

Figure 6.5 A commonly observed posture in bricklaying.

Case study 6.5 (Continued)

For this task, the materials were placed in a position where the workers did not have to bend to reach them and where they were easily handled. For example, for a right-handed worker, the bricks were placed on the left side of the cement so the worker was able to pick up a brick in his/her left hand and with his right hand pick up some cement and apply it to the brick. This layout does not require the worker to cross his/her body in order to reach these materials.

The results for the five workers retested with modified workplace layouts are displayed in Figure 6.6. The results indicated a clear difference between the initial observations and the observations made during the modified task. For the modified task, no Category 3 or 4 postures were recorded and the occurrence of Category 2 postures was only 9.2 per cent. Thus a big improvement was made to working postures.

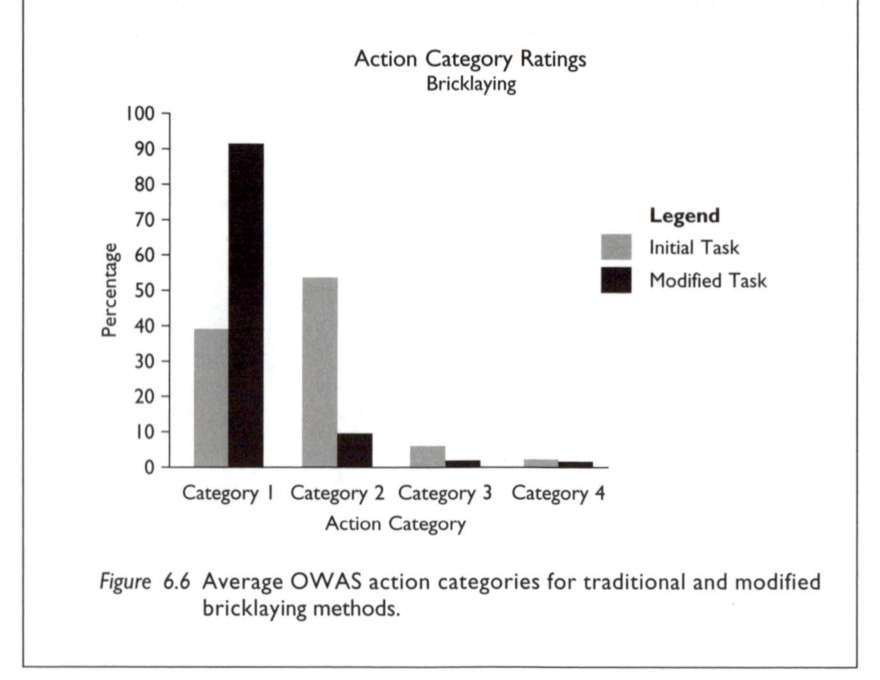

Figure 6.6 Average OWAS action categories for traditional and modified bricklaying methods.

Conclusion

Ergonomics concerns the matching of job requirements and worker capabilities. It is about fitting the job to the worker. It involves proper workplace design and work methods aimed at minimising the risk of pain and injury, particularly from MSDs.

There are a number of risk factors for MSDs which are integral parts of many construction industry jobs. Construction workers perform a great deal of manual handling, they often work in poor postures, outside their natural range of movement, in confined spaces, being exposed to various environmental conditions, vibration, noise and poor lighting.

Both the frequency and intensity of ergonomic risk factors need to be investigated. Risk factors need to be identified, assessed and control solutions developed. These solutions then need to be implemented and evaluated. This is a continuing process.

The flow of materials, sequencing and co-ordination of tasks can all directly influence ergonomic risk factors on site. Better co-ordination and communication between contractors and sub-contractors is essential for workplace planning.

Apart from the physical nature of the work itself, there are other factors that can contribute to the ergonomic risks in the construction industry. The nature of the culture itself may encourage over-exertion. Workers may feel pressure from peers and supervisors to push beyond reasonable limits. There are production pressures and work schedules which may also encourage over-exertion.

The transient nature of the construction workforce can impede efforts to identify and control the causes of injuries and illness. There is a high level of staff turnover within the industry, and sub-contractors come and go on building sites. A lot may depend on the OHS management systems of the head/principal contractor. Small sub-contractors often do not have the knowledge or resources necessary to manage OHS properly.

As demonstrated in this chapter, ergonomic interventions can go a long way towards improving workplace conditions for the construction industry worker. By identifying, assessing and controlling ergonomic risks in the workplace, tasks can become easier and place the worker at reduced risk of injury. The whole process of consulting with workers regarding their jobs/tasks can also make for a more harmonious workplace, perhaps reducing the probability of industrial disputation over safety issues. It may also generate simple, cost-effective solutions to workplace problems.

Discussion and review questions

1. How can the efficient planning and co-ordination of tasks and activities on site reduce ergonomic risk factors?
2. Is there such a thing as a 'safe weight' to lift in a particular circumstance?
3. How can a 'task analysis' help analyse ergonomic risks?

Chapter 7

The psychology of OHS

Humans possess characteristics that can give rise to accidental injury or ill health. For example, unsafe behaviours can occur when people are not aware of the hazards associated with their work, or underestimate the risks involved. Workers may feel that it is not 'macho' to follow safety procedures, such as wearing protective equipment, or may rationalise the risk, believing 'it won't happen to me'. Some people may deviate from safety procedures to gain some personal benefit, such as getting home earlier or receiving a bonus. In the context of the intense time pressures typical of construction work, people may even cut corners in the belief that they are acting in the interests of their employer in finishing the job early or on time to avoid penalties. Also, quite simply, humans are fallible and behave unpredictably, perhaps due to tiredness or a preoccupation with other issues.

It is now widely accepted that the majority of industrial accidents are in some way attributable to human as well as technical factors in the sense that actions by people either initiated or contributed to the accident or people might have acted better to avert them. For example, recent data indicate that approximately 80 per cent of industrial accidents, 50 per cent of pilot accidents and 50–70 per cent of nuclear power accidents are attributable to human error (Jensen 1982; Dougherty and Fragola 1988; Rasmussen *et al.* 1994). Though important, technological solutions to OHS are not enough. James Reason suggests that modern technology has advanced to the point at which improved safety can only be achieved through attention to human error mechanisms (Reason 1990).

No matter how automated a production process or complex a management system is, it is impossible to separate the individual from the process. People must still control production and must sometimes intervene when unplanned events occur. In fact, there is evidence to suggest that introducing safer technology can lead to more risky behaviour because people feel uncomfortable with the 'low' level of risk they experience and try to compensate for this by behaving in an unsafe manner (often called 'risk compensation').

The human element limits the effectiveness of the technical approaches to risk assessment and management described in Chapter 4, because it involves a great deal of uncertainty. Although tools for human reliability assessment are being developed, it remains difficult to model human reliability (Crossland *et al.* 1993).

Despite these difficulties, the importance of the human element means that an understanding of the influences on human behaviour is critical to the successful management of OHS. The human element is particularly important in the labour-intensive construction industry. People are fundamental to the process of constructing a structure. From commissioning through design, construction, and eventually the use of a building, bridge or dam, human decision-making and behaviour impact upon the health and safety of people who build, maintain and use that facility. It is therefore important that managers and safety practitioners have an understanding of the main theories of psychology applicable to OHS. This chapter aims to provide readers with an overview of these theories.

Psychological processes impact upon OHS at an individual level, at a group level and at the organisational level. This chapter considers the main psychological determinants of OHS behaviour at each of these three levels.

Human error

When accidents occur, investigators sometimes look no further than the identification of human error, and individuals are blamed for accidents. However, there are many *reasons* why people behave unsafely. In most cases there are more fundamental issues influencing their behaviour: issues that, unless addressed, will lead to further unsafe behaviour in the future. A human-factors approach to OHS is more appropriate and useful to guide the development of long-term solutions. Within this approach, individuals are influenced by a continuous interaction between the person, the job and the organisation. Table 7.1 lists some of the organisation, person and job characteristics that determine workers' OHS behaviour. The human-factors model is helpful because it suggests that workers' OHS behaviour, and

Table 7.1 Organisation, person and job characteristics impacting on safety behaviour

Organisational	*Personal*	*Job-related*
Policy/strategy	Recruitment	Task complexity
Planning	Person–job fit	Team work
Standards	Task/safety training	Work pace
Operating procedures	Stress	Work environment
Communications	Motivation	Man–machine interface
Monitoring/control	Personality	Conflicting goals

therefore the incidence of occupational injury and illness, can and should be controlled through managerial actions.

Reason (1990) defines human error as 'a generic term to encompass all those occasions in which a planned sequence of mental or physical activities fails to achieve its intended outcome, and when these failures cannot be attributed to the intervention of some chance agency' (p. 9).

In order to reduce the chance of human error occurring and mitigate the adverse effects of errors that do occur, it is necessary to understand *why* errors occur. There are different types of human error, and several theorists have proposed typologies for human error. For example, Miller and Swain (1987) identify the following types of human error:

- *Commission* – including an action that should not have been performed;
- *Omission* – missing something out, for example a step in a sequence of actions;
- *Selection* – choosing the wrong action from a number of options;
- *Sequence* – performing the right actions in the wrong sequence;
- *Time* – performing the right action at the wrong time; and
- *Qualitative* – not performing the action correctly.

Rasmussen identified three discrete levels on which humans perform. These are *skill-based*, *rule-based* and *knowledge-based* actions. Reason (1990) used these three levels as the basis for classifying errors into three types: skill-based errors, rule-based mistakes and knowledge-based mistakes. Mistakes are differentiated from slips and lapses because they involve doing the wrong thing in the belief that it is right (HSE 1999). The three levels of Rasmussen's human performance model are described below.

Skill-based errors

Skill-based errors occur when people are acting unconsciously or on 'automatic pilot' carrying out very familiar tasks. These tasks are subject to error due to momentary lapses of attention or distractions. Skill-based errors include slips, which are 'actions-not-as-planned', for example, performing an action at the wrong time in a sequence, omitting a step in a series of steps or turning a knob or dial in the wrong direction. Slips are more likely to occur in relation to familiar things, and can occur when unintended actions are included in a sequence of actions as a matter of habit. For example, when aspects of the existing situation are similar but not identical to features of a highly familiar environment, an established action pattern may be adopted without modification. The risk of this type of error is likely to be high in construction contexts, in which situational variables may vary slightly from location to location, even at the same worksite. For example, one of the authors knows of an incident in which a crane was involved in a routine repetitive lifting operation. The crane was moved from location to location along a roadway, and a load was being lifted outside the crane's safe working radius for use without outriggers. The lifting operation was carried out successfully several times until the crane was moved to a location in which the slope of the ground differed slightly from the location of the earlier lifts. The lifting operation was undertaken again in this location, and the crane toppled over, seriously injuring the operator. People who are highly skilled may be more likely to experience slips because they do not have to think about every minor detail involved in an operation, and may be more likely to lose concentration. Skill-based errors also include lapses of attention in which people forget to carry out a task, lose their place or forget what they intended to do. This might be the result of competing demands for attention, and lapses may be avoided by minimising distractions or providing clearly visible reminders, such as checklists for people performing tasks.

Rule-based mistakes

Rule-based mistakes occur when behaviour is based upon remembered rules or procedures (HSE 1999). These mistakes can occur when good rules are applied wrongly (Reason 1990) for example, when carrying out an inspection of an item of plant or machinery, an important safety feature is overlooked. Checklists can help to eliminate such errors. However, in emergency situations, information overload has been identified as being a problem, and in the resulting confusion, rules and procedures may not be followed correctly. Rule-based mistakes can also involve bad rules being applied. For example, where procedures are ambiguous or incomplete, rule-based errors can occur.

Knowledge-based mistakes

Knowledge-based mistakes occur when unfamiliar circumstances require problem-solving techniques. Such mistakes include misdiagnosing the problem or miscalculating the solution. Opportunities for error are considerable here, and in such situations 'groupthink' and 'tunnel vision' can block out sources of information that are inconsistent with pre-conceived diagnoses or pre-determined solutions.

Glendon and McKenna (1995) suggest that, because most forms of activity involve behaviour at all three of Rasmussen's levels, human errors can be compounded. Thus, most serious disasters involve skill-based, rule-based and knowledge-based errors. Different strategies are needed to prevent different types of error. For example, design improvements and training are the most effective means of reducing slips and lapses, while training and refresher training, duplication of information, clear labelling and colour coding can help to reduce rule-based mistakes. Knowledge-based errors can be reduced using hazard awareness programmes, through improving supervision, checking work plans and undertaking post hoc evaluation of training programmes.

Latent and active failures

Human errors can also be active or latent. *Active* failures are usually made by people in the front line, such as plant operators and drivers. These failures are immediate and can result in immediate OHS impacts. In contrast, *latent* failures are made by people removed in time and space from operational activities. Latent failures remain in the system until such time as they are triggered by an event. Latent failures often arise as a result of:

- poor design of plant and equipment;
- ineffective training;
- inadequate supervision;
- ineffective communication;
- uncertainties in roles and responsibilities; and
- management failure to provide adequate safeguards.

Latent failures are much more difficult to detect than immediate failures and, even in the aftermath of accidents, the focus of investigators is often on immediate failures.

Making sense of errors

Dekker (2002) draws a distinction between the 'old view' of human error and a 'new view'. According to this new view, human error is not regarded as an explanation for failure, but is symptomatic of trouble deeper within a system. As such, errors require an explanation, and a failure to get to this explanation can have catastrophic consequences. Furthermore, strategies for reducing human errors do not focus on the individuals but on error-producing conditions in industrial environments. While it is important to determine why people behaved as they did in the danger build-up and release phases of an accident, it is often difficult to see this behaviour from the perspective of the person involved. Dekker (2002) attributes this difficulty to hindsight bias, which causes investigators to view accidents from the position of 'retrospective outsiders', and to forget that the people involved in the accident, not having the benefit of hindsight, dealt with the situation as it unfolded, uncertain of the outcome, and often without an understanding of the entire situation.

The inability to adopt the perspective of people who made errors leads them to underestimate the complexity of accident processes and oversimplify causality. Simply pointing out what people 'should have done' is not very helpful. We need to know *why* they did not do it, and this requires an understanding of the 'local rationality' of the people who erred. In the vast majority of cases, the actions of these people made sense to them at the time, and any preventive strategies must be developed using an appreciation of why these errors made sense. In the face of complex and uncertain problems, such as exist in the onset of accidents, there is considerable scope for error in sense-making (Smith 2000). Thus, it is essential to understand human errors from the perspective of the people who made them, rather than making judgements based upon what hindsight tells us they should have done.

Dekker (2002) suggests a five-step process which can help investigators to achieve this. This is described below:

1 *Lay out the sequence of events in context-specific language.* Events influence people's mindsets, which influence decisions and behaviours. These decisions and behaviours then change the way events unfold. It is necessary to understand this dynamic relationship and organise this information temporally and spatially. From this, points at which people realised they had perceived the situation incorrectly or performed decisive actions to influence the process can be identified. Significant changes in the process can also be identified.
2 *Divide the sequence of events into episodes if necessary.* Accidents evolve over time, and it may be necessary to divide this process into separate episodes.
3 *Find out how the world looked or changed during each episode.* This involves finding out what information was available at each stage, and putting behaviour into the context of the situation at the time.
4 *Identify people's goals, focus of attention and knowledge active at the time.* Human behaviour is goal-directed and determined by knowledge. However, there are often many conflicting goals directing behaviour. It is important to consider the possibility of conflicting goals acting upon people involved in accidents. Also, bounded rationality means that people have limited knowledge about the world. The availability of knowledge, and the ease with which it can be accessed and acted upon, should be considered.
5 *Develop a conceptual description.* In this stage, the sequence of events is described in terms of human factors or psychological concepts to make sense of the errors people made in the accident process.

Human error and ergonomics

Ergonomics is concerned with issues of the human–system interface. The subject was discussed in detail in Chapter 5. However, ergonomic issues relating to the reduction of human errors is briefly considered below, because designers of equipment and machinery can have a considerable influence upon issues of operator fatigue and the likelihood of human error. It is therefore very important that the interface between human–machine systems be designed to take into consideration human limitations and reduce or eliminate the possibility of error.

The human–machine interface is as follows:

- A machine display provides information about the progress of production.
- The operator perceives this information and must be able to comprehend and evaluate it.
- The human interprets the information, using previously acquired knowledge.
- The human decides how to behave.
- This decision is communicated to the machine through the manipulation of a control.
- The machine carries out this action as programmed (Kroemer and Grandjean 2000).

Machines can move at high speed and can be very powerful. They involve complex controls, and information displayed to operators can be considerable. Thus, the design of display equipment and controls can contribute to errors, which can have serious safety consequences for operators and others.

Display equipment

Display equipment tells an operator of an item of plant of machinery how the machine is operating; for example fluctuations in temperature, speed or pressure are indicated. This information is important for safety. For example, in the cabin of a crane, information displayed indicates whether the crane is being operated within safe parameters. This information is usually displayed in one of three ways:

1. as a digital display in a 'window';
2. as a circular scale with a moving pointer, such as the speedometer in a car; or
3. as a fixed marker over a moving scale.

Different displays are suited to different purposes. If detection of change is important, a moving pointer display is best, whereas if ease of reading is

Table 7.2 Properties of different display types (Source: Kroemer and Grandjean 2000, p. 159)

Display type	*Property*		
	Ease of reading	*Detection of change*	*Setting to a reading/ controlling a process*
Moving pointer	Acceptable	Very good	Very good
Fixed marker/ moving scale	Acceptable	Acceptable	Acceptable
Digital display	Very good	Poor	Acceptable

paramount, a digital display is preferable. The properties of different displays are presented in Table 7.2.

Moving pointer displays are particularly useful when operation must be checked. Colour-coded zones, indicating safe and unsafe operation, can make it easier to check whether the machine is operating within parameters for its safe use. Glendon and McKenna (1995) suggest that using audible sounds, such as warning alarms, with visual displays can be particularly effective. For example, if a crane begins to lift a load outside its safe working radius, an alarm in the cabin can warn the crane operator of the danger.

Controls

Controls are our means of communicating our decisions to machines. The way that controls are designed also has implications for the possibility of human error. We develop in-built expectations for the way that controls operate; for example, when switching a light switch down we expect a light to come on, and we turn knobs clockwise to increase power, gas or light. Controls that act in the opposite direction are likely to cause error, particularly when people are fatigued or under stress. Kroemer and Grandjean (2000) recommend that controls be designed to take into consideration human anatomy. Thus, where quick, precise movements are needed, fingers and hands should be used, and where force is needed, the major muscle groups in arms and legs should be used. Attention should also be paid to the positioning of knobs or switches to make sure they are not too close together. Large items of plant contain many controls that need to be differentiated from each other. This can be achieved by arranging them in a particular way, for example, in accordance with their sequence of operation. Controls that need to be differentiated can also use differently shaped knobs, be made of different textured materials, be colour-coded or labelled, although the latter two methods rely upon good lighting and visual ability.

As with displays, different types of control have different properties. Some controls may be better than others for certain purposes. For example, toggle-type switches are good for speedy responses, but have a limited range, often only two positions. Where three positions are available in toggle controls, it is recommended that a movement of at least 40 degrees be provided between settings. Knobs, on the other hand, are not very good when speed of use is desirable, but have a better range and a good degree of accuracy. Foot pedals are good for speed and force, but do not permit a great deal of accuracy. See Cushman and Rosenberg (1991) for a review of appropriate ergonomic design and layout of controls.

Violations

Violations are said to occur when people deliberately break rules or deviate from procedures, for example, removing a guard-rail from a scaffold to provide easier access to a work location, or turning off the warning device in the cabin of a crane. Violations can be classified as routine, situational or exceptional (HSE 1999).

Routine violations occur when breaking rules or deviating from procedures has become a normal way of working. This may occur when rules are not enforced, or where there is the belief that rules no longer apply. Rules may also be routinely broken where they are perceived to be too restrictive, or where there is a desire to save time or cut corners. This latter situation can sometimes occur when mistaken priorities lead to a belief that saving cost or time is more important than OHS. This may be the result of management conveying conflicting goals. Routine violations may also occur when new workers start work on a job in which such violations are the norm. In this situation, the new workers may not realise that this is not the correct way of working. In the context of a construction site at which new workers will arrive on a regular basis, and different trades come onto and leave the site, it is important that site OHS rules are clearly communicated to all new workers. Interventions designed to reduce violations should focus on improving workers' risk perception ability to enable them to make more rational judgements about the likelihood and consequences of cutting corners with regard to safety. Supervision and supervisory safety leadership behaviour are also critical to reducing violations. For example, OHS motivation should be targeted through the development of group norms that support OHS and through communicating the management's commitment to OHS relative to other objectives, such as production.

Situational violations occur when the work situation makes it difficult or impossible to comply with rules or procedures, for example, where safety harnesses are provided for work at height, but no provision for fixing them is made. Comprehensive risk assessments and pre-job planning can help to overcome such violations.

Exceptional violations occur very rarely when something has gone wrong, for example, when someone enters a confined space in an attempt to rescue an unconscious or injured worker. The provision of training for emergency situations and emergency management and planning can help to avoid such violations.

Risk cognition

As the psychological models of accident occurrence described in Chapter 1 suggest, the first stage in the build-up and release phases of an accident is a failure to perceive the danger or risk (see Figure 1.3). The failure to perceive a risk or recognise its magnitude removes the possibility of appropriate preventive action, and leads to imminent danger in the danger build-up phase and injury and/or damage in the danger release phase.

Waring and Glendon (1998) propose a model of risk cognition. They suggest that humans continuously appraise risk as part of their cognitive functioning. This appraisal involves receiving information from internal and external sources and making sense of this information. Problem-solving and decision-making are influenced by cold cognitions, such as memory, learning and perception, and hot cognitions, such as moods, emotions and motivations. The outcomes of this process are goal-directed behaviours. This model recognises that behaviour in the face of risk is not entirely determined by rational thought processes. Instead, moods and emotions have an influence. The role of emotion in shaping behaviour is recognised in the law in the form of 'crimes of passion'. Emotion can also influence behaviour at work. Similarly, anger has been found to be associated with drivers' unsafe behaviour (see the section on 'Personality and OHS').

Experience is an important source of learning. For example, when someone is first learning to operate a machine or item of plant, they engage in knowledge-based learning, discovering how the machine will respond in certain operating conditions. However, once the activity has become routinised, it is carried out according to rules of thumb (or heuristics) learned during this earlier stage. The operation may become so habitual that the perceived risk associated with using the machinery or operating the plant may become lower, and habitual risk-taking may result. For example, an experienced crane operator may work according to internalised 'rules' and switch off a safety warning device, believing it to be redundant given his or her knowledge of the crane.

Sources of bias in OHS risk perception

Qualitative dimensions of risk are known to influence people's risk perception. These are discussed in Chapter 4. However, there are some other known sources of bias in the way people perceive risk. These are likely to have an influence on people's risk decision-making and behaviour.

These biases derive from a cognitive approach to motivation known as *attribution theory*. Attribution theory deals with the way in which people understand causality. Thus, in OHS, it refers to how people think that accidents or ill health are caused. This attribution of cause is developed by people who wish to predict and exert control over events (Glendon and McKenna 1995), and thus OHS behaviours may be influenced by attributional biases (DeJoy 1994). Some of these biases are described below.

Self–other bias – When bad things happen to others, people tend to attribute these things to internal factors, such as carelessness, stupidity, laziness and so on. However, when bad things happen to us, we tend to attribute these things to external, situational factors, such as an unsafe work environment.

Self-serving bias – Self-serving bias is the tendency to take credit for successes, but blame external sources for failures. For example, when an accident occurs, people may try to protect themselves by blaming others or blaming aspects of the work environment.

Severity bias – When consequences are more serious, people are held to be more responsible than when consequences are minor. Thus, attribution of responsibility is likely to be much stronger in accidents that result in multiple injuries, whereas near-miss incidents in which people were not hurt are less likely to be attributed to the actions or omissions of people, even if the potential for serious consequences was present.

Availability bias – The probability of events that are dramatic and easy to imagine is over-estimated, while the probability of more common but less dramatic events is under-estimated. Thus, experience of an accident at a worksite can increase workers' perceived probability of this type of accident.

Small numbers – Accident and incident occurrence data collected on a project level is often interpreted to indicate long-run probabilities. However, the data is statistically meaningless, and its use in this way can give a false sense of security and/or alarm.

Hindsight bias – With the benefit of hindsight, it is often easy to suggest that accidents could be foreseen. If you know that something resulted in a bad outcome, it is tempting to judge those involved as being blameworthy, although the situation may have looked quite different to those involved.

Over-confidence – People tend to be over-confident about their judgements of a situation. This situation can lead people to form the wrong conclusions and make faulty decisions in the face of risk.

Many of these sources of bias exist both at an individual level and at a company or project decision-making level. Thus, people involved in investigating incidents, undertaking risk assessments and developing risk control strategies are subject to these biases. It is very important that these issues are addressed in OHS training, OHS incident investigation, OHS risk assessment and the selection, implementation and evaluation of OHS strategies.

Personality and OHS

Personality has been defined as 'a relatively stable set of characteristics that serve to determine how a particular individual behaves in various situations' (Miner 1992, p. 145). Personality is a very broad, multi-dimensional concept and many personality traits have been identified and measured. Personality traits are similar to internal emotional states, but personality traits represent stable individual differences, while states are transient mood fluctuations (Grimshaw 1999).

There has been an ongoing debate concerning whether behaviour is influenced by personality traits or the specific situation in which a behaviour occurs. An extreme emphasis on traits would posit that people will display the same behaviours irrespective of the circumstances they find themselves in. Thus, a person who is predisposed to take risks would do so irrespective of the situation. Such an approach would lead to the conclusion that it would be useless to train such workers in safe work practices or implement safe systems of work, a conclusion which does not seem to have any merit. In contrast, the situational position views behaviour as being entirely a function of the situation. Thus, in a situation in which meeting production deadlines is viewed as being the top priority, and where these deadlines can only be met by cutting corners with respect to safety, every worker, irrespective of any personality predispositions, would adopt unsafe work practices. While this view seems more plausible than the personality-focused view, it is hard to imagine a workplace in which personality did not play any part in influencing work-related behaviour.

Thus, it seems that an interactionist view, in which personality traits and situational factors interact to determine behaviour, may be the most sensible approach (Terborg 1981; Schneider 1983). This approach does not ignore the fact that personality variables may affect people's OHS behaviour, but neither does it suggest that organisational interventions to modify behaviour will have no effect.

The interactionist approach suggests that behaviour results from a continuous interaction between an individual and his or her situation. There is a two-way interaction in which individuals shape their situations as well as being shaped by the situation. Individuals' behaviour is viewed as being driven by both internal (that is, cognitive, emotional and motivational) forces and external forces arising from the psychological meaning derived from the situation. Note also that behaviour potential is an important element of a situation that determines behaviour. The ability to behave in a certain way is an important moderator in the relationship between holding positive pre-disposition to behave in a certain way and actual behaviour (see also the discussion of attitudes and OHS behaviour).

Various personality traits have been linked to various aspects of job performance, including delinquent behaviours such as safety violations, unnecessary absenteeism and alcohol use or influence (Ashton 1998). There is some dispute as to the degree of abstraction with which personality should be considered. Thus, it is sometimes claimed that extremely broad personality traits, captured by general tests for such things as 'integrity', are sufficient to predict behaviour (Ones and Viswesvaran 1996). Others, however, recommend that a narrower view of personality traits be taken, suggesting that specific work behaviours can best be predicted by measurement of more specific traits. For example, Ashton (1998) reports that a tendency to take risks (risk-taking) and a sense of moral obligation to do something (responsibility) were stronger predictors of work-related delinquent behaviours, including safety violations, than more general personality measures. Personality is a complex and multi-level phenomenon. It has been conceptualised and measured in many ways. Several studies have explored the relationship between personality and safety and health behaviours, particularly with regard to road safety and exposure to self-imposed lifestyle-related health risks. This work may have some implications for OHS and some important results are presented below.

Personality and accidents

Researchers have linked involvement in accidents with certain personality types. For example, Shaw and Sichel (1971) report that socio-pathic extraverts and anxious neurotics are more likely to have accidents. The former are people who are self-centred, over-confident, aggressive, irresponsible, resentful, intolerant, anti-social and antagonistic to authority and the latter are people who are tension ridden, unduly sensitive to criticism, indecisive, unable to concentrate, easily fatigued, depressed, emotionally labile, easily intimidated and with feelings of inadequacy. Hansen (1989) also found that socio-pathic and neurotic attitudes predicted accidents among a sample of 362 workers in the chemical industry, when the effects of age, cognitive ability and job experience were controlled. Burgess (2000) suggests that personality can directly determine a person's behaviour and can also indirectly affect behaviour by mediating the effects of social influences designed to constrain certain personality traits relevant to safety and health behaviours. For example, people who have the traits of a socio-pathic extravert, listed above, may rebel against authority and deliberately flout safety rules.

Much of the research work investigating the influence of personality on safety behaviour has focused on road safety. However, it is probable that personality also plays a role in the occurrence of work-related accidents, though the extent of this role in construction has never been investigated. Recently, Iversen and Rundmo (2002) found that the personality traits of sensation-seeking, normlessness and anger influenced involvement in traffic accidents. Sensation seekers experience a need for varied, complex and novel sensations and are willing to take risks to satisfy this need (Zuckerman 1979). Normlessness refers to the idea that some people do not respect presumed norms, trust others to respect them or perceive consensus concerning appropriate behaviour. Normless people are therefore prepared to act in deviant ways, believing that it is acceptable to do whatever they can get away with (Kohn and Schooler 1983). Driver anger refers to frequent and intense anger while driving. Iversen and Rundmo (2002) suggest that anger can cause lapses in attention, interfere with a person's risk perception and cognition and increase the probability of risky behaviours. Furnham (1992) suggests that personality traits appear to be able to account for 10 per cent of variance in accident experience and concludes that this should not be dismissed, despite some methodological shortcomings in personality/safety research.

Accident proneness

Much of the psychology-based work in safety has examined the concept of accident proneness. Accident proneness refers to the existence of 'an enduring or stable personality characteristic that predisposes an individual towards having accidents'.

The concept originated in work undertaken by Greenwood and Woods (1919) who observed that some female workers in a British munitions factory had several accidents while the majority had none. The work of Greenwood and Woods quickly became the basis for a belief that accident experience is largely dependent upon some quality of susceptibility in the personality of individual workers. Glendon and McKenna suggest the concept of accident proneness has these two central tenets:

1. People exposed to equivalent hazards do not have equal numbers of accidents.
2. Observed differences in personal accident numbers result from enduring individual differences.

However, identifying accident repeaters is of little use unless the common characteristics of these repeaters are also known. Nichols (1997) suggests that the natural response to this concept was the scientific selection of employees. Thus, a personality profile of accident-prone individuals would be developed and used to screen out accident-prone individuals. However, as we shall see, many personality traits have been linked to accidental injury, and it has proved difficult to identify a personality type that could be called 'accident-prone'. Furthermore, there is evidence to suggest that propensity to suffer accidents is not stable over time, and that some people are more prone to accidents at certain times. The concept of accident proneness has also been criticised because it does not take into account sufficient situational and organisational factors involved in industrial accidents. It also assumes that exposure to danger is uniformly distributed, and that all individuals are equal in their propensity to report accident involvement – assumptions which are unlikely to be defensible. These problems undermine the usefulness of the concept of accident proneness.

Personality and health

Personality has been linked to health perceptions directly. For example, neurotic personality trait is associated with an increased tendency to worry about one's health and to exaggerate perceptions of the symptoms of a disease (Costa and McCrae 1985; Watson and Pennebaker 1989). An indirect relationship is also likely to exist whereby personality gives rise to engaging in risky health behaviours, such as smoking, drinking or engaging in risky sexual behaviour. Friedman and Booth-Kewley (1987) undertook a meta-analysis of 229 health studies, examining 101 of these in detail. They considered the involvement of five personality dimensions:

1 anger/hostility
2 anger/hostility/aggression
3 depression
4 extraversion
5 anxiety.

They discovered that all of the five dimensions were related to illness experience, and even suggest that it may be possible to identify personality types prone to specific illnesses. More recent research continues to identify linkages between personality and health behaviours. People high in agreeableness and conscientiousness reportedly engage in less risky behaviour than people low in these personality traits (Lemos-Giraldez and Fidalgo-Aliste 1997; Vollrath *et al.* 1999). Vollrath *et al.* also report that people high in neuroticism perceive greater susceptibility to health risks, but engage in risky behaviour less often than people low in neuroticism.

Stress and personality

Payne (1988) suggests five different avenues of investigating the relationship between personality and occupational stress. These are as follows:

1 Do individual differences play a role in causing people to choose jobs which differ in stressfulness?
2 How do individual differences relate to the development of symptoms of psychological strain, such as depression or anxiety?
3 How do individual differences relate to perceptions of stress in the environment?
4 Do individual differences alter the strength of the relationship between stress, such as role conflict, and strain outcomes, such as health complaints?
5 Do individual differences affect the way people cope with stress; that is, do they mediate the relationship between stress and strain outcomes?

Many researchers have investigated the influence of different personality traits on the stress process. Some of the findings related to four of the personality traits most commonly linked with occupational stress are described below.

Hardiness

The concept of hardiness is composed of three parts: *control*, which refers to a belief in one's ability to influence events; *commitment*, which refers to a curious nature and a sense of meaningfulness in life; and *challenge*, which is an understanding and acceptance of the inevitability of change and an appreciation that change is needed in order to develop.

Studies have revealed that hardy personalities are more resilient to physical illness and psychological distress (Kosaba *et al.* 1983). This protective effect may be because individuals high in hardiness perceive the world in a positive light, and feel that they have control over events. Thus, when faced with stressful situations, they are more likely to adopt healthy methods of coping, such as exercise regimes or relaxation techniques, to reduce the ill effects of stress.

Locus of control

Locus of control refers to the extent to which individuals perceive outcomes to be the results of their own actions. Those who believe outcomes are a result of their own actions are said to have an *internal* locus of control, whereas those who believe that outcomes are the result of external determinants, such as fate or other people, are said to have an *external* locus of control. Research has consistently linked an external locus of control with reports of ill health and distress (St-Yves *et al.* 1989) and anxiety (Parkes 1991). In contrast, people with an internal locus of control have been found to have higher levels of psychological well-being and more positive work attitudes (Koeske and Kirk 1995).

Type-A personality

Early study by Friedman and Rosenman (1974) linked type-A behavioural patterns with heart disease. Type-A characteristics include rapid speech, constant movement and restlessness, impatience, attempting to do two or more things simultaneously, possessing a chronic sense of time urgency, having aggressive feelings to other type-A people, and resorting to characteristic gestures, such as clenching the fists or banging on the table for emphasis. In contrast, Type-B personalities display none of these characteristics, and can work without agitation, urgency or impatience. Furnham (1990) reports that type-A people are more aggressive, more neurotic, more extraverted and more anxious than type-B people are.

Price (1982) suggests that the work environment is a key factor influencing type-A behaviour. Workplace pressures in construction are likely to contribute to the development of type-A behaviours. These include:

- pressure to work long hours and regular overtime;
- being outcome-oriented;
- high, sometimes unrealistic, performance standards;
- deadlines that reflect intense time urgency; and
- a sense of crisis as a consequence of time management problems.

People entering the construction industry are likely to imitate successful people who display competitive, aggressive behaviour and thus entrants 'learn' to be type-A people. In workplaces in which type-A behaviour is rewarded, considerable personal stress is experienced which can result in negative health outcomes (Price 1982).

Negative affectivity

Negative affectivity is defined as 'the tendency for a person to experience a variety of negative emotions across time and situations' (Grimshaw 1999, p. 167). It comprises a number of different negative mood states, including anger, guilt, fearfulness, depression, disgust and scorn. People high in negative affectivity report higher levels of minor ailments, such as headaches and stomach pains, even though negative affectivity appears to be unrelated to objective measures of ill health (Watson and Pennebaker 1989). Negative affectivity has also been related to job strain, anger and health symptoms. While some suggest that this is because people high in negative affectivity are more likely to exaggerate these outcomes, these linkages have been found to be significant even when the effects of negative affectivity are controlled for. Thus it seems that the correlation between negative affectivity and job strain, anger and health outcomes is not a spurious one.

Implications of personality for prevention

The implication of the linkages between personality and health and safety is that personality should not be ignored in OHS.

Personality is usually conceptualised as being relatively stable, comprising an enduring set of characteristics. Therefore, our intention is not to suggest that organisations develop interventions aimed at changing personality, because these would be unlikely to succeed and ethically dubious. Despite this, although personality is deemed relatively stable, it is acknowledged that behaviour resulting from personality traits can be changed. Personality assessment can provide employees with an understanding of their personal strengths and weaknesses and facilitate change from within (Glendon and McKenna 1995). Understanding the link between personality and safety and health is also important because personality probably influences the effect of interventions designed to improve safety and health behaviours. Iversen and Rundmo (2002) suggest that one of the reasons many road safety interventions fail is that they try to influence too large and too heterogeneous groups of drivers. The identification of groups representing different personality types associated with risky behaviour could enable the development of more effective preventive strategies for accidental injury and illness.

If personality is relatively stable and is linked to work performance, it may be possible to use personality tests to measure personality factors deemed to be relevant to certain occupations or work environments. Such tests could be used in recruitment and selection. Psychological testing was widely adopted in the USA between 1910 and 1948. In the UK, personality testing was not widely adopted in the first half of the twentieth century, but its use became more popular in the 1980s (Glendon and McKenna 1995).

Special instruments have been developed for the purpose of testing people's aptitude to meet the requirements of a particular job. However, the usefulness of such tests is not proven. Some researchers argue that personality tests do not predict job performance, because jobs involve such a wide range of activities, and job demands change over time, and thus no personality tests can adequately assess all of the different aspects of job performance requirements. Advocates of personality testing state that it can be extremely valuable provided that relevant personal qualities are identified, that personality measurement is valid and conducted using appropriate measures, and that personality-oriented job analysis is used as the basis for selecting the tests to be used. Glendon and McKenna (1995) suggest that job analysis should consider in detail the safety and risk aspects of the job, and suggest that the use of appropriate personality measures, reflecting these aspects, might be usefully deployed in selection processes.

In the construction industry, where much of the workforce is subcontracted and where a serious skills shortage exists, it is unlikely that personality testing would be practicable, except perhaps at managerial levels. However, one potentially useful application of personality tests is in the identification of training needs. For example, suitable personality measures could be linked to specific jobs, such as operating mobile plant, and occupants of these jobs tested to examine their strengths and weaknesses in relation to the relevant traits. This could then enable the identification of training needs or areas in which feedback and other motivational techniques could usefully be applied. Also, feedback of personality test results provides the opportunity to reflect on these results and consciously try to change behavioural responses in certain situations, for example suppressing anger, working less hurriedly or less impulsively.

Another objection to the use of personality tests is that behaviour is influenced only in part by personality. There are many other determinants of work behaviour, many of which arise in the work environment. Any measure of propensity to behave in a certain way must take into consideration both personality and situational influences.

Furnham estimates that 10 per cent of variance in accidental injury experience is explained by personality factors. In relation to occupational safety, while this estimate is significant, 10 per cent is far too low to justify the widespread adoption of personality testing. This estimate suggests that a much greater proportion of variance in accidental injury experience is attributable to other factors, including features of the work environment. Preventive strategies focusing on these other factors are likely to be more effective.

The role of attitudes in OHS

In many cases, unsafe work practices and accidental injury are attributed to workers' negative attitudes towards OHS. Unfortunately, this attribution is often a convenient way of transferring blame for occupational accidents onto workers. Managers often draw an overly simplified connection between a negative attitude, often deemed to be an 'attitude problem', and unsafe behaviour without having an adequate understanding of the nature and determinants of OHS attitudes, or the complex way in which these attitudes are linked to OHS behaviour.

What is an attitude?

An attitude may be defined as a predisposition to respond in a favourable or unfavourable way to objects or persons in one's environment (Steers 1981). This definition is underpinned by three assumptions about attitudes. First, it assumes that attitudes are hypothetical constructs that exist within people. Thus, attitudes cannot be observed, but their consequences can. Second, an attitude ranges on a continuum from very favourable to very unfavourable. Finally, attitudes are believed to be related to subsequent behaviour. The definition implies that people behave in a way that is consistent with how they feel.

Two leading writers on the subject of attitude define it as 'a learned tendency to act in a consistent way to a particular object or situation' (Fishbein and Ajzen 1975, p. 6). The important features of this definition are the following:

- Attitudes are learned through social interactions and experiences. People are not born with attitudes.
- Attitudes only refer to a tendency to act in a certain way. Thus, attitudes are not perfect predictors of actual behaviour.
- Attitudes are reasonably consistent.
- Attitudes are formed in relation to particular objects or situations, and cannot be generalised to other objects or situations.

Attitudes are considered to have three components: affective, cognitive and behavioural. The *affective* component of attitudes refers to how people feel about an object or situation. The *cognitive* component refers to what they think about the object or situation; and the *behavioural* component refers to the tendency to act in a certain way towards the object or situation (Triandis 1971).

Attitudes and behaviour

To the extent that attitudes influence behaviour, understanding attitudes is important in the prevention of occupational injury and illness. Figure 7.1 shows how OHS attitudes might shape OHS behaviour in construction.

Beliefs represent the information a person holds about an object. For example, an employee may describe his or her job as inherently dangerous, 'macho', profitable and exciting. These descriptions represent beliefs the individual has about the job. It is important to note that these beliefs may or may not be factual and differ between individuals. For example, an exciting job to one person may be unnerving to another. However, individuals accept their beliefs to be factual.

According to the model, these beliefs then influence the attitudes, or affective responses formed by employees. For instance, a person who believes his job to be inherently dangerous, 'macho', profitable and exciting may develop a negative attitude towards OHS rules and procedures, as these may be seen to unnecessarily interfere with the excitement or profitability of the work. This unfavourable attitude towards OHS may lead the employee to choose undesirable forms of behaviour; for example, he may choose to flout safety rules or take unnecessary risks. This conscious decision to behave unsafely is the behavioural intention. Finally, Fishbein

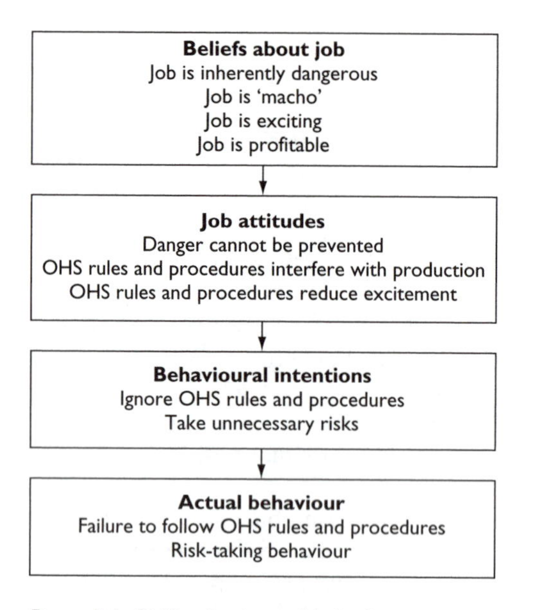

Figure 7.1 OHS attitudes and behaviour.

and Ajzen (1975) suggest that behavioural intentions become translated into actual job behaviour, such as in the occurrence of unsafe acts and risk-taking behaviour.

According to this model, actual behaviour is not determined by a person's attitudes, but is dependent upon his or her behavioural intentions. Thus, individuals with negative OHS attitudes will behave unsafely only if they make a conscious decision to do so. Similarly, if positive attitudes towards OHS exist, this will not necessarily translate into safe behaviour. It is therefore very important to understand the link between attitudes and actual behaviour in the design of any attempts to improve OHS through attitudinal change.

Glendon and McKenna (1995) identify four everyday theories about people's attitudes and the effect that these attitudes have on behaviour. These are as follows:

1 Attitudes influence behaviour, so if we know a person's attitude towards something we can predict his or her behaviour towards it.
2 Behaviour influences attitudes, thus in order to change attitudes towards something we have to force them to behave in this way. An example of this would be the implementation and enforcement of legislation relating to the use of safety belts in cars which, over time, changed people's attitudes towards the use of safety belts.

3 Attitudes and behaviour are mutually reinforcing, so change in one is likely to lead to change in another.
4 Although behaviour and attitudes are mutually consistent, it is necessary to change both using different means in a consistent way. Thus, a safety campaign may influence workers' safety behaviour with respect to a particular situation. It may also influence attitudes, for example increasing the frequency with which site supervisors remind workers to act safely. Thus, rather than attitudes influencing behaviour and vice versa, an additional factor influences both.

The implication for health and safety interventions of this latter theory is that such interventions should target both behaviour and attitudinal change. Thus, a campaign to prevent drink driving might include a media campaign about the dangers of consuming alcohol before driving, to be aired at the same time as an increase in the use of random breath testing of motorists.

Theory of reasoned action

Fishbein and Ajzen (1975) developed a more complex theory of the relationship between attitudes and behaviours, known as the theory of *reasoned action.* This theory seeks to explain what determines behavioural intentions. This is important because the observation that intentions predict behaviour does not improve our understanding of the reasons for that behaviour, and is therefore unhelpful in guiding the development of intervention strategies. The theory of reasoned behaviour is only concerned with determining what causes volitional behaviour, or behaviour that is within an individual's control. Fishbein and Ajzen (1975) suggest that intentions are a function of two basic determinants: an individual's attitude towards the behaviour, and a person's perception of social pressure to act in a certain way. The former determinant is personal in nature, and refers to an individual's positive or negative evaluation of a particular behaviour. The latter determinant is normative, and Fishbein and Ajzen (1975) refer to it as a 'subjective norm'. The theory holds that people will behave in a certain way when they evaluate this behaviour positively and perceive that important others appreciate such behaviour. This theory, depicted in Figure 7.2, has received some empirical support.

However, knowing that attitudes and subjective norms predict behavioural intention, and that behavioural intention is the most important determinant of behaviour, is still insufficient to develop appropriate intervention strategies to modify behaviour. It is important to understand determinants of attitudes and the mechanisms through which subjective norms operate. For example, what determines workers' attitudes towards OHS, and what subjective norms are at play in shaping behavioural intention with regard to workplace health and safety?

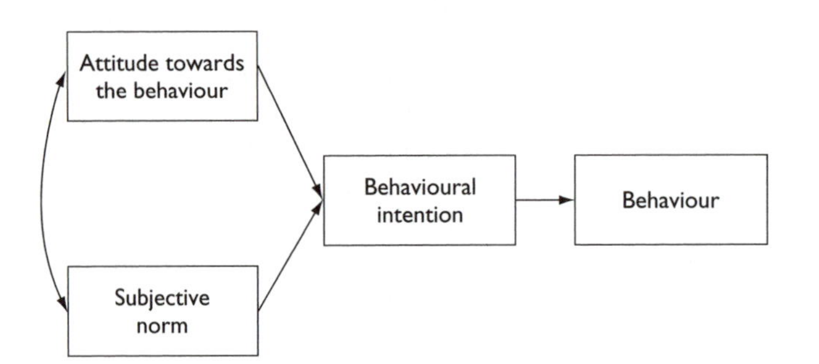

Figure 7.2 Theory of reasoned behaviour (reproduced with kind permission from Ajzen 1988, p. 118; published by Open University Press/McGraw-Hill Publishing Company).

Fishbein and Ajzen (1975) suggest that attitudes are formed on the basis of behavioural beliefs, linking a behaviour to a particular outcome, or to some other attribute such as the cost of behaving in a certain way. Thus, if it is believed that wearing personal protective equipment (a behaviour) leads to being free of a debilitating illness, and the likelihood of living a longer, healthier life, a positive attitude towards the behaviour may exist. If, on the other hand, wearing personal protective equipment is believed to cause discomfort, and to lead to a lower work rate and consequent loss of income, negative attitude towards the behaviour may develop. The strength of the belief will also influence the attitude. Thus, if the negative behavioural belief is held more strongly than the positive behavioural belief, a negative attitude is still likely to prevail. Generally, a person who believes that a certain behaviour will result in positive outcomes is more likely to hold a favourable attitude towards that behaviour, and a person who believes a behaviour will result in negative outcomes will hold an unfavourable attitude towards that behaviour.

The role of behavioural beliefs in shaping attitudes has important implications for OHS because of a common unrealistic optimism about the likelihood of being involved in an accident. Accidents are infrequent, and many occupational illnesses have long latency periods. Thus, direct personal experience of negative occupational health and safety consequences is rare. In the absence of immediate and certain rewards (good health) or punishments (pain and suffering arising from injury or illness), workers' beliefs about behaviour–outcome contingencies tend not to have a strong motivational effect upon OHS behaviour. Furthermore, information provided in OHS training can be undercut by personal experience. Workers may hold the view that 'I've done it this way for twenty years and never had an accident yet', or observe other workers who do not follow safety procedures with no apparent adverse consequences. Repeated experience of unsafe behaviour without an injury or illness can lead to systematic desensitisation and diminished fear (Job 1990). Previous research supports this evidence. For example, Weinstein (1980) discovered that a sense of 'unrealistic optimism' about the probability of being involved in an accident increased with experience. Thus OHS training programmes may need to focus on behaviour–outcome contingencies to generate positive attitudes towards safe and healthy work practices.

The second determinant of attitudes, subjective norms, refers to a person's beliefs that specific individuals or groups approve or disapprove of certain behaviour. Social pressure is exerted by people whose approval an individual values. These people, known as referents, can include family members, co-workers or others. Generally, people who believe that referents whose approval they seek value certain behaviour will feel social pressure to behave in this way. In terms of OHS, considerable normative pressures can arise in cohesive groups or work teams. The role of groups in influencing OHS is considered later in this chapter.

The role of behavioural control

The theory of reasoned action (described above) is useful in explaining behaviour that is under the control of individuals. However, in many instances, people may not be able to control their behaviour. For example, the ability to translate behavioural intentions into behaviour will depend upon non-motivational factors, such as the availability of opportunities and the resources required to behave in this way. Factors affecting behavioural control are identified by Ajzen (1988) as follows:

- *Information, skills and abilities.* Behavioural intent is thwarted when a person does not possess the necessary knowledge, skill or ability to behave in a certain way. In this regard training to deliver OHS knowledge skills and abilities is of critical importance to the effective management of OHS.
- *Emotions and compulsions.* Some acts may be beyond our control because of compulsion to behave in a certain way. A good example of this is the case of stress. Under extreme stress, people are often not held accountable for their actions.
- *Opportunity.* Sometimes external factors disrupt attempts to behave in a certain way. For example, a decision to paint a concrete structure rather than use pre-coloured concrete will result in the need to work at height, a requirement that is beyond the control of the person who must undertake the behaviour (working at height).
- *Dependence on others.* When the performance of a behaviour depends on others, it is possible that behavioural control will be diminished. In construction, in which interdependent tasks are carried out by many

different work groups, behavioural control could easily be lost through the failure of another party to fulfil their OHS obligations.

In response to the understanding that perceived behavioural control is likely to influence intention, Ajzen (1988) extended the theory of reasoned behaviour and developed a theory of planned behaviour. This is depicted in Figure 7.3.

In this model, a third determinant of behavioural intention is also included. This is the degree of perceived behavioural control a person has in relation to a particular behaviour. Past experience and anticipated obstacles contribute to a person's perception about whether certain behaviours are within their control. Constraints to behaviour have motivational implications, in that they are likely to impact upon the forming of a behavioural intention. Thus, when people believe that they do not have the resources, or will not enjoy the opportunity to behave in a certain way, they will not develop strong behavioural intentions towards this behaviour, even if they have positive attitudes towards it and believe that others would favour the behaviour. In addition, there may be a direct link between perceived behavioural control and actual behaviour to the extent that this perception accurately reflects constraints and impediments in place. According to the theory of planned behaviour, the greater the opportunity and the more resources people believe they have to behave in a certain way, the greater their perceived control over the behaviour, and the stronger their behavioural intent will be.

Some empirical evidence suggests that the theory of planned behaviour has some predictive validity with regard to OHS. For example, a study by Sheeran and Silverman (2003) revealed that attitudes, subjective norms, perceived behavioural control and intentions were all significant correlates of attendance at workplace fire safety training courses. With regard to

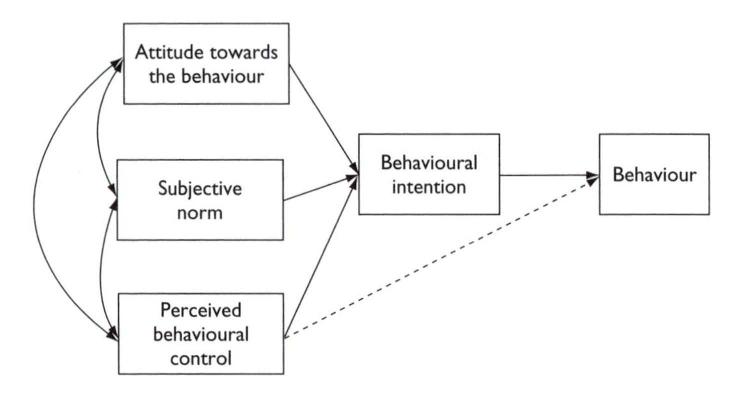

Figure 7.3 Theory of planned behaviour (reproduced with kind permission from Ajzen 1988, p. 133; published by Open University Press/McGraw-Hill Publishing Company).

OHS, the theory of planned behaviour is significant, because it implies that interventions designed to create positive attitudes towards OHS are unlikely to succeed where workers perceive they have little control over their OHS-related behaviours. For example, even if workers hold favourable attitudes towards using fall-arrest equipment when erecting steel structures, they still may not use, or even intend to use, such equipment unless appropriate fall-arrest equipment is supplied, and suitable anchorage points are available for such devices. In the absence of equipment or anchorage points, the workers' perceived control over the use of fall arrest equipment is likely to be low.

The case studies presented at the end of this chapter suggest that perceived behavioural control may be an important moderating variable in the relationship between safety motivation and behaviour.

Changing attitudes

Given that attitudes can play an important role in predicting behavioural intention, and ultimately behaviour, it may sometimes be necessary to try to change OHS-related attitudes. Attitudinal change can occur in response to many stimuli. For example, it might be prompted by safety campaigns in the mass media, or in response to a desire to be accepted into a particular work group. Glendon and McKenna (1995) suggest five criteria impacting upon the ability to change attitudes. These are described below in relation to their implications for changing OHS attitudes.

Audience

Audiences self-select, in that they listen to messages of their own choice. Thus the communication of OHS messages should be delivered on workers' 'home ground'. This may account for the success of union-led 'empowerment-based' OHS training in the USA. This approach has led to more positive OHS attitudes and behaviour change among both workers and management (Lippin *et al.* 2000).

Persuader

The communicator of OHS messages must have credibility in the eyes of the audience, and be a trusted source of information. A communicator who is similar to the audience may be particularly effective.

Personality factors

Some people may be more open to persuasion than others, and individual cognitive needs and styles should be considered. Some people may adopt defensive reasoning in justifying their existing OHS attitudes in terms of a strong belief in their personal control over events. For example, DeJoy (1989) reports that personal skill was the most important factor in beliefs about traffic accident avoidance, and asserts that a tendency to over-estimate the degree of personal control over events contributes to the belief that 'it won't happen to me'. Attempts to change OHS attitudes must overcome such obstacles.

Presentation of issues

A balanced message presenting both sides of an argument is the best way to present information. Some research has focused on whether fear arousal techniques are effective in OHS communication designed to change attitudes. The conclusion of much of this research is that arousing fear is not an effective way to change safety attitudes, because it causes people to reject the information in an attempt to build a defensive barrier. Thus, the idea that accidents and ill health are things that 'happen to others' is reinforced. Mildly arousing safety messages are more effective. Laner and Sell (1960) investigated the impact of safety posters in eliciting attitudinal change, and reports that safety messages should:

- be specific to a particular task or situation;
- give positive instruction about the correct way of doing something; and
- be on topics over which the audience has some control.

General exhortations to 'work safely' are not effective.

Persistence of change

Attitudinal and behavioural change is more likely to be long-lived where the audience participate in the delivery of the message. Thus, the use of role-play is recommended. Repeated messages are also more effective than single delivery, and safety publicity campaigns have been found to be particularly effective when they are backed up with safety training (Laner and Sell 1960).

First-aid training and OHS

Research has identified an association between traditional first-aid training and a lower incidence of workplace injuries (Miller and Agnew 1973; McKenna and Hale 1981), and also between a greater willingness to take personal responsibility for safety and a willingness to adopt safe behaviour (McKenna and Hale 1982). The effect of first-aid training on OHS behaviour was recently investigated in the Australian construction industry. This study is described in the case study presented below. The study not only assessed the impact of first-aid training on behaviour, but also explored participants' OHS attitudes. The results suggest that attitude change occurred as a result of the first-aid training intervention. Following the work of Fishbein and Ajzen, the results of this study could be interpreted as an indication that the change in attitudes resulted in a change in behavioural intent, which resulted in a positive behaviour change. Interestingly, however, performance did not improve in all categories. It was apparent that some participants perceived low behavioural control in these categories, or that prevalent subjective norms prevented any change in behavioural intent from occurring with respect to these categories of behaviour. Thus, the results presented in the case study below can be interpreted in a manner that is consistent with the theory of planned action.

Case study 7.1: Changing OHS attitudes in the Australian construction industry

A 24-week experiment was conducted to assess the effect of first-aid training on the OHS attitudes of small business construction industry employees and their occupational health and safety behaviour. The experimental sites were all small, domestic housing construction projects. The sample was made up of representatives from several different construction trades, including carpentry, bricklaying and plumbing. Participants' attitudes towards OHS were explored during in-depth interviews before and after receipt of first-aid training. In particular, the areas explored were the acceptability of risk-taking, the sense of personal vulnerability to occupational injury and ill-health, and the degree of control that participants believed they had concerning their experience of a work-related injury or illness.

Objective measurement of occupational safety and health behaviour was also conducted, by a trained researcher directly observing the workplace, before and after participants received first-aid training. Two rating scales were developed for the direct observation of behavior (Sarvela and McDermott 1993). Researchers rated the general level of safety on each site using a Global Safety Measure (GSM) and the safety performance of individual participants using an Individual Safety Measure (ISM). The GSM contained 21 items measuring four different categories of safety performance. These were

- site 'Housekeeping' (6 items),
- use of 'Personal Protective Equipment' (6 items),
- 'Use of Tools' (3 items) and
- 'Access to Heights' (6 items).

Case study 7.1 (Continued)

The ISM contained 14 items measuring four different categories of safety performance. These were

- use of 'Personal Protective Equipment' (6 items),
- 'Use of Tools' (3 items),
- 'Manual Handling' (1 item) and
- 'Access to Heights' (4 items).

Safety categories were developed to reflect the historical experience of occupational health and safety incidents in the Australian small business construction industry. For each of the performance categories, OHS codes of practice and industry guidelines were examined to identify specific safe and unsafe practices. Items were developed expressing these safe practices in behavioural terms. Each item in the scale was rated proportionally. Performance in a category was calculated by averaging the scores derived from all observable items included in that category. This figure was then expressed in percentage terms to represent an overall indication of safety performance in each category.

Following the phase one (T1) interviews and observations, all participants attended a generic emergency first-aid training course. This course provides the knowledge and skills to enable participants to provide initial assistance or treatment to a casualty in the event of an injury or sudden illness before the arrival of specialist medical assistance. The course contained both theoretical and practical components, both of which were examined at course completion. Skills taught included the performance of cardio-pulmonary resuscitation and appropriate treatment for fractures and wounds, including bandaging. The course did not contain any information specific to the construction industry. Nor did it contain any information about occupational health and safety risks or their prevention.

On completion of the first-aid training (T2), all participants were interviewed once more, and the on-site safety observations were repeated.

Behavioural scores before and after first-aid training were compared. Scores were averaged for all participants to yield a general indicator of the preventive effect of first-aid training. In the ISM, safety performance was generally highest in the 'Use of Tools' category, with scores of 94% safe and 98% safe for the pre and post-training measurements respectively. Safety performance was lowest in the 'Access to Heights' category, in which the pre-training score was only 51%. The post-training score for 'Access to Heights' rose to 93%. The use of 'Personal Protective Equipment' was found to be 65% safe before and 96% safe after participants received training. 'Manual handling' was found to be 85% safe before training and 80% safe after training.

Safety performance was generally highest in the 'Use of Tools' category, with scores of 97% safe and 96% safe for the pre and post-training measurements, respectively. Safety performance was lowest in the 'Access to Heights' category, in which the pre-training score was only 47%. The post-training score for 'Access to Heights' rose to 78%. The use of 'Personal Protective Equipment' was found to be 60% safe before and 95% safe after participants received training. Site 'Housekeeping' was found to be 79% safe before training and 85% safe after training.

Before training, OHS incidents were understood to occur as result of individual factors. The most commonly cited source of OHS risks affecting other workers was their carelessness or complacency. For example, one participant said:

> You can educate people till the cows come home. Overconfidence causes most of the problems, and there is nothing you can do about it... You can put up a handrail and provide someone with a harness, but they have to choose to use the harness and they have to operate within the boundaries of the handrail. If someone decides it is a nuisance, they will take it down.

However, although before training participants tended to believe accidents to others were attributable to a lack of care or complacency about OHS, they did not regard carelessness as being relevant to their personal experience of OHS risk. Almost half of the participants expressed the belief that accidents to themselves were attributable to factors beyond their own control, such as the negligence of others. For example, one participant said:

> There is always the risk of stepping into a puddle and finding out that someone has been negligent and dropped a power cord in there and there is a fault in the leakage switch.

Another commented:

> Well hopefully I won't [have an injury] but things can happen where it is not your fault either, I mean someone could drop a hammer and it could hit you in the head... there is nothing you can do.

Case study 7.1 (Continued)

During the T1 interviews, participants also expressed the fatalistic view that their own personal experience of occupational injury or illness was a matter of luck or chance. For example, one participant said:

> Put it this way: in ten years I've had one injury that has taken me to hospital, so that is not to be, I think.

Another commented:

> I think it is hard to say. Its your own fate.

During the T2 interviews, participants still largely attributed OHS risks to individual factors, such as complacency or carelessness. However, the perception that they could not control their own personal experience of OHS risk appears to have been reduced. Following first-aid training, most of the participants expressed the importance of taking care and concentrating to avoid occupational injury or illness. At T2, very few participants mentioned fate or the negligence of others as important influences on their personal experience of OHS risk. This indicates an increased recognition by participants that their own behaviour is also important in the prevention of occupational injury and disease.

In the T1 interviews, many participants expressed the unrealistically optimistic belief that 'it won't happen to me'. In comparison to others in their workplace, more than a third of participants indicated that others were more likely to suffer from an occupational injury or illness than themselves. For example, one participant expressed this by saying:

> You make scaffolds that aren't up to scratch – I would be the only one to walk on them because I know its safe for me, but I wouldn't want any one else doing it.

More than a third of the participants explained their ability to avoid occupational injury or illness in terms of the degree of control they exercised in the work environment. One participant expressed this by saying

> I think if you have got your wits about yourself, you can deal with anything.

At T1, another group of participants attributed their comparatively low probability of suffering a work-related injury or illness to their experience in their job. One participant expressed this by saying

> I'm probably less likely [to have an accident] because I've been doing it a long time. Not like the young guys running around madly...[they] run into things, fall off the roof and try to carry heavy weights too quickly.

At T2, two thirds of participants indicated that they had a medium to high probability of personally suffering a work-related injury or illness. Only a small minority of participants said the chance of them suffering a work-related injury or illness was low. One of these was an office-based site manager while another had just returned to work on 'light duties' having suffered a work-related back injury.

At T1, when asked whether they ever knowingly took unnecessary OHS risks at work, every participant in the sample said that they did. When asked what types of risks these were, 12 participants indicated that they were associated with working unsafely at height, for example using unsafe scaffolding, using improvised means of gaining access to height or failing to use a safety harness when required. A further five participants indicated that they occasionally took unnecessary risks using power tools and another four said they sometimes failed to use the correct personal protective equipment. There was a strong acceptance of risk-taking behaviour as 'part of the job.' Only two participants suggested that risks should not be taken or that they were concerned about taking risks. When asked why they took such risks, the most commonly cited response was 'to get the job done,' reflecting a strong production orientation among construction workers.

Following first-aid training, participants did not express such a ready acceptance of risk-taking behaviour. In the T2 interviews, only one third of the participants expressed an unreserved willingness to take OHS risks to 'get the job done'. Participants suggested that they would take OHS risks but only under certain circumstances. Five participants indicated that they had taken such risks in the past but that they were less likely to do so now. Three participants said they sometimes took risks that they recognised that they should not take. Four participants indicated that they would consider the costs and benefits before taking an OHS risk and base their behaviour on a 'calculated risk,' only taking risks where the benefits outweighed the costs and where they considered the risk to be 'worth it'.

(Source: Lingard 2002)

OHS motivation

Motivation is a critical issue in managing any aspect of organisational performance, including OHS performance.

What is motivation?

The word 'Motivation' is derived from the Latin word *movere* meaning 'to move'. Steers and Porter (1991) cite various definitions of motivation as follows:

- 'the contemporary (immediate) influences on the direction, vigour, and persistence of action' (Atkinson 1964);
- 'a process governing choice made by persons or lower organisms among alternative forms of voluntary activity' (Vroom 1964); and
- 'motivation has to do with a set of independent/dependent variable relationships that explain the direction, amplitude, and persistence of an individual's behaviour, holding constant the effects of aptitude, skill, and understanding of the task, and the constraints operating in the environment' (Campbell and Pritchard 1976).

Steers and Porter (1991) highlight three common denominators among these definitions, and state that these three components represent important factors in the understanding of behaviour in the workplace:

1 *What energises human behaviour.* This component relates to forces within an individual that drive him/her to behave in a certain way and environmental factors which trigger this drive.
2 *What directs or channels human behaviour.* This component relates to the notion of goal orientation on the part of individuals; their behaviour is directed towards something.
3 *How this behaviour is maintained or sustained.* This component reflects a systems orientation, that is, it considers the forces in the individuals and in their surrounding environments that feed back to the individuals to reinforce the intensity of their drive and the direction of their energy or to dissuade them from their course of action and redirect their efforts.

Therefore, motivation involves the arousal, direction and persistence of behaviour (Steers and Porter 1991). Using this definition, OHS motivation can be defined as the arousal, direction and persistence of behaviour that reduces the likelihood of occupational injury or illness.

Glendon and McKenna (1995) suggest that an understanding of motivation has evolved in four overlapping phases:

1 mechanistic
2 behaviourist
3 cognitive
4 applied.

These are described below.

The *mechanistic approach* held motivation to be entirely related to instinct. Thus, behaviour was understood to be driven by uncontrollable inner forces. An example of this is a survival instinct that leads us to leave a burning building.

The basis of the *behaviourist* approach to work motivation is the belief that the behaviour of an individual can be understood solely in terms of stimulus–response–reward associations. In particular, the association between response and reward is considered important. Behaviourists propose that contingent rewards lead to higher levels of performance, and thus motivation to perform arises out of response–reward contingencies. Behaviourists explain behaviour in terms of Skinner's operant conditioning approach (Skinner 1969). The operant approach theorises that learning results from the association between responses and rewards. The individual notices that a reward occurs shortly after a particular response, and tries to reproduce that response in order to enjoy the reward again. Radical behaviourists deny that such things as reasoning, judgement, creativity, concept formation or feelings of efficacy play any role in determining behaviour. Behaviour is viewed entirely as a function of its consequences. This approach is still widely used and its influence can be seen in the behavioural approaches to OHS described in Chapter 7.

Krietner and Luthans (1991) criticised the behaviourist approach because it does not account for the role of antecedents (cues that may occur prior to behaviour), which play a role in determining the course of that behaviour. Krietner and Luthans (1991) argue that 'antecedents deserve much more attention because they exercise potent feedforward control over a great deal of employee behaviour'. (*Feedforward* refers to the modification or control of a process using its anticipated results or effects.)

The *cognitive* approach to motivation is the main alternative to behaviourist interpretations. Glendon and McKenna (1995) state that the basic premise of cognitive approaches is that behaviour is purposeful, and that people seek to control their environment rather than be controlled by it (as in the behaviourist approach), or to be solely driven by uncontrollable instinct (as in the mechanistic approach). In the cognitive approach people seek to predict what will happen if they take a certain course of action and make conscious decisions as to how to behave in an attempt to take control over important events.

Finally, the *applied* tradition to motivation traverses both behaviourism and cognition. The focus is on changing or influencing behaviour. In this tradition, Krietner and Luthans (1991) advocated a move away from Skinner's radicalism towards the Social Learning theory espoused by Bandura (1986) as a more useful framework for understanding and controlling human resources in organisations. Social Learning theory takes into account both cognitive and environmental determinants of behaviour. Landy (1989) argues that behaviourism represents a technology and should not be awarded theory status. In his view, contingent reinforcement works but, in order to understand why it works, it is necessary to consider explanations offered by cognitive models, such as expectancy theory (see p. 301).

Theories of motivation

Maslow's hierarchy of needs

Maslow's need hierarchy theory is perhaps the most widely cited theory of motivation. It is based upon the concept of a hierarchical arrangement of needs ranging from basic physiological needs (such as food, water and sleep) to the need for self-actualisation or fulfilment. The theory asserts that the most basic unsatisfied need at any given time is the most important source of motivation. Thus basic needs must be satisfied before the so-called higher-order needs for love, self-esteem and self-actualisation exert any influence on behaviour. All individuals will move up the hierarchy in a systematic manner. The second most basic need in the hierarchy is the need for safety. This need relates to the self-preservation instinct inherent in human beings. Certainly this basic human need exists and exerts some influence on the way in which human beings behave at work, and elsewhere. However, research suggests that Maslow's need hierarchy theory cannot adequately explain individual behaviour (Landy 1989). Wahba and Bridwell (1976) found little support for the theory and suggest that Maslow's theory is flawed both theoretically and in practice. It is widely accepted that Maslow's need hierarchy theory is a useful model of personal development, but not a good model for predicting individual behaviour (Cherrington 1991). Thus, while recognising that a basic human need for safety exists in individuals, it is more useful to seek alternative theories of motivation for explaining individuals' safety behaviour in the workplace.

Herzberg's two-factor theory

Most theories treat motivation as a unitary concept. One alternative to this is Herzberg's two-factor theory of motivation (Herzberg 1974). Herzberg argued that the factors that motivate people at work are different and distinct from those that prevent people from becoming dissatisfied with their work. Thus, Herzberg distinguishes between *motivators* and *hygiene factors*.

Motivators influence what a person does, and can motivate improved performance. They include the following:

- a sense of achievement
- recognition
- enjoyment of job
- possibility of promotion
- responsibility
- growth opportunities
- the work itself
- advancement.

Hygiene factors have no motivational ability, but prevent people from becoming dissatisfied with their job. They include the following:

- money
- status
- relationship with boss
- company politics
- work rules
- working conditions
- supervision
- relationship with peers.

Herzberg's theory suggests that the recognition of good OHS performance would be an important motivational influence. Furthermore, the motivational influence of responsibility, the chance for growth and a sense of achievement suggest that a more participative approach to OHS management or worker empowerment might have a beneficial effect. The notion of empowerment is based on the premise that providing workers with the responsibility for their own tasks, and deciding how they will be achieved, leads to improved performance. Thus, rather than attempting to control workers, proponents of empowerment seek to give workers the responsibility for improving their own performance (Wilkinson 1998).

Empirical evidence suggests that the requisites for psychological empowerment within organisations include features closely related to Herzberg's hygiene factors. For example, Spreitzer (1995, 1996) found that access to information about the company mission and strong socio-political support from subordinates, work group, peers and supervisors or managers were required for workers to feel empowered. Spreitzer (1996) also found that low levels of role ambiguity and a participative work climate were positively linked to workers' empowerment. Koberg *et al.* (1999) also studied the antecedents of empowerment, and found that group and organisational variables were more important influences than individual differences. Thus, they conclude that workers are more likely to feel empowered if they work in a group with an approachable leader. Thus, it is likely that any motivational effect of worker empowerment in OHS will depend on features of the work environment, similar to Herzberg's hygiene factors.

Herzberg's two-factor theory of motivation has been criticised on the grounds that it assumes that everyone is motivated by the same things. The theory was developed in the USA during a period of economic growth, and doubts exist as to whether the theory would apply to depressed or developing countries or among people in low-status jobs. Neither does the theory allow for conflict among goals or motivations, for example, the tension between safety and productivity.

Vroom's expectancy theory

Vroom (1964) advanced a more sophisticated approach to work motivation, known as expectancy theory. Under the expectancy model, an individual's behaviour is determined by his or her beliefs in three areas. These are:

1. the extent to which increased effort will lead to improved performance (*expectancy*);
2. the extent to which improved performance will lead to a specified outcome (*instrumentality*); and
3. the extent to which that outcome is valued by the individual (*valence*).

Expectancy theory holds that performance is a function of skill and motivation. Skill relates to abilities, both innate and acquired (for example, through training). Motivation comprises effort expended by an individual and the knowledge of what is expected by others, for example supervisors and co-workers. Effort is determined by the value to be derived as a result of the effort and the strength of the link between effort and the rewards. This value includes both hygiene and motivators in Herzberg's theory; thus, valued outcomes could include money and status as well as a sense of achievement and personal growth. With regard to OHS, expectancy theory might lead employees to ask such questions as these:

- If I exert more effort, is the safety goal attainable?
- Will the safety goal be rewarded?
- How much do I value the reward I will receive?
- Are alternative goals likely to be rewarded more highly?

This theory more adequately reflects the complexity of work in which conflicting goals may be present, and workers must choose between rewards associated with OHS and rewards associated with other objectives, such as output and profitability. Expectancy theory also implies that effort will not be expended where workers feel that they have little control over their performance. Thus, where workers do not feel that they can control OHS risk in their work environment, offering rewards for OHS performance will not produce greater effort to meet OHS targets. This is an important implication for OHS in construction because people who are exposed to OHS risk are often removed from OHS risk management decision-making and may have limited ability to control their OHS behaviour. Case study 7.2 presents an example of the limitations of motivational management techniques in situations in which workers have little control over OHS aspects of the work environment.

Expectancy theory has some important implications for OHS motivation. First, when designing work processes or selecting risk control strategies, care must be taken to ensure that disincentives to work safety are not introduced. For example, where taking OHS precautions leads to discomfort, or where it is awkward or time-consuming to work safely, workers will be tempted to disregard safe and healthy work practices and take risks. Expectancy theory predicts that workers would weigh up the relative costs and benefits associated with taking the risk, and where the rewards associated with taking the risk outweigh the rewards expected to arise as a result of working safely, unsafe behaviour may appear more attractive. One difficulty here is that accidents are infrequent, and many occupational illnesses have long latency periods. Thus, direct personal experience of negative OHS consequences is rare. In the absence of immediate and certain rewards (good health) or punishments (pain and suffering arising from injury or illness), workers' beliefs about behaviour–outcome contingencies tend not to have a strong motivational effect upon OHS behaviour. Industries like construction often rely on employees who 'get things done', and the reward system, for both individuals and organisations, favours achievers of productivity and progress. This production orientation poses a problem for OHS and incentive schemes, such as piece-rate remuneration or bonuses, can lead to corner-cutting and unsafe work practices (Dwyer 1991). It is important that the prevailing reward structures and managerial actions do not motivate workers to improve productivity at the expense of OHS. Again, the certain, short-term reward of payment is weighed against the uncertain, long-term risk of suffering a work-related injury or illness, providing a strong motivational incentive to work unsafely.

Petersen (1989) suggests that OHS incentives often do not work because rewards are given for not being injured, an outcome subject to many determinants including chance. More effective strategies would be to link rewards to specific behaviours but, as Case study 7.2 shows, these will only be effective if the behaviours are within the volitional control of the people whose motivation they are designed to elicit. Finally, in motivating for improved OHS performance, effort may need to be devoted to changing workers' perceived relative values of rewards available.

Eliciting behaviour change

Case study 7.2: Behavioural safety management in the construction industry of Hong Kong

An experiment was undertaken on seven public housing construction projects in Hong Kong. Four categories of safety behaviour were identified and measured for a period of thirty-four weeks during the experiment. These were: housekeeping; access to heights; use of bamboo scaffolding; and personal protective equipment (PPE).

Housekeeping items related to such aspects of site safety as the storage and stacking of materials and the maintenance of clear access routes. A period of baseline measurement was undertaken using an observational safety rating instrument. After this period, motivational techniques were introduced in an attempt to change workers' safety behaviour. Meetings were held on each site and performance goals were set. Measurement continued and performance feedback was provided by means of large graphs posted at prominent positions on site. Feedback graphs were updated each week. Goal-setting and feedback were introduced to the housekeeping, access to heights and bamboo scaffolding categories at staggered intervals. PPE was maintained as a control category, and thus no goals were set or feedback was given in relation to the use of PPE.

The results of the experiment were mixed. Highly significant improvements were observed in site housekeeping with the introduction of goal-setting and feedback, coupled with significant deteriorations following the removal of the housekeeping feedback charts. This indicates that the observed improvements during the housekeeping intervention stage were attributable to the goal-setting and provision of feedback. Thus goal-setting and feedback were effective when applied to the housekeeping category. In contrast to these results, the access to heights and bamboo scaffolding interventions did not, in general, result in significant improvements in these areas. Safety performance in access to heights was found to improve on five of the sites with the introduction of goal setting and feedback in this category, but on only two of the sites was the improvement found to be statistically significant. On one site, there was found to be a highly significant deterioration in safety performance relating to access to heights at this time. Because of the presence, by week 25 of the experiment, of bamboo scaffolding on only four of the experimental sites, the bamboo scaffolding intervention could only be introduced on these four sites. No significant improvement in safety performance relating to the use of bamboo

scaffolding was observed on any of the four sites at which goal setting and feedback were introduced in the bamboo scaffolding category.
(Source: Lingard and Rowlinson 1997)

Housekeeping is a very different aspect of site safety to access to heights or bamboo scaffolding. Generally speaking, improvements can be made in housekeeping without the deployment of additional materials or equipment. Everyone on site can contribute to the improvement of site housekeeping. In contrast to this, access to heights and bamboo scaffolding items related to the work of a few specialist trades. Many of the items would require that additional materials be provided and, most significantly, that an increased time period be allowed for the job at hand. For example, if a bamboo scaffold is to be fitted with adequate, closely boarded working platforms, guard rails and toe boards, it requires that good quality timber is available to fit as platforms and toe boards. More significantly, the time required to construct a scaffold with these features is much greater than the time required to construct a bamboo scaffold without these features.

Bamboo scaffolding is typically erected by a specialist subcontractor, and the extent to which this subcontractor will incorporate working platforms, guard rails or toe boards into the scaffold will depend upon the specifications and cost agreed between the main contractor and the scaffolding subcontractor. Unlike housekeeping, improvements in bamboo scaffolding safety cannot be made by every person on the site. Furthermore, the extent to which a safe scaffold is provided for use is pre-determined by an agreement, often verbal, between the main contractor and the scaffolding firm. Under these circumstances, it is not within the control of the majority of workers on site to make improvements in this area.

A highly significant deterioration in access to heights performance occurred on one site following the introduction of the goal-setting and feedback. This finding may be because the project was close to completion and there were large numbers of finishing workers such as painters and plasterers present on site. These trades use a great deal of equipment relevant to the access to heights measurement items, including ladders, trestles and tower scaffolds. This site was also running several months behind schedule, so these finishing workers would probably have been under some pressure to complete work speedily. This pressure may have encouraged bad practices and 'corner-cutting', which would, in turn, be reflected by poor access to heights scores.

Vroom's expectancy theory of motivation may provide an insight into an individual's choice to reject or to accept and pursue a goal. It is possible that an individual's beliefs regarding expectancies, instrumentalities and valences determine whether a goal is accepted or rejected. Thus, the greater an individual's expectancy that increased effort will lead to goal attainment, the more that goal attainment is perceived to be instrumental in leading to a certain outcome, and the greater the individual values that outcome, the more likely he or she will be to accept and pursue a goal.

The results of the behavioural safety interventions presented in Case study 7.2 may be interpreted in the light of this model, in that housekeeping represents an area of safety performance in which improvements can be made relatively easily. The expectancy that increased effort will lead to improved performance and a specified outcome, that is, goal attainment, would be high. The value associated with that goal attainment may also be reasonably high and, under the goal-setting and expectancy model described above, the goal would be accepted and workers would be sufficiently motivated to attain the goal.

In contrast to this situation, organisational constraints, such as inadequate resourcing and/or time performance pressures, may impose upon workers an inability to perform work safely in the areas of access to heights or bamboo scaffolding. This can be seen as a failure by management to provide an adequate 'safety infrastructure' on site, for example, inadequate equipment, plant and personal protective equipment. This perception may cancel out any 'value' which workers may place on the outcome of goal attainment; the goal would then be rejected, and enhanced motivation would not occur.

Another factor that may affect the goal-setting/expectancy mechanism described above is the reward structure within which site workers are working. It is common for Hong Kong construction tradesmen to be paid on a piece-rate basis. Where this is not the case, a main contractor may agree on a fixed price to be paid to a subcontractor on completion of a given task. In some instances, bonuses may be paid for work completed ahead of schedule. In any of these scenarios, it will be rewarding for work to be carried out as quickly as possible. If workers perceive the value associated with the outcome of completing work quickly to be higher than the value associated with the outcome of attaining a safety performance goal, then it is likely that goal rejection will occur.

Case study 7.2 thus highlights potential problems in applying motivational techniques to OHS. Motivational techniques can only be effective where OHS behaviour is under the volitional control of workers, and care must be taken to ensure that incentives to work safely are not outweighed by incentives to take risks.

Group influences on OHS

Groups are central to the functioning of an organisation, and can exert considerable influence on the behaviour of their members. Groups provide individuals with the satisfaction of social needs, support in the achievement of personal objectives and help in the establishment of self-concept. Membership in a group can shape the attitudes an individual holds in regard to any aspect of work or non-work life, including OHS. Groups can be formal, such as OHS committees, accident investigation or problem-solving teams. Informal groups also develop spontaneously within organisations. In the construction industry, in which work teams are subcontracted, group influences on workers' attitudes and behaviour are likely to be particularly important. It is unlikely that workers who may only be involved in a construction project for a matter of days will be influenced greatly by normative pressures of a principal contractor's direct employees. Instead, acceptance by other members of the trade-based work team is likely to be a more salient influence on their behavioural intent.

Stages in group development

Group development is believed to follow a life cycle of four stages as follows:

1 *forming*, when group members come together;
2 *storming*, when initial hostility and conflict between group members is aired;
3 *norming*, when group norms and behaviours are agreed between members; and
4 *performing*, when tasks are carried out to meet group objectives.

Group norms

Group norms are a powerful predictor of risk-taking behaviour. The expectations of co-workers and supervisors in following safe work procedures and abiding by rules have a strong influence on the way we act. Group norms can be harnessed to bring into line deviants who repeatedly flout safety rules. For example, other group members may apply pressure to the deviant by verbal coercion, blacklisting or ignoring him or her until behaviour is in accordance with group norms.

The importance of group relationships in OHS is demonstrated by a study by Hinze (1981). He reports that many different aspects of worker attitudes are associated with higher levels of safety performance in a construction-related industry. He considered the feelings of county highway crew members in relation to their individual injury frequency records. His statistically significant findings were as follows:

- Workers in crews which got along better as a total unit had better safety records. From this he inferred that friction within a crew adversely affects safety.

- Workers who shared their personal problems with a co-worker or a foreman had better safety records.
- Workers with better safety records were those who worked in smaller crews. Hinze explains this by stating that smaller crews could be expected to develop more readily into cohesive, close-working units.
- Workers who were allowed to participate in management decisions, that is, their ideas were given serious consideration by management, had better safety records.
- Safer workers were found to be those who received more praise from supervisors for work well done.
- Safer workers were those who felt that their employer was genuinely concerned about their welfare.
- Safer workers were those who would select their present employer even if other similar options were available. Hinze takes this to mean that deeper loyalty to an employer results in safer performance.

Group cohesiveness

The concept of group cohesiveness is the group attribute concept most commonly investigated in the organisational behaviour literature. Traditionally, cohesion has been defined by drawing a comparison between the group and an atom. Festinger defined cohesion as 'the resultant of all forces acting on members to remain in the group' (Festinger 1950). Alternative definitions rely on the concept of 'attraction-to-group'. Lott and Lott define cohesiveness as 'that group property which is inferred from the number and strength of mutual positive attitudes among members of a group' (Lott and Lott 1965).

When groups share common values, beliefs and objectives, the mutual acceptance of ideas is a source of group cohesiveness. Members of cohesive groups share common objectives and an understanding of how best to meet those objectives. Cohesive groups are characterised by teamwork and close co-operation. The importance of group cohesion among the flight crew in civil aviation has been noted (Hawkins 1993).

However, cohesive work groups may not always be good for safety. Seashore (1967) and Greene (1989) have commented on the effects of group cohesiveness on the level of organisational effectiveness. Both writers stressed that high levels of cohesiveness do not automatically lead to greater organisational effectiveness. In relation to productivity, Seashore found that highly cohesive groups had less variation in productivity among members than did low cohesiveness groups and they differed significantly from the plant norm of productivity. However, the direction of this deviation (that is, towards higher or lower productivity) was a function of the degree to which the larger organisation was perceived by group members to provide a supportive setting for the group. Greene found strong evidence to suggest that group cohesion positively affects productivity when organisational goal acceptance is high but negatively affects it when organisational goal acceptance is low. These findings indicate that where employees feel a strong sense of belonging to a group, this group has a powerful influence over that individual's behaviour, but this influence may be either good or bad for an organisation depending on the extent to which group members identify with and accept the goals of the organisation as a whole.

The extent to which group cohesiveness has a positive effect on organisational goals (including OHS goals) will probably depend upon whether group members accept these goals. In a construction industry context the existence of highly cohesive groups (subcontractor work crews) may actively resist OHS rules imposed upon them by a head contractor to whom they have little loyalty or commitment.

Intra-group dynamics

Group decisions generally have a greater impact on the behaviour of individual members than decisions taken by a leader and communicated to other group members. Decisions made following a period of group discussion are more likely to become social norms that then guide members' behaviour. In OHS, safety committees represent a useful forum for members to improve their understanding of OHS issues and discuss ideas for improvements. The collective decision-making process adopted by a safety committee can generate shared understanding and normative pressure to behave in accordance with agreed procedures.

Communication patterns within groups, including project teams, have been described as taking different forms. Three of these forms are depicted in Figure 7.4. Cheng *et al.* (2001) suggest that construction projects traditionally adopt a linear 'chain' pattern of communication. Within this model, communication between parties to a construction project is restricted to communication between parties in direct contractual relationships concerning contractual requirements. In this model, designers who are in a contractual relationship with the client will not communicate directly with contractors, with whom they have no contractual relationship.

This restricted communication is associated with low member satisfaction, poor performance and co-ordination problems (Glendon and McKenna 1995). It is unlikely to result in inter-organisational co-operation or optimal OHS performance. Alternative communication patterns that are likely to be more effective include 'wheel' and 'free interchange' patterns. Wheel-type communication is associated with quick problem-solving, and is particularly good when groups are faced with simple tasks, but it is not well suited to complex frequently changing situations. The free interchange model of group communication is the best for solving complex problems, and has a high level of satisfaction for group members. We suggest that

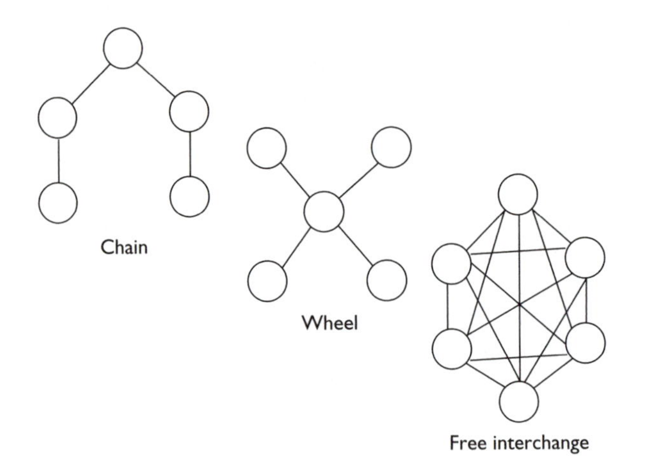

Figure 7.4 Communication patterns within groups (after Leavitt 1972).

such a decentralised communication model is appropriate for managing OHS communication in construction projects. In order for this to occur, the industry's culture of chain-type communication based upon contractual relationships must be overcome, and communication channels opened up between project participants with an interest in OHS. These may include clients, designers, suppliers, sub-contractors, employees and their trade union representatives.

Having indicated that free-exchange communication patterns lead to more integrated OHS decision-making between project team members, it must be noted that there may be circumstances when it is not the best model to adopt. For example, in crisis management situations, decisions must be made as situations rapidly change. Literally minute-by-minute developments require that life or death decisions be made under conditions of intense pressure. In such a circumstance, the free-exchange communication model is likely to be unwieldy and prone to failure. A wheel-type pattern, with a crisis management team leader at the centre, is likely to be much more effective.

Groupthink

One problem with highly cohesive work groups is that the pressures to conform are very great. Members may be concerned about appearing to be divisive and troublesome. The term *groupthink* was coined by Janis (1972) following an analysis of American military decision-making in the disastrous Bay of Pigs invasion of Cuba. Groupthink occurs when groups make decisions in isolation from any alternative views or external comment.

Alternative solutions to a problem are not considered, and the costs of alternatives are deemed to be so great that these alternatives are given no further consideration. External advice is not sought, and only expert opinion that is consistent with the group's own thinking is considered to be valid. Information that suggests that the group's preferred plan of action is not the best is dismissed as being inaccurate. Groupthink was identified as a contributing factor in the Chernobyl disaster in which the plant's operators possessed a false sense of control and invulnerability (Reason 1997).

Groupthink can prevent sound risk management decision-making, and every effort should be made to ensure that OHS decision-making does not fall victim to what Glendon and McKenna (1995) call the 'cycle of decision failure'. In such a cycle, solutions are decided prior to careful analysis and consideration of all of the issues and alternative solutions. New information is suppressed in such a cycle, leading to poor decision-making. Research also suggests that groups are subject to information-processing errors in making judgements about the level of risk inherent in a situation. This is analogous to the sources of cognitive bias in individual risk perception discussed earlier in this chapter. Houghton *et al.* (2000) report that groups are actually more susceptible to certain forms of bias in making judgements about risk than are individual group members, and therefore groups faced with a risky situation may, in some cases, be more likely to develop 'a false sense of security' than would individual members. This is more likely to be a concern in a male-dominated environment, such as construction, because research also shows that men are greater risk takers than women and, in mixed gender work groups, men's risk judgements are more influential in determining group decisions (Karakowsky and Elangovan 2001).

Shared beliefs about failure in work groups can also prevent lessons from being learned as a result of failure (Cannon and Edmonson 2001). Work groups in which members are not encouraged to discuss mistakes, and in which failure is stigmatised as incompetence, are unable to learn from failure. Such beliefs breed defensiveness and a tendency to try to cover up mistakes rather than learn from them. For example, employees may be reluctant to report near-miss incidents or minor injuries for fear of being branded incompetent. In contrast, open communication, in which conflict can be handled productively and mistakes can be a source of vicarious learning for group members, facilitates learning from failure. Schemes such as 'mistake of the month', in which work group members are encouraged to present their mistakes, can facilitate this vicarious learning and lead to group bonding.

Groupthink and decision failure can be avoided by ensuring a heterogeneity of inputs into the decision-making process. Thus, groups should contain people with different experiences, personality types and decision-making styles. The composition of effective teams is considered in detail by Belbin (1981, 1993).

The work environment

Bradley (1989) identifies the importance of the work environment in shaping workers' OHS behaviour. The creation of a work environment in which OHS is perceived to be an important and valued part of doing a job is essential to maintaining workers' OHS motivation. Everything must be done to eliminate the perceived role conflict between production and OHS, and OHS must be understood to be a non-negotiable part of working.

Management theorists have long suggested that the performance of an organisation is influenced by its culture (Likert 1967). The safety culture of an organisation has been defined as 'the product of individual and group values, attitudes, perceptions, competencies and patterns of behaviour that determine commitment to and the style and proficiency of an organisation's health and safety management' (HSC 1993). In an attempt to define what a positive safety culture means in practice, researchers have tried to identify the core features characteristic of organisations with excellent OHS performance. These features are summarised in Box 7.1.

Box 7.1 The core features of safety culture

- Genuine and committed safety leadership from senior management; visibly demonstrated by allocation of resources to safety and promoting safety personally.
- Giving safety a high priority through the development of a policy statement of high expectations which conveys a sense of optimism about what is possible.
- Long-term view of OHS as part of business strategy.
- Realistic and achievable targets and performance standards with current information available to assess performance.
- Open and ready communication concerning safety issues within and between levels of the organisation. Less formal and more frequent communications.
- Democratic, co-operative, participative and humanistic management style leading to 'ownership' of health and safety permeating all levels of the workforce.
- More and better training. Safety integrated into skills training.
- Capacity for organisational learning. Organisations responsive to structural change and results of system audits, and incident investigations.
- Line management responsibility for OHS. OHS treated as seriously as other organisational objectives.
- High job satisfaction and perception of procedural fairness in employer/employee relations.

(Source: CBI 1990; Lee 1997)

The culture of an organisation has a motivational effect because it determines how workers perceive their work environment. Schneider (1975) developed the idea that employees develop a coherent set of shared perceptions of their work environment, and termed this the 'organisational climate'. Much work has been undertaken to investigate the validity of the organisational climate concept and the antecedents and consequences of the organisational climate. It is believed that the perceptions that make up an organisation's climate are formed on the basis of cues present in the work environment, and that the climate influences employee expectations regarding behaviour–reward contingencies, acting as a frame of reference for gauging the appropriateness of behaviour. Research has led to the development of comprehensive scales for assessing organisational climates (Kozlowski and Doherty 1989; Kopelman *et al.* 1990). Several dimensions of organisational climate identified in previous research are likely to have an impact on the OHS behaviour of employees. In particular, role stress and lack of harmony, organisational goals, trust, supportiveness and resources are likely to have an impact on OHS performance. For example, where there is perceived to be tension or conflict between safety and productivity goals, OHS behaviour may be compromised.

Further support for the influence of organisational climate on OHS is derived from studies focusing specifically on safety climate. Research has indicated that, within an organisation, employees' perceptions about OHS behaviour–reward contingencies or 'safety climate' can be measured reliably (Brown and Holmes 1986; Cox and Cox 1991). Research also suggests that aspects of the safety climate are related to safety performance. For example, Zohar found that where management commitment to safety was high, employees tended to engage in safe behaviours (Zohar 1980) and, in the construction industry, Dedobbeleer and Beland (1991) found that safety performance outcomes can be predicted from safety climate scores.

More recently, Hayes *et al.* (1998) developed and tested a Work Safety Scale (WSS). The WSS measures five dimensions of the work environment: job safety; co-worker safety; supervisor safety; management safety practices; and a satisfaction with the safety programme. They report that this instrument possesses good reliability, and also predicted near-miss accidents and compliance with safety behaviours. Scores were negatively related to near-miss incidents and positively related to compliance with safety behaviours. Management safety practices were the strongest predictor of near-miss incidents, and management safety and co-worker safety were the strongest predictors of compliance with safety behaviours. This latter finding suggests the strong social and normative influence of work groups (see also the earlier section 'Group influences on OHS'). The study by Hayes *et al.* (1998) suggests that perceptions of the work environment are important indicators of OHS behaviour, and that instruments like the WSS might be useful diagnostic tools.

The use of safety climate assessment as a diagnostic tool to identify barriers to improved OHS performance was also recommended by Budworth

(1997). Aspects of the work environment can then be modified to create an environment in which workers perceive safe behaviour to be valued and rewarding. Measurement of an organisation's safety climate is usually carried out using a questionnaire. There are many different variants of safety climate questionnaires, but they usually measure a number of different dimensions, each represented by a different set of questions. Safety climate dimensions that have been measured include:

- management commitment to safety;
- perceived risk levels;
- effect of the required work pace;
- effectiveness of safety communication;
- management actions;
- attributions of blame;
- the effects of job-related stress;
- emergency preparedness;
- the effectiveness of safety training;
- the status of safety representatives and committees; and
- personal commitment to safety (Cooper and Phillips 1994).

The construction industry's 'macho' culture and production orientation have been identified as potential barriers to the development of a work environment that encourages safe behaviour (Anderson 1998b). The impact of the construction environment on the OHS motivation of construction workers needs to be considered. The assessment of the safety climate is a useful method to identify organisational, managerial or job-related issues that prompt unsafe behaviour. Once identified, these can be addressed through OHS training programmes, adoption of alternative management practices or re-structured reward and incentive schemes.

Conclusions

The improvement of construction OHS requires attention to cultural and motivational issues, as well as the organisational and systems issues. Good OHS performance relies on strong leadership and a committed and motivated project team. Workers should not be blamed for the occurrence of accidents. Instead, the pathways leading to human error should be identified and eliminated. OHS risks must no longer be accepted as 'part of the job' and psychological issues that encourage risk-taking behaviours need to be addressed.

The work environment needs to be examined to ensure that working safely is both satisfying and rewarding. This requires clear demonstration of management commitment to OHS. It is essential that the reward system for both individuals and organisations does not inadvertently reward corner-cutting in OHS.

It is likely that working in a safe and well-ordered worksite has a positive effect on workers' motivation and productivity. Therefore, construction firms should consider the benefits to be gained from demonstrating, through the provision of a safe and healthy work environment, that employees are valued. With this in mind, OHS and productivity should not be considered to be mutually exclusive goals, but part and parcel of delivering a construction project on time, within budget and without the unacceptable consequences of work-related injuries or ill-health.

Discussion and review questions

1 What is the significance of the following for OHS?
   (a) personality
   (b) attitudes
   (c) the work environment
   (d) work groups.
2 Briefly describe the main features of the following approaches to OHS motivation:
   (a) mechanistic approaches
   (b) behaviourist approaches
   (c) cognitive approaches
   (d) applied approaches.
3 Discuss the respective roles of individual screening, individual behaviour change and environmental modification in OHS.

Chapter 8

Behavioural safety management

Robin Phillips

To err is human, or why we behave unsafely

Humans are obviously fundamental to the process of manufacturing a product, from commissioning it to using it via design and manufacture. However, we have a number of characteristics that can interfere with this process, and which may be regarded as negative human reliabilities. These include the following:

1. We make mistakes.
2. We behave unpredictably.
3. We are tired, inattentive or preoccupied with other issues, for example, family problems or job security concerns.
4. We may feel safety is 'uncool' and it is not macho to follow safety procedures, such as wearing protective equipment.
5. We may not be aware of the hazards associated with our work – the possible result of poor training or lack of communication.
6. We may underestimate the risks involved.
7. We may rationalise the hazard, believing 'it won't happen to me'.
8. We may feel uncomfortable with the 'low' level of risk we are experiencing and decide to do something about it (often referred to as Risk Compensation).
9. We may cut corners or otherwise deviate from safety procedures, either because we will gain some personal benefit – getting home earlier, receiving a bonus, a favourable report at the next staff evaluation, and so on – or because we believe we are acting for the good of our employer, such as finishing the job earlier to avoid penalties (a situation known as *conflicting rewards*).

Many industries rely on companies and employees who 'get things done', and the reward system for both individuals and organisations favours 'achievers of progress and productivity'. The authors of an HSC report (HSC 1994) deduced that:

> there is a threatening trade off between production and safety. The latter, it is said, requires slower and more careful work with scrupulous following of the rules. The former can be increased by faster turn around by 'cutting corners'... A positive safety culture is one where safety not only wins out if there is a conflict, but where everything is done to remove the conflict. Well-planned and scheduled work as distinct from 'rush jobs' is the most obvious way. The conditions that make for safe operation are often those that make for a good organisational climate and hence good output.

Considerable research has been undertaken in the area of behaviour and motivation, for example, human reliability in situations where an error may have serious consequences, as is the case with train drivers, control room operators in nuclear power stations and so on. Many theories concerning human motivation exist; however, most of them, being set in the classroom, are of limited use in the 'real world'.

What is worth discussing is the change of emphasis that has taken place over the last few years in the majority of industries. Greater emphasis is now placed on the positive as opposed to the negative management of health and safety, with more attention being given to the need to involve and motivate people, as opposed to setting rules and enforcing them by constant monitoring, with penalties for any deviation from the standard, whether for legal or organisational reasons. To conclude this section, let us review the major differences between the old (negative) and the new (positive) style of health and safety management.

Old style of health and safety management

- The safety professional is perceived to be a policeman by both managers and operatives, and is often referred to as a 'safety officer', which implies someone who is reactive – setting rules, policing them and penalising infringements – and only seen when there is a problem.
- There is very little involvement of supervisory management. This is further compounded if the safety professional is referred to as a 'safety manager', because some supervisors believe it is the role of the safety manager to manage safety, while they (supervisors) manage other issues that are wrongly perceived to be unrelated to safety.
- Health and safety is viewed as a peripheral issue to the main business activity. Whether one agrees or not with the UK's Health and Safety Executive's Cost of Accidents report (HSE 1996), it provides an indicator of the potential financial damage an accident can cause an organisation.

- The management of health and safety is perceived to be setting rules and then ensuring they are obeyed.
- Health and safety inspections are only undertaken to find faults.
- Little is done to involve or motivate individuals.

To summarise, the management of health and safety, including the role of the safety professional, is segregated from the day-to-day business activity in terms of both attitude and physical presence. For example, the principal safety department of one major international construction contractor was located in 'porta-cabins' surrounded by spare and redundant plant, while other departments, such as accounts and public relations, were housed in office buildings. This gave the perception to both personnel and clients that health and safety was an unimportant issue to the contractor (though in reality this was not the case).

New style of health and safety management

- The safety professional is referred to as a 'safety adviser', which implies someone who is proactive, acting as a facilitator guiding individuals and the business through their legal responsibilities.
- The cornerstone of the style is that the principal responsibility for health and safety lies with supervisory management, ensuring an integrated approach towards all management issues.
- Health and safety are perceived as fundamental business activities, and as such are realistically resourced. The management of health and safety, including the role of the safety professional, is integrated into the day-to-day business activity. Safety is no longer perceived as a burden, but as good business practice that can win contracts and enhance profits.
- Health and safety management is seen in terms of achieving a positive goal rather than avoiding failure. To clarify this, think of the old style of management as a fire fighter, someone who reacts to a situation when needed, but most of the time remains unseen, and accepts the organisation's safety culture without question. For the new style think of an athlete, someone who has a high standard of performance and yet still tries to better it. In quality terms, think of continuous improvement.
- Audit and monitoring provide credit for what has been achieved as well as constructive criticism where standards have not been met.
- Everything is done to try to involve and motivate people.

Involvement of personnel means that management works with, rather than against, its employees. Studies in industrial psychology have shown that people are more likely to support something that they have personally contributed to – sometimes referred to as *buying into* a process. An individual's perception that he or she can make a valued contribution and change the working environment for the better is highly motivating. It is, in any case, morally right that personnel should be consulted on decisions that affect their well-being and working conditions. Personnel have detailed practical knowledge regarding their job, which is hardly surprising; however, many organisations fail to tap into this valuable source of information.

Managers can gain the involvement of their personnel in health and safety issues by:

- inviting personnel to participate in health and safety monitoring;
- encouraging membership of health and safety committees;
- creating local safety groups, perhaps on lines similar to 'quality circles';
- adopting an improvement suggestion scheme which rewards good suggestions;
- using health and safety as one aspect of team building;
- communicating health and safety issues with personnel on a regular basis;
- encouraging further reporting by acting on reports received and providing feedback to the reporter; and
- adopting an improvement strategy based on a behavioural approach.

The last concept leads us to the subject of this chapter, the behavioural approach, and how it can be used to improve health and safety.

The behavioural approach to OHS

There have been several attempts to improve a poor accident record by raising personnel's safety awareness. Informational safety campaigns (for example, posters) are generally regarded as ineffective, as demonstrated by Saarela *et al.* (1989), and fail to have a lasting impact on the accident and injury rates. Previous research (Shimmin *et al.* 1981) has shown that in a sample of accident victims, two-thirds considered their accidents to be avoidable, suggesting that personnel believe much could be done to reduce accidents. Accident causes identified by the 'victims' included many references to 'inappropriate or unsafe behaviour'.

Why focus on unsafe behaviour?

Eighty to ninety per cent of all accidents are triggered by unsafe behaviour on the part of employees – behaviours that are within the individual's control, and also within the scope of managers and superiors to effectively control. Focusing on them before an accident occurs enables managers to limit the underlying causes of accidents.

Safety behaviour in several cultures and industries has been improved through the use of psychologically based techniques (McAfee and Winn 1989) – sometimes termed *Applied Behaviour Analysis*, or more commonly, the *Behavioural Approach*. This involves systematically monitoring safety-related behaviour and providing feedback, in conjunction with goal-setting and/or another protocol (for example, training or some form of incentive scheme) to reinforce positive behaviours.

These techniques have been shown to be of value in safety (for example, Zohar and Fussfield 1981; Cooper *et al.* 1994) and also in productivity (for example, O'Brien *et al.* 1982). Research applying the techniques in construction-related industries has been conducted (for example, Komaki *et al.* 1978; Rhoton 1980; Chokkar and Wallin 1984; Mattila and Hyodynmaa 1988). In 1989 a research project entitled 'Improving Safety on Construction Sites by Changing Personnel Behaviour' commenced at the UK's University of Manchester Institute of Science and Technology (UMIST), sponsored by the HSE (Duff *et al.* 1993, 1994). A similar study was undertaken in Hong Kong (Lingard and Rowlinson 1994, 1998).

The use of the behavioural approach has increased our understanding on a number of important issues:

- why personnel undertake unsafe behaviour even though they have received safety training and know it is wrong;
- personnel's attitudes towards health and safety;
- how health and safety management systems can be improved to enable them to be used more effectively; and
- how attitudes can be modified by changing personnel's behaviour.

The final point has generated many debates that attempt to answer the question: So *what do you change – a behaviour or an attitude*? Every health and safety practitioner and psychologist has a theory about attitudes and behaviours and how they interact with each other. For the most part, our 'common-sense' theories are not in our conscious minds and therefore generally remain unarticulated. However, it is useful to identify the theory you are using to explain attitudes and behaviour, because this then makes explicit your assumptions about other people's attitudes and why they behave as they do. Typically, the everyday theories we use take one of the forms described below:

- Attitudes influence behaviour, and thus if we know a person's attitude to something (for example, using PPE) then we can predict their behaviour towards it.

- Behaviour influences attitudes, and thus if we wish to change someone's attitude towards something (for example, using PPE) then we can achieve this by obliging them to behave in a particular way (for example, by passing legislation or making a rule and enforcing it).
- Attitudes and behaviour mutually reinforce each other and thus if we change either one then this is likely to lead to a change in the other.
- While it is true that attitudes and behaviour are likely to be mutually consistent, in order to influence them, it is necessary to address both independently – that is, to influence deliberately attitudes on the one hand and behaviour on the other (in a consistent way).

How does the behavioural approach work?

The behavioural approach takes into account the following:

- how personnel think;
- how personnel behave;
- how personnel respond to situations;
- how the work environment impacts upon personnel's attitudes and behaviours.

This concept may be illustrated by the use of Bandura's reciprocal relationship between attitudes, behaviour and situation in (Figure 8.1) (Bandura 1986).

The approach is motivational in that it focuses on unsafe attitudes and behaviours in the workplace, and the interaction between health and safety performance and working environment. It is not concerned with safety attitudes alone, because people tend to alter their attitudes to fit their environment and the behaviours that they perceive to be expected from them.

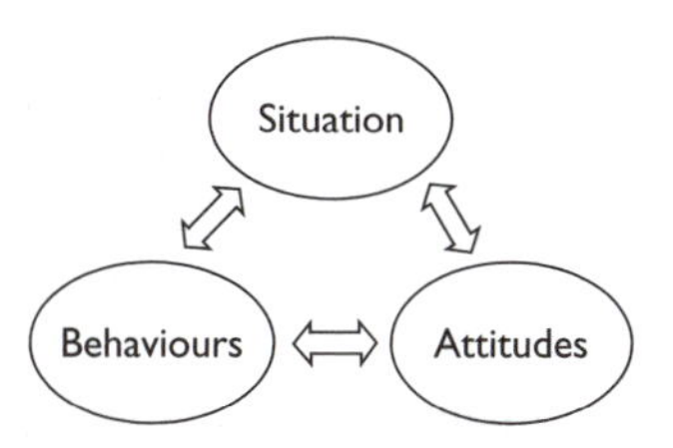

Figure 8.1 Reciprocal relationship between attitudes, behaviour and the situation.

Case study 8.1: Reducing accidents using a behavioural approach

Introduction

The study was conducted in a subsidiary of a large multinational company, on a large single-site, multi-departmental plant manufacturing cellophane film, with approximately 540 employees, located in the South West of England. Two thirds of production workers were employed on a continuous, three-shift, seven-day week, 10-day cycle, rota. The remaining production operatives are employed on a two-shift, six-day week, 10-day cycle, rota. Support staff (that is, human resources, customer services, secretarial, and administration) were employed on a 'normal' five-day, 39-hour week.

Procedure

Management briefings During the planning stages, a two hour briefing was held with line management to outline and explain the philosophy behind a behavioural approach, their role and the need for their commitment.

Developing a measure of safety performance Critical safety behaviours were identified using accident records and in-depth interviews. Measures of critical behaviours were developed for each of the plant's 14 departments.

Observing safety performance Forty eight observers were trained to observe their colleagues' safety performance and complete their measures. The measures, on average, took approximately 10 minutes to complete, and were undertaken on every shift by an observer touring their department. They were undertaken at different times during each shift, on different days, to ensure the observations reflected a true picture of safety performance. Completed measures were posted in a collection box in the main production office for the computation of results. Four weeks of data were subsequently collected from each department to provide a baseline figure against which any improvements could be compared and to enable safety performance goals to be set.

Establishing goals All personnel, including senior management, attended their respective department's goal-setting meetings. Seventy seven meetings were conducted with small groups, over a period of eight days. Each group were asked to determine a goal that was 'difficult, but achievable' for improvements in safety performance. The participative process employed encourages commitment to, and ownership of, the improvement process. Research in the UK has shown that assigned goals can de-motivate personnel.

Feedback and follow-up Following the goal-setting meetings, feedback charts were posted in the appropriate departments. Observations continued at the same rate as that during the baseline period. The results of weekly observations were posted on the departmental feedback charts every Friday morning. Additionally, information referring to the three worst-scoring items in each department was posted next to that department's feedback chart, in order to make explicit to the work force where to focus their attention the following week. During the remainder of the intervention phase, progress was monitored, and assistance was given to observers when necessary.

Results

The results indicate significant improvements in safety performance, with a corresponding reduction in the plant's accident rate.

A steady improvement in safe behaviour performance was observed across the whole plant. The plant's global safe behaviour performance levels, by week, with baseline data are illustrated in Figure 8.2a, which indicates an improvement from a 52.5 per cent average recorded over the four-week baseline period, increasing to 75.6 per cent safe at the 9–12 week point, however, a drop in safety levels occurred during the last four weeks (12–16) of the intervention period to an average of 70 per cent safe, which coincided with a significant increase in the overall accident rate during this time period (Figure 8.2b), due to

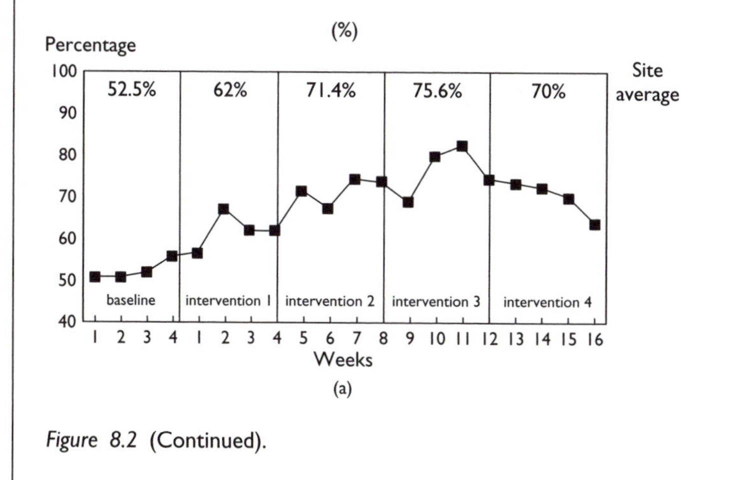

Figure 8.2 (Continued).

Case study 8.1 (Continued)

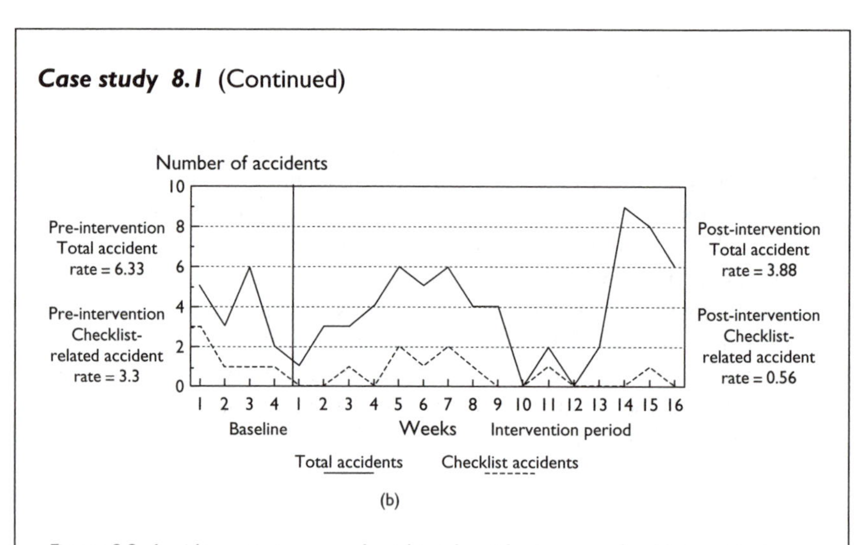

Figure 8.2 Accident rates per week and total accident versus checklist accident rate.

maintenance being undertaken in a haphazard manner, in one of the departments.

In order to examine the effects of the approach on the plant's accident rate, it is necessary to examine past accident performance. It is also important to make a distinction between the number of accidents related to the measures and the total number of accidents. During the year 1991, 307 minor injuries, and 22 lost-time accidents were reported. Of these, 172 (52.3%) were related to the subsequent measures, while during 1992, the year of introduction, in the 20 weeks prior to the intervention (including the baseline period) 77 minor, and two lost-time accidents occurred, of which 35 (44%) were related to the measures. During the 16-week intervention period, 61 minor accidents and two lost-time accidents were recorded, nine (14.3%) of which were related to unsafe behaviours monitored on the measures.

A comparison between the 1992 20-week pre- and the 16-week post-period indicates a 21% decrease in the plant's overall accident rate, and a 74% reduction in accident occurrences related to the departmental measures. Figure 8.2b illustrates the accident rate, by week, for both the pre and post periods. The total accident rate prior to the intervention was 6.33, reducing to 3.88 by the end of the intervention, while those relating to the measures, the measure accident rate prior was 3.3, reducing to 0.56 by the end. On the basis of these figures, it can be deduced that a safety related behavioural intervention can have a positive effect on a plant's accident rate, further supporting the hypothesis that behavioural programmes focus attention and action.

Conclusion

In summary, the above study demonstrates that the application of the behavioural approach to safety, utilising a participative bottom-up approach within manufacturing industries, has considerable merit. Positive effects upon safe behaviour, methods of working, communications and industrial relations, in addition to reductions in accident occurrence and related costs, were found. Perhaps even more importantly, the study's results demonstrated that the relationship between safety performance and accident rates is mediated by organisational variables not normally associated with safety.

It is also worth noting that this plant had its first experience of a behavioural programme during 1992, it is still operational, despite the plant being sold twice. More importantly, approximately 90% of personnel have undergone observer training and undertaken at least one period of observations. The management of occupational safety and health have been given a higher prominence within the organisation, which has led to an improved safety climate, that is, management demonstrating a higher level of commitment as perceived by shop-floor personnel.

The introduction of the programme has influenced other management issues, for example the management of quality, which has resulted in the plant being awarded quality certification.

Edited version of Cooper, M. D., Phillips, R. A., Sutherland, V. J. and Makin, P. J. (1994). 'Reducing accidents using goal-setting and feedback: A field study'. *Journal of Occupational and Organisational Psychology*, 67, pp. 219–240.

Benefits and problems

Let us consider the benefits and the problems that may arise from the use of a behavioural approach.

Potential benefits include:

- improvements in personnel's safety behaviour;
- a reduction in unsafe situations;
- fewer accidents;
- better safety management systems;
- improved personnel attitudes towards health and safety;
- improvements in communications, involvement and co-operation between operatives and management;
- less damage to plant, equipment or work in progress;

- improvements and more continuous productivity – for example, fewer delays in the production process; and finally,
- a reduction in the costs normally associated with accidents.

However, the carrot-and-stick approach has the potential to be unreliable, simply because human beings are individualist and emotional organisms. Punishment for mistakes may well, for example, lead to personnel behaving in the same way again, but making greater efforts not to be caught, thus driving unsafe behaviour underground. On top of this, there is always the possibility that such a strategy will cause resentment, frustration and hostility amongst personnel, which may affect morale and performance. We will be considering the potential problems experienced when introducing a behavioural approach by viewing the approach through the eyes of operatives and managers later in this section.

The key to a successful introduction of the behavioural approach appears to lie in personnel's commitment to the organisation and work group. The motivational effects of goal-setting and feedback (constituent parts of the behavioural approach) may be limited if employees do not identify with, and feel part of, the organisation (Lingard and Rowlinson 1994).

As previously discussed, research has shown that an effective way of reducing unsafe behaviour is to employ social recognition and praise for those who behave safely. However, we have already seen that this is something that is rarely done because both management and operatives feel uncomfortable about the giving and receiving of praise. The behavioural approach helps to create a climate that is 'psychologically safe', which uses and fosters personnel involvement by allowing them to accept or provide praise and encouragement.

Implementing the behavioural approach

Only one type of behavioural intervention will be discussed here: participative goal-setting and performance feedback – the intervention that had the greatest impact upon levels of posted safety performance in the Duff *et al.* (1993) study.

For the successful implementation of such a strategy a number of actions are needed:

- developing a measure of safety performance;
- monitoring safety performance by an observer to ascertain levels of performance prior to the introduction (a baseline);
- the goal-setting and feedback introduction;
- continuous monitoring of safety performance;
- providing feedback and praising success; and
- correcting unsafe behaviour.

The development of a safety performance measure

A safety performance measure should be capable of fulfilling several functions, such as safety monitoring, control or evaluation of safety improvement strategies. However, to fulfil these functions adequately, it must be capable of measuring safety in a valid, reliable and sensitive manner.

Quantifiable measurement

A good measure should allow the opportunity to identify real changes since, like other measurements involving human behaviours, safety performance will be prone to large random variation.

Sensitive

The measure should be sensitive enough to detect all significant changes in safety performance.

Reliable

The measure must be reliable in that it should be capable of recording the same result, given the same situation, regardless of who is taking the observation.

Valid

The measure must be valid in that it genuinely represents the safety condition.

Understandable

The measure should be easily understandable by those who have to use it, and by those whose performance is to be measured, if it is to be useful and credible.

Efficient

The cost of obtaining and using the data must be less than the benefit to be gained; therefore, the measure should be capable of producing data without disrupting production operations.

Universally applicable

The measure should be capable of measuring safety performance levels on a variety of construction sites regardless of the size and type of building being constructed, or the stage at which the measure is used.

The seven characteristics of a safety performance measure

Methods that rely on accident injury rates as the primary measure are flawed. Accidents, though they are the direct result of unsafe behaviours, or occur indirectly through unsafe situations, are to some degree chance events. This makes them unreliable measures of safety performance, other than over the very long term. Measures that focus on accident or injury rates reveal very little about the antecedent behaviour and situational malfunctions that are the real causes of accidents. Ideally, safety measures should not only provide an indicator of safety performance, but should help to prevent accidents. In this regard, they should not only indicate when and where to expect a safety problem, but should also direct and guide those responsible for safety towards possible solutions.

A subsidiary function of a measure is to evaluate and determine the effectiveness of any safety improvement strategy. The measure must be capable of showing the magnitude and direction of the consequences of the strategy. It is recognised that accidents can be a function of the environment as well as behaviour. However, an unsafe environment is often the result of an unsafe behaviour.

Identifying unsafe behaviours or situations

Identifying unsafe behaviours or situations is achieved by considering:

- knowledge and experience of the work undertaken – knowing the hazards and risks and how often problems have occurred as a result of undertaking the work;
- reviewing the organisation's historical accident and near-miss records so as to deduce which behaviours are related to the majority of incidents; and
- reviewing the risk assessments and method statements of hazardous activities that are likely to be undertaken on a regular basis.

The identified unsafe behaviours and situations should be discussed with personnel, allowing them the opportunity to discuss and review the issues, thus gaining their commitment to, and ownership of, the finalised safety performance measure. In addition, such a process increases the face validity of the measure, and identifies issues that may be of concern to personnel and that have not appeared in any accident report or risk assessment. (After all, the personnel who undertake the task are in an ideal position to pass comment on it.)

Developing a safety performance measure

A safety performance measure will consist of a mixture of:

- specific safety behaviours; and
- unsafe situations that result from an unsafe behaviour.

The golden rule is that each unsafe behaviour or situation must be easily observable and written in a clear, specific and unambiguous manner to avoid confusion for the observer. For each issue, there need to be clear instructions for the observer, informing them of good practice or legal requirement. For some issues it will be necessary to concentrate on the situation, rather than the behaviour, because typically the situation is observable for longer periods. Most unsafe behaviours are of relatively short duration, and difficult to capture with a safety measure. For example, the unsafe erection of a ladder, or failing to correctly lash it, may take only a couple of minutes; however, the incorrectly lashed ladder – the evidence of the unsafe behaviour – will be present for some time.

The role of the observer will be to:

- regularly and systematically measure safety performance using the developed measure; and
- provide feedback on a weekly basis to the whole site, and to any personnel who request it.

The observer should be credible – that is, considered conscientious and safe by his or her peers – and, ideally, knowledgeable about the work being performed, with good verbal and interpersonal skills.

To aid with credibility and worker ownership of the intervention, any observers should include non-management workers, for example, workers, union representatives, charge hands and so on, in addition to the usual representation of line managers and safety advisors. The number of observers will depend upon the size and the complexity of the site.

The observers need to be trained in observation techniques, the use of the developed safety performance measure and giving verbal feedback to personnel.

Establishing a safety performance baseline

The safety performance baseline is the 'percentage safe' score calculated from the developed measure that reflects, over a period of time, the site's safety performance levels. It is used as a benchmark to which the ongoing safety performance can be compared. The baseline is produced from observations made over a couple of weeks, with at least three observations per week, preferably once a day. It is important that no feedback about levels of safety performance is given to personnel during this period, to ensure that the baseline is a true reflection of health and safety performance.

At the completion of the baseline period, the performance baseline is calculated and plotted on large graphical feedback charts that are publicly displayed around the site (Figure 8.3).

What to do with the baseline information

Before discussing how the baseline information is used to improve safety performance levels, we need to briefly review:

- participative goal-setting; and
- feedback, sometimes referred to as 'knowledge of results'.

Goal-setting

The literature on goal-setting as a procedure for influencing behaviour is substantial. For example, Wood *et al.* (1987) reviewed approximately 200 studies. Goal-setting theory states that goals are the immediate, though not sole, regulators of human action, and that performance will improve when the goals are hard, specific and accepted by the individual or group. Goal-setting is said to affect performance by directing the attention and actions of the individual or group, mobilising effort and increasing motivation (Locke and Latham 1990). For goals to be effective, they should:

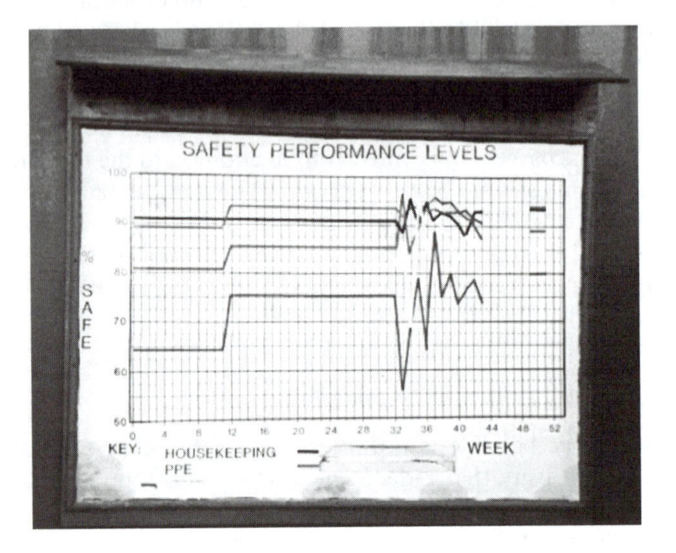

Figure 8.3 Sample graphical feedback chart: HSE phase II (Robertson *et al.* 1995).

- be ambitious but reasonably achievable;
- be set by those who will work towards them, at all levels;
- be specific and measurable; and
- represent long-term improvement, not just short-term gain.

Research has shown that a major moderator of any goal-based behavioural intervention is the commitment given to the goal by personnel.

Factors that have been found to enhance commitment fall into two broad categories: those that convince people that achieving the goal is *possible*, and those that convince people that achieving the goal is *important*. Managers can play an important role in facilitating commitment to goals by persuading their personnel that the desired goals are both achievable and important.

Allowing employees to participate in the goal-setting process can also considerably improve commitment to a goal. Research suggests that for many people, a goal set and delegated by others serves as a disincentive, and may be rejected for that reason (Naylor *et al.* 1980).

A comparison was made between assigned and participatively set goals in the behavioural study undertaken in the UK construction industry (Duff *et al.* 1993). The results indicated that on sites where workers participated in the goal-setting process, there was better performance, compared with sites where goals were assigned to workers (Cooper 1992).

Encouraging participation and acceptance of goals does not mean that workers should be given complete freedom to set their own goals, but that targets should be arrived at by a process of sensible and realistic discussion. Taking part in discussions to decide the target level to be achieved not only makes it clear to workers how hard and specific the goal should be, but also clarifies, for all parties concerned, the best strategy to adopt, and what resources may be needed to attain the goal.

Further benefits derived from participation include 'workers' ownership' of the improvement process, better working relationships between managers and workers, improved job satisfaction, and the reduction of perceived conflicts between competing organisational goals, for example the perceived conflict between safety and productivity.

Some managers and safety professionals feel uncomfortable with a genuinely participatively set goal. Setting goals participatively allows personnel true ownership. It also has an important symbolic function – it introduces the concepts of consultation and communication that are so vital to a successful intervention. In addition, it is a typical reaction for management to feel uncomfortable with a goal 'less than 100 per cent'. Personal experience has shown that all organisations (good and bad) have a 'zero accident goal' – but that the better ones are working towards it realistically one step at a time. This problem is often overcome with explicit reference to 'short' or 'medium' term goals.

Feedback

The literature on the role of feedback in determining performance effectiveness clearly indicates the positive effect of knowledge of the results of one's behaviour. Reviews of feedback research, for example Algera (1990), indicate that performance is enhanced when management provides clear feedback of performance-related information as can be seen on the sample feedback chart (Figure 8.4).

In practical terms, generating high intervention commitment, and subsequent performance, during the process may best be achieved by following certain procedures, as specified by Cooper in Robertson *et al.* (1992):

- Displaying results of the baseline measure on a feedback chart of some description.
- Giving explanations as to why improvements in the particular area of health and safety performance are necessary, how the measure was obtained and developed and how it is used.
- Emphasis by the goal setter, usually the observer acting as a facilitator, that the baseline is the current performance level and that performance can and should be increased, and also upon the benefits of reaching the goals to the individuals themselves, as well as to the company.

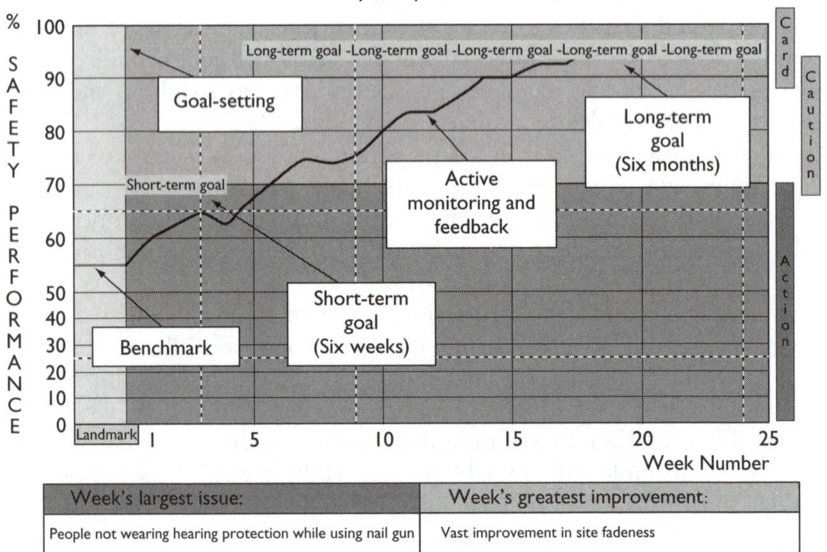

Figure 8.4 Sample feedback chart: start of the intervention process.

- Stressing the importance of personnel participating in setting a difficult, specific but realistic goal to improve performance. This encourages commitment and acceptance of goals and minimises the likelihood of resistance.
- Emphasising that some goals are unrealistically high and therefore not expected. (Goal-setting works only if the goals are perceived as realistic and attainable.)
- Stressing that no sanctions will be applied for not reaching the goal. This ensures and encourages positive actions. Previous research has shown that personnel are likely to test this. If sanctions are applied, it is probable that personnel will reject the goals, with subsequent detrimental effects upon performance.
- Emphasising the necessary actions that will enable the personnel to reach the goal. This helps to clarify the particular behaviours that are required.
- Asking the personnel to set a difficult, specific goal that the majority agree with. If the goal appears to be too easy, it will be necessary for the goal setter to suggest a more difficult level. Essentially, this is a period of negotiation between management and personnel, the outcome of which should be satisfactory to both parties.
- Explaining to personnel that as performance is monitored, the results will be posted at regular intervals on the feedback chart, which will be placed in a prominent position in their workplace.
- Ensuring that an authority figure who is supportive, and who will exert reasonable pressure on subordinates to reach the goals set, is present during the goal-setting session. Management commitment to the intervention is of extreme importance. Many organisational interventions have floundered because managerial commitment was lacking (Duff *et al.* 1993, 1994).
- Regularly monitoring safety performance. Once the goals have been determined, the observers continue to take observations, preferably each day, until it is deemed unnecessary to have such a high profile health and safety intervention.
- Regularly providing feedback and related issues. At the end of every week of the intervention, the weekly safety performance is calculated and posted on the publicly displayed feedback chart. Some organisations believe it is enough just to do this, believing personnel will seek out the information; however, this is rather naïve. Others use it as the catalyst for their on-the-job training programme by using short meetings similar to a 'tool-box talk', where the previous week's safety performance, and any issues arising from it, are discussed – for example, behaviours or situations causing the greatest concern and any remedial action that has been taken. On the basis of the feedback, the personnel can take the appropriate corrective action over the course of the next week so that their safety performance improves.

As previously discussed, management has a vested interest in improving health and safety because it reduces costs and lost production time. It is, however, important that personnel are encouraged and given praise to take the necessary actions for them to improve their performance. If it is perceived that managers or supervisors are trying to impose their will, personnel will lose confidence in the behavioural approach and in their manager's commitment to health and safety.

Some people may feel that any behavioural intervention would stand a better chance of success if some form of incentive were used. However, research suggests that incentives may be of only short-term benefit, as rewards may be seen as ends in themselves, and may actually hinder the internalisation of safe attitudes in the long term. Linking safety performance to rewards may compromise objectivity of scoring; and *should* personnel be paid to be safe?

Summary of the behavioural approach

After reading this chapter, you may feel the behavioural approach is somewhat naïve, and be sceptical of it. Nevertheless, if it is used correctly, the improvements can be dramatic, as can be seen in the studies referenced in this chapter.

Goal-setting and feedback can be used to produce large improvements in safety performance. Available evidence suggests these improvements may be caused by any of the following:

- The existence of site observers monitoring aspects of safety on a regular basis, using a detailed checklist, serves to highlight areas of health and safety requiring management input – areas that might otherwise have gone unnoticed.
- The presence of safety performance goals and feedback serves to motivate site personnel to prioritise safety.
- The goal-setting and feedback activities improve the level of communication on site, and encourage greater discussion of health and safety issues in a manner akin to Total Quality Management.
- The goal-setting and feedback sessions permit direct communication between the top of the management hierarchy and the workers, which has the effect of improving senior management's understanding of day-to-day site health and safety issues.
- Health and safety provides a relatively neutral platform for the debate on safety- and production-related issues, which might not have otherwise become apparent. This can have other benefits in the form of improvements in safety, productivity and quality.
- Commitment (supportiveness) of site management enhances the effectiveness of the goal-setting and feedback approach.

The behavioural approach promotes a positive safety culture, using the 'four Cs' of OHS (G 65):

- *control* – proactive rather than reactive safety management;
- *co-operation* – stimulate teamwork by involving personnel in planning and implementation;
- *communication* – improvement in communications across all issues; and
- *competence* – improvement in safety awareness.

However:

The behavioural approach is *not* a quick fix as some managers and safety professionals believe. It must work in harmony with existing health and safety systems, and not as a separate entity and requires a culture change in health and safety management style from reactive to proactive. In order to succeed, the behavioural approach requires a safety culture that allows the concepts of consultation, co-operation and communication to flourish and requires that management be able to recognise and praise good safety performance, rather than focus on attributing blame for poor performance.

Discussion and review questions

1. The management of health and safety is perceived by some to be a continuous fight with human nature. Discuss how it is possible to move away from this perception and identify motivational factors that can be changed to encourage safe behaviour.
2. What do you change first, an individual's behaviours or attitudes? What influences this decision?
3. Discuss how a 'typical' construction organisation could make the move from traditional safety management (reactive) to a behavioural continuous improvement approach (proactive). What could prevent this move?

Chapter 9

Innovation and IT in OHS management

Introduction

Innovation is a key element in continuous improvement. In the business world of the construction industry, there are many opportunities for innovation in process: for example, in project delivery strategies or OHS management systems. Hence, this chapter brings examples of innovation, many in the context of the IT platform, as a mechanism for driving continuous improvement. Innovations in terms of product are also highlighted, and the chapter makes use of a series of example case studies. The reasoning behind this mode of presentation is that innovation is a somewhat intangible concept and that the use of practical examples of innovation provides a sound basis for understanding.

Technology and innovation

It seems the vast resources put into the development of technology and into the driving of innovation are almost invariably directed at the production process itself rather than the OHS management system. This chapter attempts to address how innovation and technology, and specifically IT, can assist in improving the OHS management process. The 'bottom line' in this argument is that innovation and technology implementation need to be seen to be worthwhile. How the impact of an innovation is measured has been a problem taxing accountants and managers for many years. A similar problem has dogged progress in OHS management systems in the construction industry for many years also.

The five key issues addressed in this chapter are:

1 design
2 technology
3 information technology

4 training and
5 decision-making.

Innovation is viewed as both product and process innovation. One might suggest that this is a very diverse set of elements to consider in terms of innovation and technology management. However, they are all crucial to the effective and efficient running of an organisation. As such, they are vitally important for the implementation of an effective OHS management system.

A good example of a process innovation, requiring technological changes, is shown in the following case study (source: Innovation project 43, http://www.m4i.org.uk). The developer was Slough Estates (www.sloughestates.com), a very experienced UK developer with its own contracting capabilities, and Cirus Ltd (www.cirus.org.uk) was the trade contractor. In this project, the designers and contractors got together before construction commenced and decided that productivity and safety could be improved by minimising external scaffolding, which not only provides access for workers, but can also restrict access for materials, particularly large panels. In order to ensure that workers could get access safely, a completely new piece of plant was designed. The outcome is summarised below:

Case study 9.1: Example of innovative construction process improving safety and productivity

Elimination of external scaffolding

Construction was undertaken with no standing external scaffolding. This was implemented by employing a specialist cladding detailer, a cladding specifier and then procuring all the materials using in-house buyers, from lists prepared by the detailer. Materials were fixed by labour-only sub-contractors who were dedicated to this work and so 'learnt' quickly, and productivity improved dramatically.

Results

1 Accidents were reduced compared to the norm for similar projects.
2 The speed of construction was increased because areas of external works were not 'sterilised' by standing scaffolding.
3 To erect the large glazing units, the curtain walling contractor designed and built a specialist piece of plant which is safe and is now used on other projects.

Case study 9.1 (Continued)

4 Less supervision of fixers was needed.
5 Continuous improvement was achieved via feedback from fixers.
6 Costs were significantly reduced (approximately 15% below usual industry rates).

Lessons learnt

The structure should be designed to fit the cladding. Integration of the cladding designer with the in-house architects provides a route of continuous improvement in this area of construction. Cladding panels had to be redesigned to enable positioning by mobile access platform. Access points, vulnerable to damage by mobile access platforms had to be installed last. Prior to the commencement of construction, the site had to be made flat and level in order to use the mobile access platforms.

Transferability

This method can be implemented on any project if the design (especially of the cladding) is thought through in terms of construction methods but only if the client, designer and contractors are willing to review their process together. Early consideration of construction methods is essential.

OHS management as a business driver

In the general management field, it is possible to devise a model that deals with innovation and strategic technology management. Hampson (1997) adapted the model of Rosenbloom and Burgelman (1989) in order to undertake a strategic analysis of the technology management and innovation process within construction (Figure 9.1). This particular model had one of its foundations in the work of Porter (1985) and his *five forces* model. Effectively, this model provides a holistic view of the strategic management of technology and innovation, and is the basis of a SWOT analysis of any organisation and its products. (A SWOT analysis investigates and analyses a company's strengths, weaknesses, opportunities and threats, focusing both internally and externally. For a good guide to conducting a SWOT analysis see Khurana 2003.)

The external forces acting upon the organisation can be identified using Porter's five forces, and the extent to which innovation is managed and integrated into the organisation can be assessed using the model. In addition, one might use other tools, such as the Value Creation Index (developed by The Cap Gemini Ernst & Young [CGE&Y] Center for Business

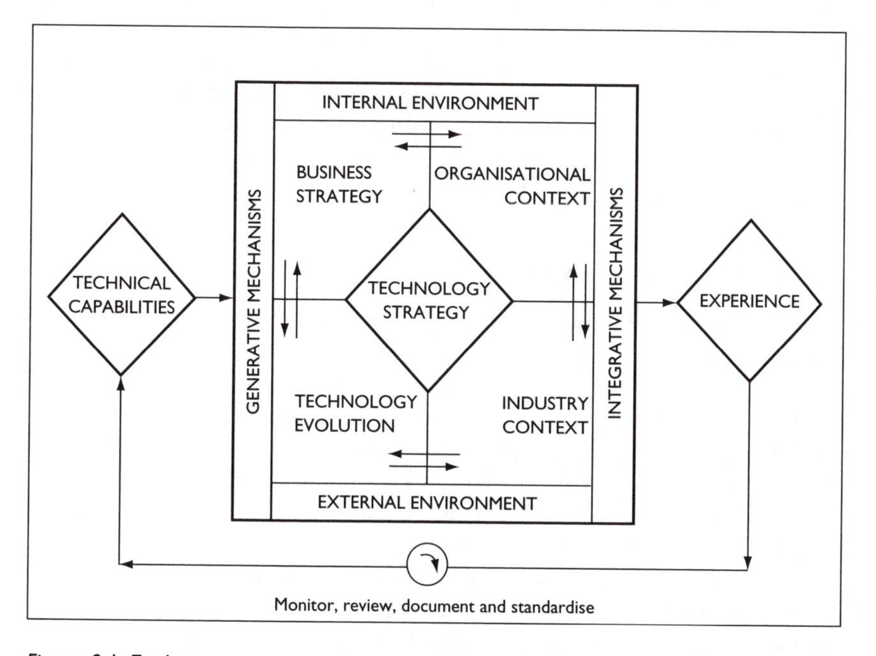

Figure 9.1 Evolutionary process framework for technology strategy (Rosenbloom and Burgelman 1989; adapted from Hampson 1997).

Innovation [CBI]) or the Intangible Assets Monitor (Sveiby 1997), to assess how readily innovation and technology improvement in OHS can be generated and accepted in companies involved in construction. Business focus in North America and Europe has shifted over the past decade from financials and productivity to valuing intangible assets – investigating how an organisation's non-financial performance can be valued in the market. Some of the intangible assets in a company are:

- employee talent
- management quality
- organisational culture
- staffing stability (low turnover and ratio of seniors to juniors)
- technological capabilities
- strategic alliances
- workplace environment, and
- company reputation.

These drivers are important for all aspects of business. However, they become essential in a new paradigm of business performance assessment that is strongly based on intangible drivers that stimulate innovation and ideas, and lead to reputation – not just the conventional measures of earning and assets base. This emphasis on the human capital of the business – its knowledge, knowledge management and reputation as an employer – should bring OHS to the forefront as an asset that can be valued. This being the case, OHS management can become a driver of the business, and can be seen as adding value to the business' valuation; this will be a very positive change for the construction industry.

Role of the design engineer

The technology management and innovation process take place within the general context of the construction industry and, in particular, in an industry where professional engineering practice and professional construction management are regarded as two separate disciplines. The changing paradigms of engineering practice and their impact on OHS management lead us to the view that the role of the design engineer is integral to positive OHS outcomes in the workplace. It has been noted in a number of studies of engineering education that the study of ergonomics within this education system is very basic, if not non-existent (Smallwood in Rowlinson ed. 2003). It seems now that there should be strong support for the teaching of ergonomic principles in the undergraduate engineering curriculum if we are to bring the design process into line with the construction process in terms of OHS. The issue of integrating OHS into design was dealt with in detail in Chapter 4. Whilst emphasising the importance of ergonomics in OHS. management, it should be made clear that the problem cannot be fixed merely by looking at the ergonomics of the issue but also by implementing a process whereby work practices are changed (administrative controls) and technology is introduced (engineering controls). Thus, the application of the scientific principles of ergonomics must go hand in hand with proper administrative organisation and the implementation of effective engineering controls if innovation in, say, hazard analysis is to be effective. This is the strength of the Hampson model, in that it draws all of the business activities together to provide a mechanism for stimulating, developing and implementing innovations, be these innovations in processes or innovations in products.

Barriers to innovation

Again, however, the contextual situation within which innovations and new technology are implemented must be considered. Lingard and Holmes (2001) discuss the barriers to the implementation of technological controls for OHS risks in small business. They suggest that, in small business, too much emphasis is placed on workers' behaviour and that the adaptation of technology to reduce OHS risk is often overlooked or believed to be too costly to implement. These perceptions are likely to impede innovation in the control of OHS risks, and create an environment in which workers are resigned to living with risks as 'part of the job'.

Information systems

Whole-life issues are also important in relation to OHS in construction. For example, with regard to as-built drawings, Tang (2001) states: 'The lack of accurate as-built records of some underground utilities and the prolonged process for obtaining excavation permits have an adverse impact on project delivery.'

This has OHS implications because inadequate information about underground services and facilities can pose a serious risk to those performing excavation works. Tang recommends that 'public sector construction clients should take a lead in developing an efficient information system on underground utilities and streamlining the existing procedures for processing excavation permits'.

Tang discusses OHS issues in the construction industry but as-built drawings are an OHS issue in themselves. By carefully mapping buildings, and particularly building services, risks, such as electrocution and gas explosions during excavation, refurbishment and demolition can be more effectively managed.

If the utilities are mapped using 3DCAD systems then the process of OHS management will become even more efficient (a discussion on 3DCAD will follow later in the chapter).

Tracking performance

Incident information systems are an essential part of OHS management and provide the data on which analysis of accident trends and management failings can be based. Such systems need to be robust and provide rapid feedback on current trends. IT has a very important role to play in this respect. By collecting incident data remotely, say by using an intranet-based data capture system, incident data can be analysed and tracked almost in real time. The following section looks at accident investigation in detail and indicates where IT can contribute to improvements.

Accident analysis

Accident investigation is a key element in accident prevention. It is important to learn from experience and to be able to identify commonly recurring 'themes' in accidents in order to be able to devise new prevention methods. Many organisations have recognised this fact and have developed detailed and sophisticated incident reporting procedures. In most developed countries, detailed incident reporting is a legal requirement.

Analysis of safety performance does not have to depend solely on the analysis of data from accidents that have happened. More positive assessments of safety performance can be undertaken, and these are discussed later in this chapter; however, a brief discussion of accident analysis will be given here.

Accident records fall into two types: those required by statute and those required by good management practice. One can categorise records into two main types: *reactive records* and *proactive records* (HASTAM 1999). Reactive data records are incident reports involving near misses or dangerous occurrences, such as:

- accident reports involving injury;
- accident reports involving damage;
- prosecutions under government ordinances and improvement notices; and
- insurance claims.

Proactive records include:

- hazard reports;
- personal protective equipment issue records;
- training records; and
- OHS inspection records.

All of the records maintained on site (and at head office) should be used to assess whether the OHS performance of the site or company is improving or deteriorating. Obviously, there is no point in maintaining records if these records are not adequately analysed. However, as can be seen above, there is a wide range of data that can be collected and analysed. By developing an IT package that can deal with all of the above inputs (including data capture, data aggregation, data analysis and report generation), management can be informed very quickly of current trends in accidents and incidents, and should be able to interrogate the database in order to formulate new OHS initiatives and monitor their effectiveness.

The strength of an intranet is that it can give access to the same data to the Managing Director, the section head, the OHS officer, the project manager and the site foreman, without making the information available to the public. By making these data immediately available to all levels in the organisation, OHS can be brought to the forefront of people's minds and, with appropriate encouragement, innovative OHS solutions can be developed.

Objectives of reporting systems

The objective of an accident reporting system is:

- to monitor accident rates;
- to identify accident causes;
- to monitor the effect of on-site OHS initiatives; and
- to estimate the costs of accidents.

Again, by providing access to the OHS statistics across the organisation in 'real time', the immediacy of OHS management issues can be emphasised and a proactive approach facilitated by the structured use of IT.

Benchmarking

Benchmarking is an external focus on internal functions. Benchmarking can be used to assess a company's performance as far as its OHS management system goes compared with industry best practice and compared with competitors. Benchmarking can be undertaken as an input or an output analysis. Atkin *et al.* (unpublished 2000) have developed an OHS management benchmarking tool which assesses contractors' OHS management systems inputs. In Australia, Trethewy *et al.* (2000) have developed a benchmarking tool which assesses a contractor's OHS inputs. By combining these two with conventional measures of safety performance such as accident rates and audit results, a contractor can get an overview of its position in both output and input terms.

Figure 9.2 shows the typical output from a benchmarking exercise for a large contracting organization using the instrument developed by Atkin *et al.*

It can be seen that the system developed by Atkin *et al.* uses four key dimensions of OHS management

1. human resource management;
2. implementation of OHS management procedures;
3. specific OHS-related project objectives; and
4. organisational management for OHS.

This tool, based on an Excel spreadsheet, can give almost instantaneous feedback on the current state of the OHS management system in a company or even on a particular site. This is a quick and powerful IT tool for focusing attention on key dimensions of OHS management and giving pointers for improvement. Such a system can be run on a laptop or a PDA (personal digital assistant, such as a Palm®) and so is set to become a powerful, IT-enabled OHS management tool.

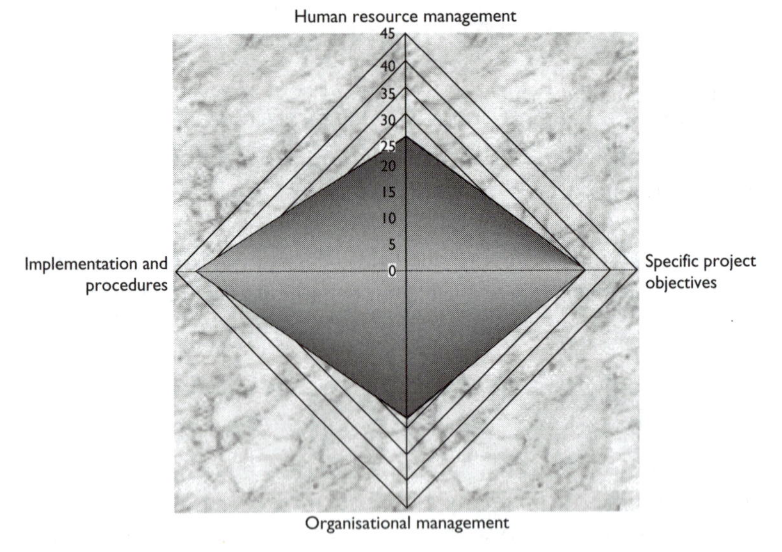

Figure 9.2 Typical output from an OHS management system input audit.

Other audit tools, such as construction CHASE and SABRE also use Personal Digital Assistants (PDAs) for data collection and analysis, and it is certain that much of our data collection on site will be done almost automatically using similar devices. (Undoubtedly, the PDAs that we are writing about now will be superseded by PC tablets and other technologies within only two to three years.) Technologies such as bar-coding will also facilitate this data collection process, as will voice recognition systems which will allow managers and supervisors to dictate reports of situations directly into their IT on the spot.

It is essential that company-wide OHS management systems are developed to take advantage of such technologies; the data collected need to be integrated and analysed virtually automatically in order to produce regular, weekly or even daily, exception reports in order to stimulate the process of continuous improvement in OHS. In this role, IT can be a driver for continuous improvement by facilitating the provision of rapid, focused feedback company-wide.

Examples of a benchmarking approach

In the UK, a benchmarking approach was used in the construction of the Greater London Authority headquarters. The use of a scorecard or 'league' of the best and worst OHS performers on site enabled the main contractor, Mace Ltd, to pinpoint in its supply chain specific areas in which OHS required improvement. The demonstration of this method of improving OHS highlights the system itself and the savings made from its implementation (source: http://www.m4i.org.uk, project 219). Mace Ltd (http://www.mace.co.uk) have taken advantage of their interest and expertise in OHS to the extent that they offer services in this area, such as:

- management and workforce training;
- audit, gap analysis and gap bridging;
- full policy and management system compilation;
- independent construction project auditing; and
- independent site inspections for contractors.

Thus, a recognition of the importance of OHS has turned into a business opportunity for this company.

Mohamed (2003) gives a detailed account of how a balanced scorecard approach can be used in construction OHS. He argues that the tool he has developed has the potential to translate an organisation's safety policy into a set of goals across what he terms as four perspectives: *management*, *operational*, *customer*, and *learning*. These goals should then be converted into performance measures intended to produce a strategic focus on safety across the entire organisation. The perspectives represent all stakeholders, and so ensures a holistic view of safety. The tool is intended to be used for strategic reflection and implementation.

Mohamed argues that by selecting and evaluating appropriate measures in each perspective, functional requirements can be identified and actions taken to attain goals identified in the process. The balanced scorecard approach has the potential to enable construction organisations to pursue incremental safety performance improvements through the use of a 'strategy focussed organisational paradigm' (Kaplan and Norton 1996), which values the company's intangible assets. Hence, the movement towards more open evaluation of a company's worth through valuing its intangible assets is, in the long term, likely to lead to major and sustainable improvements in construction OHS. After all, construction companies have little of value on their books other than their human capital.

Communication and information

Mobile phone messaging technology, SMS, has a great potential to be used as part of an organisation's OHS management system, as it can instantaneously deliver important and urgent OHS information, as well as delivering regular reminders, hints and tips concerning OHS. This concept is premised on the possession by virtually all workers on site of a mobile phone, and the difficulty of locating or contacting most workers on a construction site during the course of a day at work. As a consequence, SMS provides a unique and powerful opportunity to be used as a cheap OHS management tool. How should SMS be used in this context, and in what areas can it be most effective?

The focus of such an innovation is likely to be construction foremen and gangers, the node from which much site information flows. Recent work in small-world theory postulates that information is disseminated through an organisation by means of 'well-connected' nodes (Barabasi 2002) in a network. Networks may take many forms, such as the Internet, the old-boy network, a company's supply chain and so on; each is a network, but with differing properties. In the situation discussed here, the network is an information network and the node, the foreman, can be used more effectively through prompting using SMS.

SMS could be used to deliver messages to workers in the following categories:

- emergency scenarios, such as fire, rainstorm or typhoon warnings;
- warning scenarios, such as a planned power outage in a particular building section;
- scheduling meetings, such as reminders concerning scheduled OHS meetings and tool-box talks;
- reinforcement of OHS initiatives, such as broadcasting general OHS information or delivering topical slogans;
- trade-specific scenarios, such as broadcasting OHS reminders to workers in high-risk situations (for example, working at height, trenching works);
- OHS bulletins publicity broadcasts; and
- target setting and reporting of OHS performance.

Such an approach is currently being investigated in a number of countries, but the major impediments to implementation are worker attitudes and their 'wariness' of management intentions. Thus, the human element will always be a factor moderating the uptake of innovative uses of new and old technologies.

Mobile phone technology provides a range of opportunities for construction as it develops apace. Another potential innovation, for tunnelling or other high-risk projects, is the use of biometrics for monitoring access and egress to the site. Palm scanners are already in place in order to monitor entry and egress to Hong Kong sites, such as the Cyberport project, and such systems are much more secure than, say, the old tally system for tunnelling.

Using databases for risk reduction

'Knowledge is power' and 'lies, damn lies and statistics' are important concepts in OHS management. A well-constructed, up-to-date database of incident data that can be readily interrogated at all levels of the organisation is a powerful OHS management tool that can enable continuous OHS improvement.

However, we need to be aware of the limitations of our dataset, and be sure that we are drawing valid conclusions from our analysis. Hence, the use of a powerful, up-to-date intranet accessible throughout the organisation, with data capture and analysis capabilities, is a prerequisite for modern management. This applies to both contractors and clients – the following example of the Hong Kong Housing Authority system (reported by Lingard and Rowlinson 1997) illustrates this point.

Example of the Hong Kong Housing Authority system

In the early 1980s the Hong Kong Housing Authority decided, through one of its principal directors, that safety performance on its sites was well below acceptable standards. However, when the Authority attempted to analyse causes of accidents based on reports returned to the Hong Kong Labour Department, it found that the report data were of insufficient detail for it to be able to analyse the nature and underlying causes of accidents on its sites. Hence, a consultancy was commissioned from Hong Kong University to devise a bespoke reporting system for the Housing Authority, and the completion of accident details was made compulsory in the Housing Authority contract. Lingard and Rowlinson devised a reporting system based on an epidemiological approach to accident analysis. This system allowed for a detailed analysis of the multiple causes of accidents, and identified the underlying management actions and other issues which contributed to the accident. This system was implemented for all new work constructed by the Housing Authority and an example of the reports produced can be seen at the following Web page: http://www.hkusury2.hku.hk/steve/.

Based on the analysis conducted at Hong Kong University, a number of areas of weakness in terms of safety performance were identified. The most significant of these were access to heights internally, the use of bamboo scaffolding, steel flexing and bending operations and use of hand tools. These issues were then investigated in more detail by interrogating the database which was built up over the months by the research team. By undertaking this research, the Housing Authority were able to pinpoint changes in their contract and its specification which were required in order to achieve an improvement in safety performance.

The Authority ran a regular series of seminars where the results of the analysis were reported back to all of their main contractors, and the contractors had the opportunity to quiz the research team about the exact nature and details of the problems identified. Because of the impact of this research, the Authority made certain changes to its ways of operating, including:

- implementation of double row (as compared to single row) bamboo scaffolding on all of its contracts;
- implementation of a Green Card system which ensured that all operatives were adequately trained when they first started on a construction site; and
- other initiatives which cumulatively led to a continuous decrease in the accident rate on the Authority's construction sites.

The newly devised accident reporting system was seen to be a major success from the Housing Authority's viewpoint, in that it assisted in determining new initiatives to implement on its sites, and from the contractors' viewpoint, in that they were able to learn from the experience of other contractors, pinpoint unsafe practices on their sites and make appropriate improvements to their safety management system. In addition, the accident reporting system led to the Labour Department reviewing its data collection procedures, and to the adoption of a completely new accident report form, based very closely on the accident report form developed for the Housing Authority.

Thus, real-time information was being fed back via this reporting system to the Authority and its contractors and, indirectly, this led to the achievement of continuous improvement in safety performance. Unfortunately, although an online version of the accident report form was developed, at the time few contractors had managed to implement Internet access on their construction sites, and so the downloading and uploading of reports at site level was never fully implemented. However, given the current situation in the industry, such an online uploading and downloading system could now be effectively implemented in most countries around the world. This would facilitate the rapid analysis of results, with feedback to all concerned, and the highlighting of any dangers and problems in the feedback report.

Role of method statements

Method statements indicate how a particular element in the building or structure will be put together. Method statements must take into account construction feasibility, use of plant and equipment and the nature of the materials being used. Construction method statements provide an ideal opportunity to review the construction process and to determine what is the most appropriate and safest method of construction. If an attitude of *design for OHS* is embodied in the production of the method statement then method statements should enable safer construction to take place. However, it is important that the method statement so devised is actually followed in its implementation. Hence, one might demand that, in future, method statements deal with all the items of personal protective equipment, the method of access, equipment to be used and other items that directly impinge on the OHS of workers.

Method statements are a pre-requisite for any construction project. They indicate in detail how each element of the works comprising the project will be constructed, and can go so far as to provide detailed layouts of temporary works and associated calculations. OHS should be a prime consideration in devising such method statements; they provide an opportunity for hazard identification, hazard management and scope for building safe practice into the design. They should be devised with the contents of codes of practice in mind.

Risk assessments and method statements are intimately linked, and by building up a common database of standard method statements and risk assessments the contractor (or designer, or client's representative for that matter) is in a position to quickly assess the 'lie of the land' in relation to risks and then get down to a more detailed analysis specific to the particular project in hand. Thus, the OHS risks can be value engineered (VE) at the same time as the design and construction processes. The establishment of a database that hyperlinks method statements and risk assessments will allow the VE process and the risk assessment process to move forward together in an efficient and co-ordinated manner. If the costings for the method are linked into this database, as well as the costings for risk amelioration measures, then a true picture of the costs of safety and efficient production can be developed. It will often be the case that efficient, well-engineered construction methods will be safer and cheaper. However, current practice tends to separate all of the elements discussed above, mainly because of the sentient differences between the professions, the adversarial nature of the traditional construction process and the reliance on individual expertise to translate two-dimensional drawings into three- and four-dimensional models. The use of 4DCAD can overcome many of these shortcomings, and at the same time incorporate the databases mentioned above into one powerful visualisation of the project. This concept is discussed in detail later in this chapter.

For often-repeated processes, such as flying form erection, piling operations and caisson construction, it is possible to devise and circulate a standard method statement to be adopted on each and every project. However, each project should be judged on its own merits, and the site, surroundings and pertinent variables, such as water levels, should be carefully checked and analysed and amendments made to the standard method statements before they are used. It is just as important if such statements are to be effectively used that workers are periodically sent on refresher courses to remind them of the details of the processes and checks they are expected to conduct, that audits are conducted to check that procedures are being followed and that the procedures are themselves regularly reviewed and improved in the light of new technologies. An example of the importance of method statements and communication is given below from innovation project 355 on the m4i website.

Sharing specialist subcontractor expertise

The following case study shows how IT can be used as a tool to stimulate the capture and retention of knowledge and information through a virtually shared resource, the spreadsheet.

Case study 9.2: Asbestos removal contract

1. The contract was let on the basis of best practice from a previous contract, with method statements, particularly in relation to OHS, overriding the agreement of the price. Continual improvement was built into the review procedure. The objective was to ensure an excellent level of OHS performance from all individuals and teams and increased tenant satisfaction.
2. A Key Performance Indicator, relating to the working time within each dwelling, was agreed, in addition to the Indicators of contract, time, cost and OHS on site. The objective was to improve performance by providing tenants with a high level of certainty as to the amount of disruption that they were likely to experience and to minimise it, thereby reducing contractor's access difficulties.
3. All parties agreed that as far as possible they would use the same staff as on the previous contract and to maintain them until the end of the contract. The objective was to aid the smooth running of the project through better teamwork, greater trust, closer liaison and improved communication.
4. All properties were surveyed in the light of experience gained from the previous contract. This information was scheduled on a computer spreadsheet against the description of works in the specification. The spreadsheet was to be shared with (and could be e-mailed to) all parties in the development team. The objective was to anticipate problems on site, to assist cost planning, programming, purchasing, monitoring and valuations and to speed up agreement of the final account.

This case study raises an important issue: that of knowledge management, and creating a learning organisation. The goal of knowledge capture and retention is central to good business management, but has proven to be an elusive goal in industries that, like construction, are project based and characterised by the transience of team members.

It should be noted that the innovation here is not just brought about by a product, the virtual knowledge base, but also a process change – the way in which people worked together in a collaborative, partnering manner – and how emphasis was laid on continuity of working relationships. These latter two issues are commonly overlooked, but are particularly relevant to OHS: by fostering a collaborative, sharing atmosphere amongst project team members, a proactive attitude towards all problem-solving, including OHS, is stimulated. This is the soft infrastructure that must be developed for project teams from different organisations to work together effectively.

In this instance, IT was a driver that enhanced this collaborative atmosphere. In fact, this collaborative sharing of information on best practice is the basis behind 'Rethinking Construction', which promotes the m4i website.

Inspections and audits

Inspections of worksites should take place continuously. Inspections are required in order to keep track of progress but there is no reason why such inspections should not always be carried out to check on OHS. This is a process in which every supervisor, and worker, can participate on a daily, and even hourly, basis. If a culture of OHS were engendered throughout an organisation, such inspections would be a perfectly natural part of the construction process. Planned inspections also have their place, particularly when dealing with items of plant and equipment such as scaffolding, hoists and cranes. Auditing is a different process, the objective being to monitor the performance of OHS systems. The scope and timing of audits will vary according to need and they may or may not be conducted by independent agents; whichever, they should always be conducted objectively. The output from them should be a report of current levels of conformance and effectiveness and a series of recommendations for future improvements and initiatives. Immediate action might also be called for in extreme circumstance. Such a situation would indicate a major problem with the OHS system. PDAs are an ideal IT gadget to enable this process to take place efficiently and effectively. By loading a simple tick box checklist onto the PDA, a structured and consistent measure of OHS performance can be produced at every inspection and it can be analysed and distributed electronically almost immediately.

Site OHS audits have become common practice nowadays. These are valuable elements of any OHS management system.

The role of audit

The role of the OHS management audit is to assess whether the OHS management system, as described in the documentation, is actually being implemented in practice. The audit attempts to measure performance in terms of implementation in a whole series of areas crucial to the OHS management system. A properly designed audit will deal with all of the components of the OHS management system. An example of how the implementation of an innovative audit system can improve OHS and productivity is given below (source: http://www.m4i.org.uk, innovation case 155, contractor www.willmottdixon.co.uk).

Non such High School for Girls, new classroom block: Health and OHS system

The project demonstrates how a structured approach to OHS can produce quantified evidence for comparison of sites. This information forms the basis for rewarding best practice and penalising indifference. On the Non Such project, there were no reportable accidents at all in some 32,000 hours worked on site over ten months. The site was awarded third place out of about 120 Willmott Dixon projects nationally. They maintained a safe environment for controlled open days and work experience for students.

Healthy productivity demonstrates that a well-managed OHS regime does not inhibit productivity and may well improve it. Productivity figures measured using the construction industry key performance indicators rank the Non Such project (KPI=80%) well ahead of the industry median (KPI=50%).

The OHS Inspection Scheme covers the inspector's role, inspection activities, awards, fines and penalties. It also includes a model inspection report and a model prohibition notice. A company OHS officer who is looking at the general appearance of the site and evidence of best practice inspects Willmott Dixon sites every 4–6 weeks. Sites are audited every six months against a rigorous checklist with 20 sections and 180 items. From this data, sites are ranked. Sites and individuals worthy of an award are identified, along with failing sites.

Penalties can be imposed on sites that score less than 75% grading in the six-monthly report, ranging from £100 for marginal cases to £2,000 fine for infringement of a prohibition notice. In effect this is an adjustment to the project's contribution but it is also taken into account in individual staff performance appraisals. Financial sanctions do not extend to individuals.

The complete package of incentives includes monetary rewards for site managers, based on six months' performance (£500 to the top five site managers and £350 to the next five). The OHS inspector can also recommend awards for effective work on difficult, high-risk sites. Top performing subcontractors are recognised with 'best subcontractor' certificates to the foreman and a letter to the managing director. The Non Such site manager received two awards, each £750, for achieving the 3rd best Willmott Dixon site nationally (January–June 2000) and also the best site manager in Hitchin business unit for the year 2000.

The nature of the audit

Audits can be either internal or external. An internal audit will be undertaken by the company's own staff, and will attempt to be as objective as possible in assessing the performance of the OHS management system. However, if a totally objective audit is required, then an independent auditor may be brought in to conduct an extensive, independent audit of the company's OHS management system. In reality, it is essential for companies to undertake both types of audit. The independent audit should be undertaken on a regular, perhaps annual, basis, and the company-based audit should be undertaken at regular intervals, say every six months or less. The whole aim of conducting an audit is to assess where weaknesses lie in OHS management systems and so aim for continuous improvement, the same principle which applies to Total Quality Management systems.

Training of auditors

One cannot undertake an audit without properly trained auditors. These auditors must have undertaken an auditing course, and must work with experienced auditors before qualifying as fully-fledged auditors themselves. In order for such a system to work effectively, it is essential therefore that the qualification and the training of auditors is undertaken before OHS management systems are implemented. Unfortunately, this is often a chicken-and-egg situation, and in Hong Kong, in 1999, there were only 16 properly qualified auditors for hundreds of construction sites. One can imagine the problems this brings about in terms of undertaking meaningful audits.

An example of IT in OHS audit: SABRE

SABRE is a new site OHS audit tool that was launched on 23 May 2002, and underwent trials on construction sites belonging to UK-based supermarket giant, ASDA. With the success of the trials, ASDA plans to use SABRE on all of its sites.

SABRE is designed to make construction site audits faster, simpler and consistently accurate. 'It is a proactive, hazard spotting tool that provides the sort of radical preventative approach needed to improve OHS on construction sites,' says BRE's Tony McKernan. 'I believe that it could help to reduce the rate of injuries and deaths on UK sites.'

SABRE's launch followed extensive field trials of the tool by FaberMaunsell, HBG, Laing, Carillion, Pearce Retail and others. It was developed with government sponsorship by BRE, working closely with industry and the Health and Safety Executive, and is supported by Sypol, an OHS specialist company.

The OHS tool is being operated on ASDA sites by FaberMaunsell in its role as planning supervisor and client's agent under CDM, with responsibility for CDM and OHS site inspections, for all ASDA stores.

SABRE combines a checklist, recording device and analysis tool in one easily transportable and robust piece of equipment, says Ashley Potts of FaberMaunsell. 'Having undertaken a number of inspections using the tool we feel that it provides a realistic measure of performance. I hope that it will soon provide the basis for national benchmarking populated by real users' results.' The system can be used on all construction sites, from minor building works to major civil engineering projects. It enables anyone with basic OHS training (such as that offered by CITB OHS courses) to gather data on working conditions, and instantly allocate simple scores to their findings.

A hand-held computer presents the user with a series of questions on all aspects of site OHS and covers, among other areas, scaffolding, cranes and lifting gear, welfare and regulations such as COSHH. Examples of these simple questions include: Is work adequately lit? Do ladders have level and firm footing? Are the materials covered by COSHH regulations used and stored correctly? The system allocates a score to each question, calculates the overall score for the site, or area under scrutiny, and generates immediate action notes.

SABRE's underlying principle is to encourage improvement by introducing an element of competition into site OHS. Sites that score poorly require immediate action to improve conditions. But more important is SABRE's allocation of pass scores ranging from 'acceptable' through 'well managed' to 'excellent'. With regular use, the scores can be used to demonstrate improvements, or identify areas of weakness.

(Source: http://www.bre.co.uk/services/SABRE.html)

Knowledge-based and expert systems

Recently, interest has turned to the use of knowledge-based or expert systems for industrial safety purposes. The purpose of knowledge-based or expert systems is to capture an expert's knowledge and re-produce the expert so that a novice can benefit from this expertise without having to learn by experience, which is a slow and error-prone process (Gaines 1990). Robertson and Fox (2000) suggest that knowledge-based or expert systems can be used to improve industrial safety in a number of ways. Some of these applications are described below.

The provision of regulatory advice. Knowledge-based or expert systems can be used to aid a user's navigation through regulations or safety guidelines, for example assisting the user to check conformance. Expert systems of this type have been used by the US Department of Labor Occupational Safety and Health Administration, to advise people of their OHS responsibilities. For example, a knowledge-based system has been used to assist people to determine which types of spaces are classed as confined spaces and whether they require a permit to enter.

Hazard analysis and avoidance. Knowledge-based or expert systems can also be used to provide advice about the extent of risks and the most effective ways to control identified risks. These systems are typically prescriptive and the quality and currency of expert information supporting them is therefore critical. In collaboration with Australia's CSIRO, the company 3M has developed a knowledge-based or expert system to assist people to identify the appropriate type of respiratory protective device (from among 100 products) for people working with one of 1700 chemicals.

Post-accident analysis and corporate knowledge. Knowledge-based or expert systems can be used to automate standard accident investigation procedures, ensuring that data is captured in a consistent way and stored for re-use. For example, the US Federal Highway Administration has developed a hand-held computer crash data collection system that uses expert investigators knowledge to advise investigators as to the best form of questioning relevant to given features of a crash. The Internet has made large-scale knowledge collection and sharing possible. For example, accident data can now be collected by hand-held devices and transmitted to central knowledge databases which can then generate reports and present these on the Internet. This is a very powerful means for sharing information, since it permits analysts to identify patterns and trends in incident occurrence and to become aware of new hazards sooner.

Recently, Davison (2003) developed a prototype knowledge-based system to assist construction designers to identify hazards, evaluate risks and specify suitable risk controls in their designs. The prototype provides health and safety information, which can be delivered in combination with computer-aided design (CAD) tools. The knowledge-based system enables building features, such as fragile rooflights, to be included in design documentation and checks designs to determine whether there are any OHS risks inherent in them. If risks are identified, the designer is alerted and provided with information about how to remove or reduce the risk. Initial tests found that designers found having easy-to-access OHS information was helpful and plans are now in place for the full implementation of the system.

The benefits of knowledge-based or expert systems are becoming apparent. As the construction design example described above demonstrates, such systems permit more thorough and easier checking of compliance against regulations and can deliver domain-specific expertise in risk analysis and control to people who may not possess this expertise. Knowledge-based or expert systems can also provide decision support, enabling rapid and timely OHS decision-making as well as facilitating the capture and analysis of incident information which can then be stored and retrieved for future use. It is likely that the use of knowledge-based or expert systems in OHS will grow in the future, bringing many benefits.

Visualisation

Taking technological solutions to present-day limits, the use of virtual reality (VR) or visualisation, is becoming very important as a future mechanism for improving construction site safety. Hadikusumo and Rowlinson (2002) and Rowlinson (2003) discuss the use of VR systems to assist in construction site layout and safety analysis. They address the issue as follows: 'Visualisation allows the representation of virtual product and process data.'

By *product* they mean the finished building and the components that are needed to make the building grow, such as formwork for concrete casting and pre-cast concrete elements.

By *process* they mean the sequence of construction and plant and materials needed for the building to grow. Thus, the VR model produces on a computer screen what the planner, estimator and safety manager visualise in their heads from the two-dimensional architectural drawings and their experiences of working on a construction site.

In this way, visualisation supports the user's goal of embedding the *design-for-safety* process (DFSP) into the overall design by using theories of accident causation, as well as axioms derived from OHS best practice and regulations to identify hazards. This gives virtual opportunities related to the construction process that include the following:

- The safety engineer can do a virtual walk-through in order to analyse the proper design, for example, of scaffolding, safety nets and other temporary safety protection.

- The collision-detection facility offered in VR can be used to evaluate the planned construction process model from a number of perspectives, such as access space, ladders or other vertical access means and fall protection for materials.
- An illumination facility in the virtual model can be used to design proper lighting during the construction process.

Site layout

Hadikusumo and Rowlinson (2002) go on to discuss the use of visualisation in what they term the DFSP approach. Visualisation allows the user to interact with the data by a virtual walk-through, which is very useful for presentation of product and process data. The user can see the product from any position or location. The user can also walk inside the virtual product (building), which cannot be done by other means, such as rapid prototyping of a miniature of the building. Therefore, many members from different organisations can walk through the virtual product model and discuss the production process whilst the building's design is still to be finalised, in a DFSP mode.

Conventionally, a safe system of work is devised through six essential steps:

1 assess the task;
2 identify the hazards;
3 define safe methods;
4 implement the system;
5 monitor the system; and
6 review the system.

Unfortunately, despite legislative and contractual requirements, the reality is that in the construction industry most OHS hazard identifications are conducted at the site. In other words, they are conducted by visual site inspections at the construction stage. This pressure on the management team is really a product of the competitive tendering system. Often tender periods are as short as four weeks for complex projects and then, once tenders have been adjudicated, work starts very quickly on site, and planners and safety managers have inadequate time to conduct risk assessment studies thoroughly. This aspect of OHS management can be improved if OHS hazard assessment can be conducted as soon as possible. For this reason, a system has been developed to conduct OHS hazard identification before the construction stage; that is, at the completion of design stage.

The importance of OHS hazard identification in preventing an accident is stressed in the theories of accident causation and the OHS management systems used in construction site OHS practice today.

OHS hazard identification methods

There are several established methods of OHS hazard identification including:

- Preliminary Hazard Analysis (PHA)
- Failure Modes and Effects Analysis (FMEA)
- Failure Mode, Effects and Criticality Analysis (FMECA)
- Hazard and Operability Study (HAZOP) and
- Master Logic Diagram (MLD).

Some of these are covered in Chapter 5.

Summarising, the key issue in construction then is the medium – two-dimensional architectural drawings – used in the hazard identification task. This method has severe limitations because:

- it is difficult to interpret two-dimensional drawings as three-dimensional mental objects; and
- two-dimensional drawings only represent static information of a project design; in other words, the dynamics of construction processes are not represented.

This is a serious problem because OHS hazards are also inherited within construction processes. As the two-dimensional drawings are normally not delivered until just prior to the start of construction, the OHS team often does not have adequate time to develop a sound, comprehensive OHS plan.

Visualisation for OHS hazard identification

In order to solve these problems, the potential of visualisation technologies has been utilised in a trial project, based on a Hong Kong Housing Authority standard block. This technology has three benefits: interaction, immersion and imagination (Young 1996).

The interactive feature enables a user to modify a virtual world instantaneously. This feature can support a virtual site inspection in which a user can inspect a virtually real construction object (product model) and identify any OHS hazard inherited within it.

The immersion feature allows the user to see, as well as touch and feel, a realistic-looking world. Although the degree of immersiveness may vary from one VR system to another, a simple VR system can be used to represent a three-dimensional object that supports a 'what-you-see-is-what-you-get' (WYSIWYG) environment. This advantage can be used to eliminate the problem of interpreting two-dimensional drawings as three-dimensional mental objects, since all construction objects can be represented as three-dimensional objects in a VR world.

The imagination feature enables a developer of VR to create an application that can solve a particular problem. This feature provides significant benefits, since a VR developer can create a VR application to suit a specific project need. For example, this feature is used to represent the dynamic of construction processes that cannot be represented in two-dimensional drawings.

As a result of exploiting the benefits of VR, several significant advantages can be achieved in OHS hazards identification:

- Users (that is, safety managers or teams) can see virtually real construction objects easily since they are represented as three-dimensional objects; this condition enhances the safety hazard identification process by removing the difficult mental interpretation process of imagining two-dimensional drawings as three-dimensional objects.
- The virtually real construction project can be designed to represent the static and dynamic components of a construction project. This can support the hazard identification task since hazards occur within both the construction components and the construction processes.
- Virtual site inspection can be conducted even before the project design has been completed, allowing significant contributions to be made to the project's safety planning.

Hence, VR can provide a complete and ongoing evaluation of the construction process even as the project design evolves. This is a concept similar to simulation of a project, but its sophistication lies in its ability to present life-like construction components and processes and its facility for allowing the user to 'walk through' this environment as if it were real.

In order to develop a VR system that allows for visual site inspection to identify OHS hazards, two basic components are required:

- an OHS database of hazards; and
- VR functions in a software package.

OHS database

The main functions of the OHS database are:

- to assist a user to identify OHS hazard(s) inherited within the visualisation; and
- to assist a user to assign accident precaution(s) to prevent accidents.

The VR functions are the necessary behaviours and tools needed in the safety hazard identification task.

Keywords and checklists offer the most flexible way to identify OHS hazards, and in this system the construction components are the keywords. These are used to retrieve the possible OHS hazards from the database, for example a slab (a keyword) may inherit several OHS hazards, if:

- there is an elevation break in the slab, a worker might trip; and
- there is an unprotected opening within a slab, a worker might fall.

In the system, by clicking on the slab with the mouse, the user can search the database for possible OHS hazards related to the slab. The possible hazards are then presented as checklists, which can supplement the user's knowledge and experience and ensure all hazards are identified. This function also has a potential application in training workers to identify hazards for themselves on site.

The OHS database is designed based on a Construction Components/ Possible OHS Hazards/Safety Precautions set of relationships. One construction component can have many possible OHS hazards, and one possible OHS hazard can have many safety precautions. The advantage of using this relationship is that OHS hazard information related to a construction component and its process of installation can be attributed to the construction component, becoming obvious as the user walks through the visualisation.

VR functions

VR functions are needed to support the virtual site inspection. This is conducted according to the following sequence:

- The user walks through the virtual project.
- The user identifies OHS hazards related to components or processes.
- The OHS database lists possible OHS hazards.
- The user checks which possible OHS hazards are pertinent and adds any further hazards.
- The OHS database lists a range of accident precautions.
- The user chooses suitable safety precautions, and adds any additional ones.
- The safety precautions selected are documented in the risk assessment and in the project safety plan.

Thus, DFSP visualisation has the potential to produce, in a semi-automatic manner, risk assessments, a project safety plan and method statements.

There are certain technical issues that must be addressed when designing a VR system. The user must be able to walk like a normal human being and not float on air or walk on water. This problem can be solved by providing a terrain-following mechanism, a VR function, for doing the walk-through. Nor can the user be allowed to 'ghost' through walls and other solid objects. This can be achieved by providing a collision-detection mechanism, a VR function, to block a user from walking through a solid object. If a user finds that a situation encountered on the walk-through may present a safety hazard, they can click the mouse on the virtually real construction component in order to retrieve possible OHS hazards data from the OHS database. For this, a geometry-picking function must be created. The user might also need to take a measurement; such as if there is an unprotected edge in a slab and there is a potential for a worker to fall, it may be necessary to measure the height of the drop, so a VR tape measure is needed.

These four essential VR functions, then, must be created in order to support the virtual site inspection to identify OHS hazards:

1 collision detection
2 terrain following
3 geometry picking and
4 VR tape measurement.

nDCAD

Although the discussion above has focused on what is loosely termed 3DCAD, what we are really focusing on is nDCAD, that is, multi-dimensional CAD. The system presented above has a fourth dimension of time, in that we can see the building grow (see a demonstration at www.hkusury2.hku.hk/steve). However, each component in the building also has attributes, such as time, weight, cost, safety hazards associated with it and so on. We will shortly be able to move into a virtual world which is capable of automatically generating our project schedule, our bill of quantities, our method statement, our safety plan and so on. The opportunity for collaborative design and design for safety (in construction and use) is very close to becoming a reality.

A similar case study has been reported by BovisLendLease on the 'Rethinking Construction' website. In this case, three-dimensional graphics were used rather than a full visualisation (Figures 9.3 and 9.4). However, the ability to see beyond the normal two-dimensional drawings enabled careful consideration of engineering and construction issues. By drawing the team together, even in a virtual scenario, and presenting a three-dimensional model, a safe and effective solution was devised. In fact, BovisLendLease use what they call *iKonnect* as a virtual resource, where clients and others can post questions concerning difficult problems, and BovisLendLease can proffer solutions. This may well be a marketing tool, but it is an effective use of IT in dealing with problems of all types, not just OHS issues.

Figure 9.3 Virtually real construction components of the harmony type of Hong Kong housing authority's standard block.

Figure 9.4 Virtually real construction processes of the harmony type of Hong Kong housing authority's standard block.

Innovation: IT – 3D modelling case study

Computer modelling was used to plan and execute difficult bespoke earthwork support and concrete installation. This innovation enabled the project team to deliver a difficult earthworks package on time and to cost by modelling the working procedure using three-dimensional graphics. This enabled the project team at all levels to understand the process and procedure required for the works, and to execute them safely and without undue delay.

(Source: http://www.m4i.org.uk/innovation case 143)

Robotics

The word *robot* was first used by Czech playwright Karl Čapek in 1920, and is taken from the Czech word *robota* meaning forced labour. Robotics is an important area of OHS improvement. When dangerous substances have to be sprayed or when hazardous environmental operations have to be undertaken, a robot or remotely controlled machine can be used rather than exposing a human being to the risk. Where processes are tediously repetitive, and the problem of boredom and fatigue comes into play, a robot will perform much better than a human. There are many advances in robotics but most of these have been specifically directed at manufacturing industries where the process is much more static than in the construction industry. However, as robotics develop and artificial intelligence systems improve, the scope for use of robotics on construction sites to deal with dangerous and repetitive tasks becomes increasingly a reality. Areas in which robotics might prove beneficial include:

- demolition;
- excavation and earthmoving;
- paving;
- tunnelling;
- concrete slab screeding and finishing;
- operation of cranes and autonomous trucks;
- welding and positioning of structural steel members;
- fire resisting and paint spraying; and
- inspection and maintenance.

Areas where robotics are now making an impact on safety are developing rapidly. Earthmoving machines and other plant can be run remotely; see, for example, the Mechatronics group at Lancaster University in collaboration with JCB, the plant manufacturer. However, there is always a danger that a worker or structure will be hit by such automatic vehicles. In order to address this issue, work has been done on analysing the images provided by cameras mounted on the machines so that they can 'recognise' humans and structures, and avoid collisions.

As reported by the Mechatronics group:

> Safety is a crucial issue when large mobile robots are required to work alongside people. At the heart of our research is the concept of a safety manager which is conceived as an independent entity whose job it is to monitor the environment, and give permission for all behaviour which has a safety critical component. It is argued that this is a viable approach for complex non-deterministic systems. The work is divided into six work packages: safety analysis, safety requirements specification, hazard partitioning, information requirements, system architecture design and prototype development. The last stage involves a trial implemented on our robot excavator.
>
> (Source: http://www.comp.lancs.ac.uk/engineering/research/mechatronics/mobile.html)

At a lower level of technology, crane drivers and plant operators now have equipment fitted with cameras and small VDUs (Visual Display Units) in their cabs thus enabling them to see previous 'blind spots'. Nowadays, this is a very basic technology that we take for granted, but such innovations have the potential to reduce accidents significantly and make workplaces safer. Similarly, there is now great potential for using GPS (Global Positioning System) to direct automatic plant.

As a postscript, it seems that the best progress being made on construction robots is by NASA in its Jet Propulsion Laboratory. Recently, it reported the successful testing of two 'rovers' that were to construct a solar power station on Mars:

> The Robotic Work Crew can traverse uneven, hazardous terrain. The crew visually detects and tracks its goal, identifies nearby objects in its path and works collectively to avoid obstacles. Throughout this process, the robots constantly update each other about payload forces and motions as felt at their respective grippers. If the beam is slipping, the rovers collectively sense the problem and compensate. The robot team robustly fuses this information into a bigger picture, coming up with a best cooperative control solution. The JPL researchers say the rovers function much like a construction crew without a foreman. They note that once the system has been programmed with basic behaviours and coordination models, it is a truly distributed and autonomous intelligence across the robot team that gets the job done, responding to situations of the minute.
>
> (Source: www.spacedaily.com)

Discussion and review questions

1. There appears to be a basic assumption in the construction industry that innovation is a good thing and should be encouraged. Discuss the extent to which innovation is the key to improving the management of OHS in the industry.
2. To what extent can visualisation revolutionise the construction procurement process and redefine the roles and responsibilities of the various construction professions.
3. Consider the advantages (and disadvantages) of using the Internet, mobile phone technology and hand/palm-held computers in managing OHS on construction sites.

Chapter 10

Conclusions

One of the main lessons to have come from the writing of this book is that OHS problems are complicated. There are no easy answers to the difficulties we face in the construction industry, and to look for easy solutions would be a foolish and unrewarding task. Thus, in this concluding chapter we try to indicate what we believe is the direction for the better management of OHS in the construction industry. In so doing, we highlight what we consider to be some of the most important issues to be tackled; whether these be positive steps, such as the incorporation of knowledge management into organisations, or the pointing out of flaws in current approaches to the management of OHS in projects. We offer no glib solutions. Hence, the issues that we discuss in this concluding chapter are predicated on the fact that it will take considerable time and effort to improve OHS, through systematic management, in some sectors of the construction industry. As a consequence, one of the themes of this chapter is that participants in the construction industry should look inwards at how they organise their operations but, at the same time, be prepared to consider the external environment in which their organisations operate. Failure to maintain this strategic external focus in the management of OHS puts organisations at risk of legal liability and business failure, through public condemnation and an inability to recruit and retain skilled employees in the future.

This chapter identifies some new directions for the management of OHS in the construction industry. In particular, we comment upon the need to change the culture of the construction industry from one in which risks are regarded as an inherent part of the job to one in which employees at all levels actively care about not only their own OHS, but also the health and safety of others. We argue not only that senior management must demonstrate commitment to OHS, but also that middle managers and supervisors play a key role in creating a safe and healthy work environment. The critical role of supervisors' leadership in OHS means that OHS management systems should not be centralised and bureaucratic. In the decentralised project-based construction industry, it is particularly important that OHS leadership be demonstrated at a local, site and work crew level. We suggest that traditional emphasis on occupational injury and, more recently, illnesses recognised by employees' compensation schemes will be insufficient. Rather, in the demanding, high-pressure construction industry, excessive workloads and longer-than-average hours pose a threat to workers' safety, and can also negatively impact upon workers' overall health and well-being. We suggest that, in the future, construction organisations will need to consider the impact of job demands on workers' mental and physical well-being. Lastly, we suggest that the emphasis on corporate social responsibility and ethical conduct in business will increase the pressure upon organisations to treat workers with respect. There is a moral imperative to manage OHS effectively, and failure to do so will increasingly threaten a firm's 'license to operate'. High-calibre existing and potential employees will be reluctant to work for organisations whose OHS record is poor and socially responsible clients and shareholders will similarly avoid doing business with or investing in these organisations. Thus, we conclude that, in the future, the management of OHS will play an important role in the management of construction projects and OHS will be an issue of strategic importance within construction firms seeking to maintain a competitive edge through attracting and retaining a highly skilled and motivated workforce.

Organisational learning

At the start of this book, we described how the construction industry fails to learn from its experience of occupational injuries and illnesses. We described how incidents of a similar type occur with alarming regularity, causing injury and illness to construction workers. Furthermore, there is considerable similarity between OHS incidents occurring in construction industries throughout the world. The need for the construction industry to collectively learn from its experiences and use incident information to prevent future injuries or illnesses is clear, yet the construction industry appears to be unable to do this.

Dale (1994) suggests that mistakes and setbacks are elemental features of learning and development. The way that organisations respond to the changing business environment and learn from experiences, both good and bad, is the ability that defines 'learning organisations'.

Learning organisations have been described as those in which people at all levels, individually and collectively, continuously increase their capability to create desirable outcomes for the organisation as a whole (Garvin 1998). Pedler *et al.* (1988) defined a learning company as an organisation that facilitates the learning of all of its members and continuously transforms itself. While different authors single out different issues as being central to becoming a learning organisation, Argyris (1999) identifies some common features. These are notions of organisational adaptability, flexibility, avoidance of stability traps, propensity to experiment, readiness to re-think means and ends, an inquiry orientation, the realisation of human potential for learning in the service of the organisation and the creation of an organisational environment that supports human development.

The concept of organisational learning is underpinned by the notion that organisations are more than a collective of individuals. Instead, organisations are viewed as entities that have unique identities manifested in their organisational cultures, which develop and change over time as a result of organisational experiences (Dale 1994). Organisational learning refers to the development of collective skills and knowledge, shared assumptions and values between members.

We argue that the concept of organisational learning is one that is critical to the construction industry's ability to improve its OHS performance and suggest that, with regard to OHS, construction organisations need to develop the ability to learn.

The ability to identify and learn from one's mistakes is an important part of incident management. In order to achieve this, organisations must have incident information systems, enabling reliable incident data to be captured, analysed and deployed in prevention efforts. An analysis of present performance is an essential part of any continuous improvement effort. However, incident information systems will not be sufficient in themselves. Senge (1994) suggests that effective learning requires not just a good understanding of 'current reality'. He suggests that organisations also need a strong sense of where they want to be in relation to current reality, or a *vision*. When organisations understand where they are and where they want to be, the resultant 'creative tension' allows organisations to generate the energy required to improve their performance.

Senge (1994) distinguishes between improvement arising from creative tension and improvement achieved through traditional problem-solving. When organisational learning occurs as a result of creative tension, the impetus for change comes from vision juxtaposed with an understanding of current reality. However, in problem-solving, the energy for change comes from reacting to an undesirable situation which has become bad enough to necessitate change. In organisational learning, change is driven from within; in problem-solving the impetus for change is external. As such, organisational learning is sustainable, while problem-solving usually runs out of steam. It is therefore important that construction organisations seeking to improve their OHS performance do so by analysing their 'current reality', that is how they are currently performing, and develop a vision of where they would like to be with regard to OHS. Fostering a shared understanding of the gap between these can then provide the 'creative tension' needed to drive performance improvement.

Learning can occur at different levels. Argyris (1999) uses an electrical engineering analogy to describe the difference between single-loop and double-loop learning. This is depicted in Figure 10.1. A thermostat is an

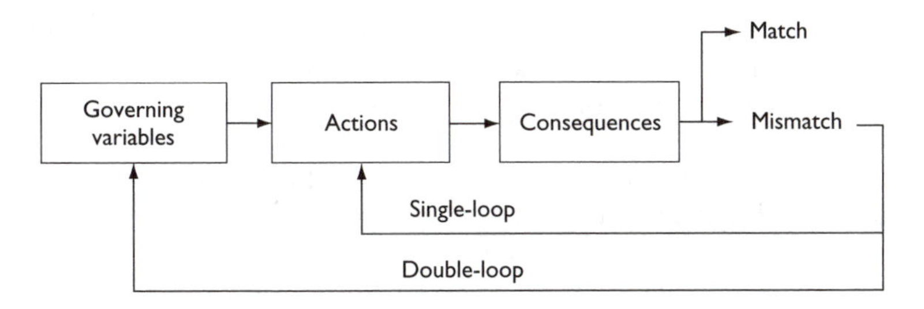

Figure 10.1 Single- and double-loop learning (adapted from Argyris 1999, p. 68).

example of a single loop learner because it is programmed to identify states of 'too hot' and 'too cold'. When it detects a mismatch between the actual and desired temperatures, it adjusts itself, engaging in single-loop learning. Single-loop learning is therefore a lower level of learning focusing on the symptoms of a problem. Fiol and Lyles (1985) suggest that single-loop learning can lead to some improvement in organisational performance because problems are rectified as they become apparent, albeit at a superficial level.

In contrast, double-loop learning occurs when mismatches between desired and actual outcomes are examined by consideration of governing variables. Double-loop learning focuses on the root causes of problems. Argyris (1999) suggests an example of double-loop learning would occur if the thermostat questioned why it was set on a particular temperature in the first place. In relation to OHS the distinction between single- and double-loop learning might be demonstrated by the difference between identification of immediate causes of incidents and their underlying systemic causes. Only by addressing the latter can long-term preventive strategies be developed.

There are many impediments to organisational learning. These include the adoption of defensive routines, mixed messages, the use of deception, organisational camouflage and taboos concerning the discussability of some issues. OHS problems are often embarrassing and threatening and can produce defensive routines in which people try to prevent embarrassment or threat. Hofmann and Stetzer (1998) suggest that attribution errors are a form of defensive routines that inhibit the ability to identify the true causes of OHS incidents and therefore limit the ability to identify appropriate strategies to prevent similar incidents in the future. Two types of attribution error are identified. One of them is the *fundamental attribution error*, in which people have a tendency to over-estimate personal factors and under-estimate situational factors in incidents involving others. Thus, workers' contributions to workplace incidents are often over-estimated by investigation teams. Another attribution bias is termed *defensive attribution bias*. This occurs when people, who perceive themselves to be personally and situationally similar to the victim of an incident, make external attributions about its cause and exaggerate situational factors. These attribution biases may explain why managers often blame workers for incidents, while workers blame features of the work process or work context (Prussia *et al.* 2003).

In this context, it is difficult to identify and reduce the causes of embarrassment and threat and the result is an inability to learn from experiences. Defensive routines act to cover up errors and people engage in these routines because they believe that they are necessary for individuals and the organisation to survive (Argyris 1999). Furthermore, these routines are often engaged in without participants' awareness that they are behaving defensively. Argyris (1999) describes this behaviour as Model I or theory-in-use behaviour. He suggests that people are socialised to behave in accordance with four governing values:

- Achieve your intended purpose.
- Maximise winning and minimise losing.
- Suppress negative feelings.
- Behave according to what you consider to be rational.

The most prevalent action strategies these principles produce are as follows:

- Advocate your position.
- Evaluate your thoughts and actions of others.
- Attribute causes for whatever you are trying to understand.

The way in which participants' governing values are satisfied is through remaining in control, evaluating events and attributing their causes in a way that reduces negative feelings, embarrassment and threat and supports the individuals' position. This behaviour is, according to Argyris (1999), the antithesis of learning and is often self-serving and supportive of maintaining the status quo. What is even more problematic is that people are unaware that they are behaving defensively and therefore it is almost impossible for individuals to overcome these barriers to double-loop learning.

However, organisations can develop mechanisms to facilitate organisational double-loop learning. These can overcome the problems associated with the use of mixed or inconsistent messages that prevent the discussion of important OHS issues in the workplace. For example, a manager who says, 'Jim, finish the job as quickly as possible, but be safe' sends an inconsistent message but behaves as though the message is not inconsistent. The manager then makes the apparent tension between productivity and safety undiscussable and makes the undiscussability of this tension undiscussable, thereby defending him or herself against embarrassment.

Collective retrospection provides opportunities for organisational learning to occur. For example, Busby (1999) describes the processes by which collective retrospection enabled defensive routines to be overcome in several engineering design projects. The use of group techniques, such as dialectics, in which one individual would state a thesis and another would argue its antithesis before the first or a third participant identified a synthesis, was observed in a post-project analysis of performance. Antitheses were often signalled as tentative by the use of pre-fixes, such as 'Let me play devil's advocate.' Busby (1999) suggests that this type of interaction encourages a vigorous search for evidence to support positions taken and can facilitate learning. In addition, the use of probing interaction styles in the post-project reviews facilitated double-loop learning though Busby (1999) cautions that this technique could also have negative outcomes because it sometimes challenges the self-efficacy of participants.

Reviews in which the pattern of events was re-constructed were particularly helpful, and group interactions often facilitated the identification of the way in which events were interconnected. Thus, collective retrospection might present a useful strategy in enabling learning from OHS incidents and experiences. However, it is likely that this learning will be most effective when input is gathered from individuals within the organisation who hold different perspectives. In the construction context, this may include representatives of different trades, supervisors, foremen, subcontractors, health and safety representatives, union representatives and others.

In order to overcome defensive reasoning in incident investigation or during collective retrospection, it is important to develop a positive safety culture and create an environment in which OHS issues are communicated freely and openly. Hofmann and Stetzer (1998) report that workers in work groups in which safety-related issues were openly discussed were more likely to make internal attributions concerning incidents involving other workers, therefore showing a reduced tendency towards the defensive attribution error. However, where OHS communication was not open, workers were less willing to recognise that the fellow worker contributed to the incident. A positive OHS climate is also reported to increase agreement between workers and managers about OHS issues, including workers' safety-related behaviour, and reduce the extent to which managers make fundamental attribution errors (Prussia *et al.* 2003). These results suggest that, in order for organisations to effectively learn from OHS incidents, the organisational context must be supportive of this learning. OHS climates are discussed later in this chapter.

An important feature of leadership in learning organisations is that leaders no longer act as authoritarian charismatic decision-makers. Instead, they design organisations that determine the future behaviour of the organisations they lead. This involves establishing a shared sense of purpose and core values (Senge 1994). Senge (1994) cites the 1982 Johnson & Johnson

Tylenol issue as a case in which acting in accordance with the organisation's stated values saved the company from a major crisis. When bottles of the best-selling Tylenol were tampered with, resulting in several deaths, the corporation immediately removed all Tylenol from retail outlets, destroying 31 million capsules even though they were tested and found safe. Johnson & Johnson's vision, written forty years earlier, read that 'service to customers comes first'. Johnson & Johnson's speedy and decisive action averted a crisis of confidence in the company.

Thus, leaders in learning organisations must also be prepared to expose for scrutiny mental models, assumptions or worldviews that are held in the company. With regard to OHS, assumptions that injuries or illness are inevitable parts of work prevail in some construction organisations. These assumptions must be exposed, scrutinised and dispelled. Leaders need to encourage employees at all levels to question assumptions about limitations and conditions imposed by the way that work is organised and seek better, safer and healthier ways of working.

Unfortunately, engineering cultures, typically found in many construction organisations, militate against organisational learning (Ford *et al.* 2000). Engineering cultures tend to be driven by optimism based on science and available technology. While the human factor is recognised, equipment and processes are often designed to make things as automatic as possible. This is anathema to systems thinking, which focuses on the inter-connected nature of complex socio-technical systems, and it is important that other cultural influences are strengthened in support of organisational learning in construction organisations.

There are many ways by which organisations can develop the capacity to learn. Kululanga *et al.* (1999) explored the way that construction contractors in the UK currently learn and report that while some mechanisms of learning are widely used, others are not. For example, many companies adopted partnering, joint venturing or corporate mentoring as mechanisms for learning. However, other techniques, such as research, employee-based networks and benchmarking, were seldom used. These results suggest that a wider range of organisational learning strategies focusing on continuous improvement in OHS could be used by construction organisations.

Dale (1994) identifies a number of conditions that help to create an environment in which organisations can learn effectively. These include:

- developing a learning strategy;
- participative policy-making;
- informating (the use of IT to inform and empower people to ask questions and base decisions on data);
- formative accounting (control systems that are structured to assist learning from decisions);
- internal exchange;

- reward flexibility;
- enabling structures;
- front-line workers as environmental scanners;
- inter-company learning;
- learning climate; and
- self-development of all.

While these are general conditions, the focus on employee participation and empowerment is particularly relevant to the field of OHS in which employees' interests are of paramount importance. We advocate that learning from OHS experiences become a core part of all operations within construction organisations and suggest that until these organisations actively strive to learn from their mistakes, the industry's OHS performance will remain poor.

The ability to learn effectively requires a work context in which OHS issues are not covered up, dismissed or treated as undiscussable. Instead, OHS communication needs to be open and honest, and active involvement in improving OHS needs to be encouraged, recognised and rewarded. Sadly, this is often not the case in construction organisations in which a culture of blame prevails, OHS issues are not discussed and, if they are discussed, are dealt within the context of a hostile industrial relations climate.

Overcoming these problems will be difficult and will require considerable cultural change. This will not occur automatically but will ultimately require that managers within the construction industry demonstrate strong OHS leadership, to lead by example and communicate the importance of OHS through recognising and rewarding those who work safely and do not take risks, even in the face of tight schedules and looming deadlines. The notions of safety culture, climate and leadership are key issues in enabling construction organisations to learn from their mistakes. These issues are discussed below.

OHS culture, climate and leadership

The *safety culture* and *safety climate* concepts are often used interchangeably. Both are based upon an extensive body of research into the concepts of organisational culture and climate. In this body of research, culture is understood to embody values, beliefs and underlying assumptions in organisations, while climate is a descriptive measure of employees' shared perceptions of the organisation's atmosphere. Methodological approaches to investigating organisational climate and culture usually differ in that climate is usually measured using quantitative, psychometric questionnaire studies, while culture is explored by qualitative anthropological studies. There has been a long-standing debate about whether organisational culture or organisational climate should be considered more important.

The 'culture or climate' debate is now being mirrored in the OHS research field. Safety culture is distinguished from safety climate in that the former refers to underlying core organisational beliefs, while the latter represents employees' attitudes and perceptions of OHS at a given point in time (Flin *et al.* 2000). Given this interpretation, an organisation's safety culture is expressed through its safety climate (Guldenmund 2000). If this interpretation is accepted, then the development of a positive safety culture should be the most important aim for those who wish to improve OHS performance, while the measurement of the safety climate can be viewed as a useful diagnostic tool and method for measuring the safety culture. Safety climate surveys can therefore be used as a snapshot assessment of the state of the safety culture within an organisation or at a particular site. Safety climate assessment can also be used to identify problem areas that can then become the focus for programmes for change (Diaz and Cabrera 1997).

Different researchers have presented different sets of indicators said to reflect a positive safety culture. Some of these were discussed in Chapter 7. Questions remain about whether these indicators are general features of a 'good' safety culture in all industries or countries or whether some safety culture indicators may be unique to certain sectors or national cultures.

Guldenmund (2000) suggests the study of safety culture is hampered by the lack of a unifying theoretical model. In particular, models of safety culture do not embody a causal chain but, instead, specify desirable attributes believed to be associated with excellent OHS performance. The way that safety culture impacts upon OHS behaviour is therefore unclear. Pidgeon (1991) characterises a positive safety culture as comprising:

- norms and rules for dealing with hazards;
- positive attitudes towards safety; and
- reflexivity on safety practices.

This suggests that good safety cultures establish acceptable norms, which can be expressed simply as 'the way we do things round here', generate enthusiasm and a belief in the importance of safety and encompass the capacity for retrospective analysis and organisational learning discussed earlier in this chapter.

Safety culture is a relatively new concept and, although questions remain about what it is and the mechanisms by which it operates, it is potentially a valuable concept for improving OHS performance in the construction industry. However, it is important that the concept of safety culture be investigated in regard to the construction industry to determine what constitutes a good safety culture in this context. Examples of best practice are particularly helpful in this regard and construction organisations should be encouraged to share their success stories.

Unfortunately, examples of excellence in OHS are not commonly found in construction. We suggest that insufficient attention has been paid to the cultural impediments to improving OHS in the construction industry, and we turn to some of these impediments next. It is critical that these cultural issues be addressed if the construction industry is to improve its OHS performance.

Cultural impediments to safe working in construction

A genuine commitment to safety from senior management is one of the most commonly cited features of a 'good' safety culture. This includes establishing the company's direction through a company's OHS policy, communicating the importance of OHS in all senior management's actions and adequately resourcing the OHS programme. Good safety cultures are said to take a long-term view of OHS, which is viewed as part of business strategy. Unfortunately, the structure and characteristics of the construction industry militate against the presence of these features. Anderson (1998b) suggests that a 'task culture' prevails in construction, in which the ultimate focus is on the production of an end product, and effective OHS management therefore has to fit with this culture.

Most participants in the construction industry are usually focused on short-term project goals and often spend time reacting to unforeseen events, making it difficult to take a long-term strategic view of the organisations' values and business objectives. Furthermore, organisations operating in a highly uncertain environment, like the construction industry, may find it difficult to formulate business strategy, particularly as it relates to an issue that is often deemed not to be central to core business activity.

Unfortunately, many construction managers still regard OHS as an 'add on' rather than a central part of the management of a construction project. Furthermore, the pressures imposed by competitive tendering and the resulting need to cut project costs to a minimum means that OHS is sometimes overlooked in construction estimating and planning. The structure of construction organisations and the organisation of work also present difficulties for the growth of a good safety culture.

Projects operating as cost centres are unlikely to be able to resource the implementation of a full-blown OHS management system, and it is often difficult to demonstrate commitment to OHS through the implementation of a systematic approach, unless the organisation possesses an established central OHS department. This is particularly difficult for small- to medium-sized firms lacking the resources to establish such a department. However, at the same time, the presence of a centralised OHS department can lead project or site managers to perceive that OHS is not an integral part of their job, but is something to be managed and policed by the OHS specialist. This is contrary to the requirement of a good safety culture – that OHS is an important responsibility for everyone, particularly managers. The danger of project or construction managers not embracing their OHS responsibilities is heightened in the decentralised construction industry, where managers, at site level, have considerable decision-making autonomy. This makes the importance of supervisory OHS behaviours even more important in construction than in most other industries. The role of supervision in safety leadership will be discussed below.

'Good' safety cultures also support open and ready communication concerning safety issues within and between levels of the organisation. Organisations with good safety cultures do not seek to attribute blame when incidents happen. However, an unfortunate feature of the construction industry is that conflicts and disputes are commonplace. When problems arise, it is common for project participants to seek to lay the blame at the door of another party. This culture of blame was recently highlighted in the Australian 'Cole' Royal Commission into the Building and Construction Industry in which construction employers blamed the trade unions for holding up building work and using restrictive OHS practices to gain control of projects (*Sydney Morning Herald*, 10 February 2003, p. 7). The unions lambasted employers for failing to meet their OHS obligations, and called the Royal Commission a 'disgraceful sham' because it ignored breaches of OHS regulations by employers (*Australian Financial Review*, 31 March 2003, p. 3). A culture of conflict and blame is unlikely to encourage open communication with regard to OHS and is likely to foster defensiveness and even attempts to cover up OHS problems. As we have already suggested, this behaviour prevents organisations from learning effectively from their mistakes.

Organisations with good safety cultures also invest in employee training and integrate OHS into skills training. As noted in Chapter 4, construction organisations are often reluctant to train a workforce that is highly transient, or comprised mainly of subcontractors. In addition, the tight time and cost constraints involved in construction projects make it difficult to resource training or release employees to undergo training.

A sense of optimism that all incidents can be prevented also pervades good organisational safety cultures. Unfortunately, as Case study 4.1 (in Chapter 4) revealed, many construction workers accept injuries as an inevitable part of their job. This perception is exacerbated by the emphasis placed by managers and supervisors on construction project timelines and the avoidance of delays. The strong focus on cost and time performance may in some cases pressure workers to take unnecessary risks, or convey the message that OHS is less important than production and that risk-taking is sometimes necessary to achieve scheduled production targets. The OHS implications arising as a result of unrealistic deadlines that require workers to work considerable overtime are considered later in this chapter. The understanding that excessive overtime and fatigue present considerable health and safety risks has recently raised these issues to the fore in construction OHS.

A final important impediment to enhancing the safety culture of construction organisations is the fact that OHS is not well integrated into the education of construction professionals and managers. A recent study of thirty-one UK higher education institutions revealed that while there were some excellent examples of OHS in the curriculum, these initiatives were driven by individuals who tended to be few in number (Carpenter *et al.* 2001). Furthermore, in a significant number of the courses, the incorporation of OHS into the curriculum was not actively supported by the head of department. This may, in part, be because construction industry professional associations tend not to require clear and specific OHS competencies to satisfy accreditation requirements. In many cases, OHS requirements are too vague to be interpreted consistently.

We argue that in order to achieve cultural change, architects, engineers, surveyors and construction managers who will lead the industry in the future need to be well versed in the ethos of OHS and the principles of risk management. Educational institutions need to recognise that OHS pervades all aspects of a construction project and is a complex and stimulating field of study. As the previous chapters of this book have demonstrated, OHS is underpinned by different theoretical frameworks, including risk management, organisation theory and psychology. It is therefore an area of practical relevance and intellectual merit. Furthermore, OHS is now acknowledged to be one part of an organisation's broader risk management and corporate governance imperatives. All managers and professionals should understand OHS in order to exercise professional skill and judgement in their decision-making. However, while OHS is not integrated into the curriculum, aspiring construction professionals are not being inculcated into the correct culture, and will not be equipped to lead the cultural change needed when they reach positions of authority and influence.

Nurturing a good safety culture

The Advisory Committee on the Safety of Nuclear Installations (ACSNI – Human Factors Study Group) prepared a report for the UK HSC in 1993. This report contained some suggestions as to how organisations could go about creating a good safety culture (HSC 1993). The Committee suggested that in the first instance the existing safety culture should be reviewed. Following this review, aspects of the culture which needed to be changed should be prioritised and actions to promote cultural change in these areas should be decided and implemented. This process was described as one of continuous change, and thus these steps are to be repeated indefinitely.

Several methods for reviewing organisational safety cultures have been developed. For example, Ludborzs (1995) recommends the use of a safety culture audit, in which key individuals and employees are interviewed and observed, as is common practice in safety management systems audits. This allows both documented and lived aspects of the OHS management system and operations to be analysed, and enables shortcomings to be identified, including disparities between what people say and what they do. The consistency between stated beliefs and behaviours is a key facet of safety culture (Anderson 1998b). The safety culture audit method assesses ten broad areas, and includes detailed checklists of indicators. One advantage of such an approach is that it allows varying results that can be used as evidence of the existence of sub-cultures or counter-cultures. This is particularly useful in decentralised organisational structures such as those in construction organisations because while documented safety may be uniform across an organisation, 'lived safety' may vary considerably from project to project.

Kennedy (1997) also presents a modified version of a Hazard and Operability (HAZOP) study methodology, described in Chapter 5, to measure safety culture. Like HAZOP studies, the Safety Culture HAZOP (SCHAZOP) is undertaken as a group, and uses brainstorming techniques prompted by keywords or guidewords and property words. For example, guidewords include 'missing' and 'mistimed'. Property words include 'person', 'action' and 'procedure'. Issues identified are then refined in more structured discussions, and vulnerabilities in the management of safety are identified.

Another method widely used as a diagnostic tool is the use of a safety climate survey. This involves conducting a large-scale questionnaire survey among employees. Climate questionnaires seek to measure employees' perceptions concerning key features of safety culture. Different surveys vary in length and include different categories of safety. One instrument, used in the construction industry, comprised nine items and measured management commitment and worker involvement in OHS (Dedobbeleer and Beland 1991). Another instrument, used in manufacturing, comprised 40 items and measured safety training, management attitudes, promotion, levels of risk, work pace, safety officer status, social status and perceptions of the safety committee. Flin *et al.* (2000) provide a comprehensive review of safety climate measurement instruments.

Coyle *et al.* (1995) suggest that no one set of safety climate factors has universal significance, but, they argue, this does not mean it is not helpful to measure safety climate and compare climate measurements between organisations or units within an organisation. In fact, the identification of different climate factors can be used to identify where cultural change might be needed. Safety climate factor scores can also be compared between groups of employees, for example plant operators and other trades, males and females, foremen and labourers, to determine whether any of these groups differ significantly in their perceptions of OHS within an organisation. These differences can then inform change programmes. For example, if management think an OHS training programme is effective but workers consider it a waste of time, understanding this difference in perception of the training programme is critical to improving it.

Safety climate and OHS outcomes

Safety climate is widely held to have an impact upon organisational behaviours, such as communication, decision-making, problem-solving, conflict resolution, motivation and safety-related behaviour. Research has demonstrated a link between safety climate and OHS outcomes (Zohar 1980; Diaz and Cabrera 1997; Varonen and Mattila 2000). These studies suggest that safety climate can predict incident occurrence, and also be used to discriminate between organisations with good or bad safety performance. The results of safety climate survey studies indicate that global perceptions employees form about their work environment are related to management values and organisational policies and practices and that these values, policies and practices act through the safety climate to shape the OHS behaviour of workers.

Griffin and Neal (2000) tested a causal model of safety climate and OHS behaviour in the Australian manufacturing and mining industries. They undertook path analysis to identify the nature of the direct and indirect relationships between organisational values, policies and practices, the global safety climate, safety knowledge and motivation and workers' safety behaviour. Their results suggest that organisational policies, including safety practices and safety equipment, together with management values, are directly related to safety climate. However, they report that safety climate is *indirectly* rather than directly related to workers' participation in safety activities of the organisation, and compliance with safety requirements. The relationship between safety climate and these desirable safety behaviours is mediated by workers' safety knowledge and motivation. The implication of this finding is that, though safety climate is an important determinant of OHS behaviour, it acts through workers' safety awareness, reflecting the critical importance of OHS training and the dissemination of information about the extent and consequences of OHS risk exposure to workers. Unless fully informed, workers may not actively participate in the safety process and may resist the adoption of necessary risk control measures, even if organisational policies and practices are supportive of OHS.

Understanding the nature of relationships between organisational practices, policies and values, individuals' perceptions of the work environment and

OHS behaviour is important if safety climate is to be used effectively to diagnose vulnerabilities and develop appropriate intervention strategies. A better understanding of these linking mechanisms in the construction industry should be sought.

Multi-level safety climates in construction

One feature of the construction industry that may limit the extent to which shared perceptions of the safety climate exist in construction organisations is the decentralised structure of these organisations. Productive work in construction occurs in projects geographically remote from the organisations' corporate head offices. Multi-disciplinary teams typically manage projects, with considerable autonomy in making project-related decisions. OHS policies that determine an organisation's strategic OHS goals and standard operating procedures are often consistent between projects within a single organisation. However, 'lived' OHS practices and management values may vary considerably between projects. Consequently, workers' perceptions of the prevailing safety climate may vary considerably from project to project.

This possibility suggests that a multi-level safety climate model may be appropriate, in which workers are influenced by their perceptions of expected behaviours at both an organisational and project level. Work-group level safety climates have been investigated in a manufacturing organisation by Zohar (2000), whose results revealed that the safety climate varied significantly between units within the same organisation. Furthermore, Zohar (2000) reports that climate scores predicted the safety performance of sub-units in the months following the climate assessment, indicating that those sub-units with more positive safety climates experience fewer incidents. Zohar (2000) also found that workers discriminated between perceptions of the organisation's safety climate and the sub-unit safety climate, demonstrating that it is possible to perceive the organisation takes safety seriously but that sub-unit leadership does not value safety. In particular, group-level safety climate related to patterns of supervisory safety practices, or ways in which company level policies were implemented within each sub-unit. This finding has significant implications for the construction industry in which site-based managers and supervisors act as a conduit, through which organisational policies and objectives are communicated to the workforce. It is highly likely that the prevailing project safety climate will be the strongest predictor of workers' OHS behaviour, because project-level supervisory behaviour has the most immediate effect on the recognition and rewards enjoyed by workers.

The existence of workgroup level climates is particularly important in the project-based construction industry. In this industrial context, the role played by project-level managers and site supervisors in defining the safety climate will be critical. The measurement of project-level safety climates is also potentially beneficial because it could provide the basis for identifying supervisory behaviours and attributes associated with positive safety climates. This information could then be drawn upon in the training and selection of site supervisors and managers.

Supervisory safety leadership

Many authors have written about the importance of top management commitment to OHS. While top management define organisational OHS goals and undoubtedly play an important part in defining and communicating organisational values relating to OHS, the role of first-line supervisors is also critical.

Research suggests that supervisory practices play a key role in shaping workers' understanding of what is expected of them, by communicating the priority of safety in the workgroup or the importance of acting safely in a particular job. First-line supervisors are typically positioned at the intersection between management and the workforce and, as such, their actions are likely to influence the way in which workers interpret top managements' expectations. They act as a conduit between management and workers, monitor compliance with management's directives and provide managers with information about workers' compliance or negligence with regard to OHS (Niskanen 1994). Supervisors provide important feedback to workers concerning the appropriateness of their behaviour. Consequently, supervisors' behaviour and expression of views will have considerable influence on the development of workers' beliefs about management policies and priorities. Leather (1987) points out that intentions communicated by supervisors as what management 'really wants' is not always consistent with the contents of formal policy statements or plans.

Clarke (1999) suggests that one barrier to the development of a strong and positive OHS culture is the lack of mutual inter-group perceptions of OHS. Different groups within organisations will have perceptions of what the other group believes or how its members will act. Where these perceptions are inaccurate or biased, shared values and norms cannot be achieved. Clarke (1999) examined drivers', supervisors' and senior managers' understanding of the OHS attitudes of each group in a large rail company and found considerable bias. For example, drivers estimated that supervisors and managers would have less awareness of important safety issues than themselves, when in fact supervisors and managers shared drivers' concern about safe working conditions. The negative stereotyping of senior managers' safety concerns was reflected in the fact that supervisors also believed senior managers to be less concerned than they were, suggesting that supervisors may contribute to workers' perception that senior management is unconcerned about safe working conditions. These biases will not only prevent the development of a shared understanding of OHS issues, but will also hinder staff-management communication, confidence in management and the development of trust (Clarke 1999).

A multi-level management model of workers' safety behaviour was developed by Simard and Marchand, who suggest that supervisors can influence accident prevention in two ways:

1 By being personally involved in accident prevention, for example by undertaking site inspections, being involved in incident investigations, training new employees or by analysing hazards in job safety analyses; and
2 By encouraging workers' participation in these safety activities (Simard and Marchand 1994).

Simard and Marchand investigated the extent to which a series of macro- and micro-organisational factors influenced workers' safety behaviour and found that supervisory participative practices were the strongest predictor of workgroups' propensity to take safety initiatives (Simard and Marchand 1995) and to comply with safety rules (Simard and Marchand 1997). The effect of work group and supervisory practices were considerably higher than macro-level factors, such as top management commitment, the appointment of a full-time safety professional and the organisation's implementation of consultative processes. Interestingly, Simard and Marchand (1995) found that these macro-level factors did have a positive effect on workers' safety behaviours but that this effect was an indirect one in the sense that macro-level factors positively influenced the participative management of workgroup safety, which in turn led to enhanced safety behaviour. Again, supervisory level practices were crucial. Simard and Marchand (1994, 1995, 1997) conclude that senior management should adopt a decentralised approach to OHS management, rather than a centralised, bureaucratic approach. In particular, the importance of worker participation in the safety process should be noted. Training supervisors in the benefits and importance of adopting participative management in OHS is one strategy that should be adopted.

The relationship between leadership and OHS has recently been explored. For example, Hofmann and Morgeson (1999) report that group leaders who have high-quality relationships with their own supervisors are more likely to feel free to raise safety concerns. This safety-related communication, in turn, is related to safety commitment and improved safety performance in the work-groups they lead. Their findings are consistent with those of Simard and Marchand in that they suggest that senior management support for OHS is important because it conveys, to supervisory employees, that the organisation values its workers. These supervisors then behave accordingly. Also, positive exchanges between supervisors and subordinates are more likely to foster an environment in which safety concerns are raised and in which the organisation can learn from its mistakes.

While employee involvement in OHS is universally held to be an essential feature of effective OHS management, the nature of this involvement is also important. Geller (2001) suggests that workers' involvement in OHS should extend beyond looking after one's own health and safety. Geller (2001) writes of a Total Safety Culture in which everyone goes beyond the call of duty for the safety of themselves and acts to benefit the safety of others, for example by giving co-workers feedback about their safe and unsafe behaviours. Geller (1991) refers to this as 'actively caring' for safety.

Different leadership approaches are likely to impact upon workers' propensity to actively care for safety, and Geller (2001) suggests that workers' propensity to actively care for safety can be increased by strategies designed to enhance workers' self-esteem, belonging and empowerment in OHS. Giving people personal control to enable them to develop ownership of OHS is a characteristic of transformational leadership in OHS, which transcends the 'rules and rewards' approach of transactional leadership. Transformational leadership is characterised by value-based interactions underpinned by trust, loyalty, openness and reciprocity. In contrast, transactional leadership is based to a lesser extent on these values and is more focused on hierarchical, rather than egalitarian values. Transactional leadership can be sub-divided into the following three dimensions:

1 *Constructive leadership* – in which leaders identify employees' needs and expectations and try to motivate them by offering suitable rewards for performance.
2 *Corrective leadership* – in which leaders monitor subordinates' actions in relation to certain standards and detect and correct errors.
3 *Laissez-faire leadership* – in which leaders disown their supervisory responsibility.

Geller *et al.* (1996) suggest that a top-down, rule-enforcement perception of OHS, such as the corrective type of leadership, is unlikely to encourage workers to actively care for the safety of others. Zohar (2002a) examined the role of leadership style on safety climate and OHS performance, and reports that transformational and constructive leadership predicted the injury rates in workgroups, and that this effect occurred indirectly through participative safety management. This suggests that certain leadership styles, in which supervisors develop close and individualised working relationships with subordinates and show a concern for their health and safety, are associated with participative supervisory practices and improved OHS performance.

Vassie and Lucas (2001) differentiate between teams managed using supervisors, team leaders and self-managed teams. In a study of manufacturing organisations, firms using team leaders and self-managed groups were found to have higher management involvement in setting and monitoring OHS objectives than those using traditional supervisor-managed work teams, indicating greater management accountability for OHS in these firms. Where traditional supervisor-managed teams were used, management relegated responsibility for OHS to supervisors and did not get directly involved in monitoring OHS performance (Vassie and Lucas 2001). Companies using traditional supervisors also showed the lowest levels of employee empowerment to determine OHS practices, with OHS decision-making being undertaken largely by supervisors. While in construction work is often performed by semi-autonomous teams of tradesmen, the traditional supervisory model still prevails. The results of Vassie and Lucas suggest that encouraging greater worker participation in decision-making at workgroup level might present an opportunity to improve OHS.

The study of OHS and leadership has only recently commenced and therefore we cannot conclude that construction site supervisors should adopt a particular leadership style. Indeed, the construction industry presents difficulties for transformational leadership because the prevalence of subcontracting means that value-based, individualised relationships are difficult to foster. However, the results clearly indicate that participative supervision is important and, suggest that, the application of OHS rules and a corrective approach, without participation, will not be effective.

Work, safety and well-being

The needs and experiences of project-based workers are a source of particular concern in the construction industry. One aspect of work that is increasingly recognised to be an OHS issue is the amount of time workers spend working and the physical, mental and even social hardships that result from excessive workloads.

Work hours

Hours in construction are notoriously long. For example, Merlino *et al.* (2003) report that, among 996 construction apprentices, the average hours worked per week was 45.1. Dong (2002) also reports construction workers work longer hours per day than workers in other industries. The amount of time that people are working and the pressures under which they are working have been investigated by one of the authors in the Australian construction industry. In Australia, the Construction Forestry, Energy and Mining Union undertook a survey of 800 of its members and reported that 80 per cent of those surveyed nominated excessive working hours as the main reason for workplace injuries and accidents and 30 per cent nominated work hours as a key cause of their personal relationships breaking down. This has led the union to focus on improving members' quality of life (*Sydney Morning Herald*, 16 November 2002, p. 11).

The 1995 Australian Workplace Industrial Relations Survey revealed that in Australia, work hours and work intensity had risen (Moorehead *et al.* 1997). Thus, work has become 'greedier', demanding more time and energy from employees. Glezer and Wolcott (2000) report that in late 1996, 66 per cent of Australian men and 23 per cent of Australian women were working more than 41 hours a week. Of these, half of employed men and 46 per cent of women felt work interfered with home life compared to less than a quarter of women and the small proportion of men who worked less than 30 hours per week.

The construction industry is schedule-driven and pressures to meet tight deadlines often result in the need to work extended work hours and regular overtime. The issue of excessive work hours is experienced by blue collar workers and professional staff alike. The results of two separate studies of Australian construction professionals revealed that, although there is a significant difference between hours worked by different professional groups, all professional groups in construction work in excess of 41 hours a week. In a sample of women working in the Victorian construction industry, architects reported working 42 hours per week, project/construction managers reported working 49 hours per week and site/project engineers reported working 57 hours per week (Lingard and Lin 2003). In a study of employees of a large Australian construction firm, site-based employees reported working longer, more irregular hours than office-based employees. The average number of hours worked each week was 63 among site-based respondents in direct construction activity, 56 among respondents who work mostly in a site office and 49 among respondents in the head or regional office.

Work hours have been the subject of considerable debate in the Australian construction industry. In the year 2000, in the Australian State of Victoria, the construction trade union (the CFMEU) successfully negotiated a standard 36-hour working week for construction workers. The extension of the 36-hour week to other states has been a contentious issue and employer organisations are deeply unhappy about the prospect of 36-hour weeks, claiming that builders tendering for projects worth AU$10–50 million would lose work if they agreed to a 36-hour week because it would add 5.5 per cent to labour costs (*Sydney Morning Herald*, 16 November 2002, p. 11).

Prior to the change, the basic working week was 38 hours but, in reality, work hours have not changed despite reducing this standard to 36 hours because the extra two hours a week are accumulated into five extra rostered days off and added to public holidays. High rates of overtime are still the norm and the 'normal' working week in construction is 56 hours, comprising four days of ten hours and two days of eight hours. Work undertaken over and above the 36 hour standard is deemed to be overtime and paid at a higher rate creating a financial incentive for workers to exceed the 36-hour week.

There is considerable international variation in working hours but a typical week of 56 hours is long in comparison to international averages. For example, in the UK, the average hours worked per week is 44.7 compared to 39.9 in Germany, and 39 in Denmark and the Netherlands. Lingard (2003) reports long work hours were found to be the most significant predictor of burnout among engineers in the Australian construction industry.

Work stressors

Not only do construction industry employees work long hours, the nature of the work is also inherently stressful and, in the case of blue-collar workers, physically demanding, increasing the likelihood of accumulated fatigue. Common stressors in the physical work environment (the construction site) include noise, vibration, restricted workspaces, inadequate lighting and extremes of temperature. Task-related or organisational stressors commonly found in construction include time and cost pressures, excessive workloads, inter-personal conflict and conflicting roles. These stressors can impact upon both blue-collar workers and professional/managerial employees. For example, a recent survey confirmed that Australian engineers experience considerable time-related work pressures (APESMA 2000). The survey found that professional engineers work long hours, including significant amounts of regular unpaid overtime. Over the twelve-month period studied, engineers reported that the amount of work to be done had increased (63 per cent), the pace of work had increased (62 per cent) and the amount of stress had increased (52 per cent). APESMA also reports that more than a quarter of respondents believed there had been an increase in health problems as a result of their working lives. The most common ailments they identified were those related to excessive workloads, such as continual tiredness (66 per cent) and stress (70 per cent). Bacharach *et al.* (1991) also suggest that for engineers, role conflict (for example, conflict between professional standards and budget constraints) may be strongly associated with severe 'life and death' consequences, increasing the levels of stress.

Human error

Long hours and work stress can have an impact upon OHS and workers' well-being in several respects. For example, there is emerging evidence to suggest that long hours negatively impact on workers' performance and the propensity for human error (Akerstedt 1995; Williamson and Feyer 1995). Human error is widely acknowledged to be a contributing factor in the majority of accidents in the workplace (HSE 1999). However, Reason (1997) argues that human errors are the consequences, not the causes of accidents, in the sense that errors are caused by personal, task-related, situational and organisational factors, or 'latent' error-producing conditions.

Rosa (1995) argues that the extension of the working day can cause the accumulation of fatigue, which, in turn, can have a deleterious effect on the ability to concentrate and perform effectively (Rosa 1995). Personnel fatigue leading to faulty decision-making has been implicated in major disasters including the Challenger space shuttle disaster in 1986 (HSE 1999). Laboratory studies have also revealed that sleepiness and fatigue are significant contributory factors to the risk of human error and accidents. Dinges (1995) and Lilley *et al.* (2002) report that near-miss injury events are significantly more common among forestry workers reporting a high level of fatigue at work. In a study of the effects of overtime, in construction, Dong (2002) found that overtime increased the risk of injury. Lowery *et al.* (1998) also report that construction contracts with large overtime payrolls experience higher lost-time injury frequency rates, again suggesting that overtime is linked to accidents.

It has also been suggested that longer work hours reduce the opportunity for recovery through sleep, which has a negative effect on alertness (Rosa and Colligan 1988). Sleep deprivation, arising from disrupted sleep patterns occurring as a result of extended work hours and shift work, may be more common than we imagine. Alarmingly, one US-based study suggests that 50 per cent of the American adult population is sleep deprived and sleep deprivation is so widespread as to be labelled an epidemic (Atkinson 1999). Harrington (2001) also reports that sleep problems are a risk factor for increased worker error, suggesting sleep-deprived workers may be more prone to injury. The results of a recent prospective cohort study also indicate that sleeping difficulties are significantly related to occupational fatalities, although the causes of these fatalities are not reported (Akerstedt *et al.* 2002). Sleep deprivation does not only affect the safety of workers while on the job, but the ability to drive home safely after working a very long day can be impaired. In a recent US court case, a construction worker who was permanently disabled after he crashed while driving home after a 36-hour day was deemed to have lost the ability to judge whether he was fit enough to drive (Goldenhar 2003). As Dong (2002) notes, construction workers often drive longer distances to and from work, increasing the risk of journey injuries.

In Europe, policy makers have acknowledged the link between work hours and OHS. In 1996, members of the European Union agreed upon the European Community Directive on Working Time. This Directive gives employees the legal right to refuse to work more than 48 hours a week and requires that employees be given a daily rest period of 11 consecutive hours in each 24-hour period and a minimum weekly rest period of one day.

Human error is also more likely to occur under circumstances in which work environment stressors are present, such as extreme temperatures, humidity, noise, vibration, poor lighting and restricted workspace (HSE 1999). Construction site environments are typically dirty and dangerous and one or more of these work environment stressors are very likely to be present. Construction is also characterised by social and organisational stressors, including excessive workloads, time pressures, peer pressure, conflict with co-workers, other tradesmen or contractors and conflicting goals, such as productivity, cost and safety. The chaotic work practices, and job and organisational conditions, are likely to increase the likelihood of human error in construction work.

In forestry, it has been demonstrated that injuries peak just before the first break in the day's work (Slappendel *et al.* 1993). Lilley *et al.* (2002) suggest that this is due to a decrease in workers' energy levels and accumulating fatigue at this point during the day. It is therefore important to ensure that workers are able to take regular rest breaks to avoid fatigue.

Psychological well-being

Studies also show that work can cause mental health problems and impact upon workers' well-being and quality of life, both at work and in other life domains. One mental health outcome that has received much research attention is the phenomenon known as burnout. The most widely accepted definition of burnout conceptualises the phenomenon as 'a syndrome of emotional exhaustion, depersonalisation and reduced personal accomplishment' (Maslach *et al.* 1996). *Emotional exhaustion* describes feelings of depleted emotional resources and a lack of energy. In such a state, employees feel unable to 'give of themselves' at a psychological level. *Depersonalisation* is characterised by a cynical attitude and an exaggerated distancing from one's work. *Diminished personal accomplishment* refers to a situation in which employees tend to evaluate themselves negatively and become dissatisfied with their accomplishments at work.

Research evidence suggests that burnout is associated with negative outcomes for both individuals and organisations. At an individual level, burnout has been associated with the experience of psychological distress, anxiety, depression, reduced self-esteem and substance abuse (Maslach *et al.* 2001). Some studies also suggest a link between burnout and coronary heart disease (Appels and Schouten 1991; Tennant 1996). This suggests that burnout should be regarded as an occupational health issue.

Researchers have identified many job demands (stressors) to be associated with employees' experience of burnout (Lee and Ashforth 1996; Gmelch and Gates 1998). These include subjective overload, responsibility, role clarity or role conflict, various aspects of job satisfaction and control over one's work (Schaufeli and Enzmann 1998; Maslach *et al.* 2001). In a recent study of practising civil engineers in the Australian construction industry, work hours, a subjective sense of overload (or having too much to do in the time available) and conflicting roles within one's job were all significantly linked to burnout (Lingard 2003).

Work-life balance

Another rapidly growing body of research investigates the interface between work and non-work life. Obviously, excessive numbers of hours spent at work impact upon aspects of workers' lives outside work because time is finite and involvement in one activity limits time available to do other things. The notion of work-life imbalance is beginning to receive attention in construction.

In a recent pilot study, Lingard and Francis (2002) report on qualitative data collected from professional and managerial employees of a large construction company in Australia. The most common theme that emerged in the analysis of participants' comments related to job schedule issues and the availability of respite from work. The comments made by respondents suggest that many employees feel that long hours, including weekend work, and the inability to take time off severely compromise their ability to balance work and non-work life. As one respondent said: 'My main concern is the amount of hours we work each week. On most projects [we work] over 100 hours each week.'

This theme had a higher priority among male respondents than among female respondents, which probably reflects the higher proportion of men in site-based roles requiring weekend work. The following comment made by one respondent who moved from a site-based position to the head or regional office reflects this difference well. He wrote:

> My quality of life has vastly improved now that I don't need/have to work weekends. My hours have dropped from over 60 hours per week to 50 hours. Twelve months ago I was burnt out, ready to resign, and exhausted and angry. Now I get enough sleep, I'm not stressed out all the time. Although I terribly miss all the action, chaos, teamwork and instant gratification achieved on site, I feel my stress levels and resting periods are where they should be.

While these comments were made during a pilot study, the results of which cannot be generalised to the construction industry as a whole, they do suggest that construction industry employees struggle to balance their work and non-work lives.

Prior research has consistently demonstrated the ill effects of work-life imbalance on employees' quality of life and psychological well-being. For example, work-to-family conflict is reported to be inversely correlated with job satisfaction (Kossek and Ozeki 1998; Bruck *et al.* 2002) and psychological well-being in both men and women (Bedeian *et al.* 1988; Frone *et al.* 1992). In Finnish dual-earner couples, work-to-family conflict is reported to have an indirect negative effect on marital function, through job exhaustion and psychosomatic symptoms (Mauno and Kinnunen 1999), and Lingard and Sublet (2002) report long hours of work predict marital conflict among engineers in the Australian construction industry. Negative spillover from work into family life has also been reported to increase the odds of problem drinking among employed, midlife adults (Grzywacz and Marks 2000).

Preventive strategies

A study of overtime in construction revealed that working overtime is accepted as part of the culture of the construction industry, and a mechanism workers use to cope financially with fluctuations in demand for their labour (Goldenhar 2003). However, in the same study, participants raised several OHS issues associated with working overtime. These were sleep deprivation, the increased likelihood of injury, fatigue and stress, both at work and outside work. It seems inevitable that overtime will continue in construction. Given the nature of construction work and the excessive hours expected of construction industry employees, the issues of burnout and work-life balance are likely to increase in importance. As links between work conditions and health outcomes become clearer, these issues may well become critical OHS issues in the future. Although at present work-life balance and mental health issues remain poorly understood, and their work-relatedness is often questioned, construction organisations that are serious about the health, safety and well-being of their employees and contractors should examine the impact of work schedule expectations and job-related stressors and ensure that construction programmes are realistic, projects are adequately resourced and employees are able to rest and recover from work as necessary. This may mean that instead of relying on overtime, as is currently the practice, it may be better to hire more workers. Pre-planning construction work more carefully can avoid the need for unscheduled overtime and, in some cases, management may need to limit the number of consecutive days or hours worked. Allowing workers some control over their work scheduling and enabling them to refuse overtime when they need to can also counter the negative health effects of long work hours (Sparks *et al.* 1997). Lastly, it is important that appropriate facilities are provided in which workers can rest. Site sheds should be provided with a supply of cold drinking water and workers should be encouraged to take regular short rest breaks, at least every two hours, remain hydrated and have food available for snacks.

The scientific evaluation of OHS interventions

While much is now known about organisational, situational and individual factors associated with negative OHS outcomes, there has been surprisingly little scientific evaluation of preventive strategies implemented as a result of this knowledge. OHS interventions are actions designed and implemented to improve OHS performance. They can include a whole range of different activities, including workplace modification, the implementation of OHS committees, changing job procedures, providing OHS training or implementing behavioural safety programmes. OHS interventions are often recommended by well-meaning OHS practitioners and others. Sometimes these interventions are resourced and implemented, but rarely are they systematically evaluated. For example, Vojtecky and Schmitz (1986) suggest that OHS training interventions are usually assumed to be having the desired effect and there is little systematic assessment of whether this is actually the case. We suggest that expensive and time-consuming OHS improvement measures should be tested, rigorously and scientifically, on a small but representative sample before they are extended to the population as a whole. This could prevent the wholesale adoption of unproven and ineffective measures which may cost companies, clients or governing bodies a great deal of money while yielding unsatisfactory results. It cannot simply be assumed that OHS improvement measures are having the desired effect and, for this reason, evaluation issues are extremely important to OHS researchers and practitioners alike.

Evaluating the effects of an OHS intervention in a work setting is not simple. The simplest way to evaluate the effectiveness of an OHS intervention is to examine injury rates following the implementation of an intervention. Unfortunately, this is not a very good method of evaluation because injury rates are not a very reliable indicator of safety performance, particularly in small worksites. Also, injury rates are subject to many influences and it is possible that an improvement in injury rate observed following an OHS intervention was actually caused by some unrelated factor, for example a change in the government's policy regarding enforcement of OHS legislation. If this change coincided with the intervention, spurious conclusions could be drawn as to the effectiveness of the intervention. It is important that evaluation of OHS interventions be carefully designed. Furthermore, these evaluations should be designed at the same time as the intervention itself, so that OHS performance data can be gathered before the implementation of the intervention if necessary.

Robson *et al.* (2001) identify a number of different evaluation designs that can be used for evaluating OHS interventions. They suggest that different evaluation designs have different strengths in that they provide differing degrees of confidence with which evaluators can say that a specific intervention *caused* a desired outcome. They identify three types of evaluation design. These are:

1 experimental designs
2 quasi-experimental designs and
3 non-experimental designs.

Experimental designs

Experimental designs provide the strongest evidence of a causal link between an intervention and the observed outcome, for example an improvement in OHS performance. Traditional experimental designs use a control group, which does not participate in the intervention. This control group is then compared with an experimental group that experiences the intervention. Thus, one group of workers who receive a training programme (the experimental group) is compared with another that does not experience the training programme (the control group). If the experimental group's OHS performance differs significantly from the control group's, then this is taken to indicate a causal link. However, causation can only be inferred if evaluators can be sure that workers in the experimental group and the control group did not differ in a systematic way before the intervention. Therefore, in true experimental designs, people or work groups must be assigned to intervention and control groups randomly. This is often difficult to achieve in work settings in which pre-established work groups cannot be changed. Thus, in this case, quasi-experimental designs might be used instead.

Quasi-experimental designs

Quasi-experimental designs offer a compromise between rigour in evaluation and the practical constraints of the workplace. They often include a control group with which comparisons are made, although this control group is often generated by a non-random process, for example comparing the OHS performance of one construction site at which an OHS intervention is implemented (the experimental site) with the OHS performance of a comparable site at which the intervention was not implemented (the control group). While it is always possible that any improvement at the experimental site was the result of systematic differences that already existed between the sites, rather than being caused by the OHS intervention, quasi-experimental designs are better than non-experimental evaluations.

Non-experimental designs

Non-experimental designs provide the weakest evidence of a cause and effect relationship between the intervention and any change in OHS performance. A simple 'before and after' comparison is a type of non-experimental design. 'Before and after' comparisons are sometimes the only way in which an intervention can be evaluated. However, a simple comparison of pre- and post-intervention data from a person or work group is not sufficient to infer a causal relationship. This is because factors, such as the passage of time, maturation of participants or statistical regression can bring about change in performance that is unrelated to the intervention (Campbell and Stanley 1963). 'Before and after' comparisons should only be used when time and circumstances do not permit the use of a more sophisticated evaluation design, and the results of 'before and after' comparisons should be treated cautiously. However, while 'before and after' comparisons provide only weak evidence that an intervention has had its desired effect, Robson *et al.* (2001) suggest they are still better than no evaluation at all.

Single-case experimental designs

As described above, the traditional experimental design involves assigning people or workplaces, by a random process, into a control group and an experimental group, to expose only the experimental group to the intervention and then to compare performance between the two groups. However, in work settings, random assignment is very difficult to arrange, making this type of design extremely difficult to implement.

An alternative to the control group design, which still permits evaluators to determine with confidence whether an OHS intervention has been effective or not, is to draw comparisons within the same group of subjects, in a so-called single-case experimental design (Barlow and Hersen 1984; Komaki and Jensen 1986). Single-case experimental designs have been widely used by researchers evaluating the effectiveness of behavioural safety management interventions. Two types of single-case design are commonly used. These are described below.

The withdrawal design

In its simplest form, this design is denoted as ABA. The A-phase of the evaluation refers to the baseline during which OHS performance is measured to ascertain its natural level. The B-phase of the study refers to the intervention period. An OHS intervention is introduced and maintained for the whole of the B-phase. The withdrawal refers to the removal of the OHS intervention after a pre-set length of time has elapsed, and a return to baseline conditions. If the intervention was responsible for any change in OHS performance, then this change would occur concurrently with the introduction of the intervention, and the trend would be reversed when the intervention is withdrawn and conditions returned to the baseline conditions.

The disadvantage of withdrawal experimental design is that it may be impossible to return to baseline conditions if people have learned new information as a result of the treatment variable, for example in the case of OHS training interventions. In this situation people cannot be expected to un-learn information for the purposes of the evaluation. The withdrawal of the treatment variable may also be undesirable, which is the case in OHS interventions. For example, if an intervention has been introduced that improves OHS behaviour or provides a safer working environment, it would be unethical to remove it for evaluation purposes.

The multiple baseline design

An alternative to the withdrawal design is the multiple baseline experimental design. A multiple baseline design can be carried out across groups, people or behaviours. Data are collected for two or more baselines at the same time, and then the OHS intervention is introduced to the groups, people or behaviours at staggered intervals. The evaluator then examines whether OHS performance changes after the intervention is introduced in the first group, person or behaviour. The evaluator then assesses whether other groups, people or behaviours that have not yet received the intervention continue to perform at their baseline levels. Finally, the evaluator examines whether the same change from baseline to intervention performance is replicated when the intervention is applied to other groups, people or behaviours. If the same OHS improvement occurs at different times for the different groups, people or behaviours, then the evaluator concludes that the intervention is the cause of the improvement.

Ethical considerations in evaluating OHS interventions

In designing evaluations of OHS interventions, there are certain ethical considerations to be made. First, if an OHS intervention is expected to bring about an improvement in OHS performance, for example a reduction in risk to workers performing a particular task, then the potential for harm implied in an evaluation design must be carefully weighed against the likely benefits to be achieved. Thus, if serious hand injuries have been experienced using a particular type of machine, for example a circular saw, and a new type of guard has been purchased for this machine, it would be undesirable to purposely provide this guard to workers at one site but deny it to workers at a control site to evaluate the effectiveness of the guard. In this case, a 'before and after' design may be preferable. The delayed delivery of an intervention to individuals or groups of workers is also an ethical concern in multiple baseline evaluation designs.

A second ethical issue in intervention evaluation is the consent of individuals involved. In Western societies, there is an emphasis on fully informed consent from people in evaluation studies and any evaluations undertaken within universities of government agencies usually require approval by ethics committees, especially when health records or employees' opinions are involved. Therefore, the issue of informed consent should be considered when evaluating OHS interventions.

OHS, social responsibility and business ethics

As the above consideration of the ethical implications of evaluating OHS interventions indicates, OHS has a moral dimension. The results of managerial actions have extended consequences. These consequences are often experienced by people who have no control over the actions that caused them and consequently, there is an argument that these consequences should be considered when decisions are made. If decisions can hurt or harm people in ways that are outside their individual control, then the issue is a moral one, which requires some ethical analysis.

Unsafe systems of work damage individual lives and pollution harms environmental health. A shocking example of the impacts of faulty managerial decision-making on the community is the Union Carbide disaster at Bhopal. The plant at Bhopal was a major manufacturing facility producing pesticide chemicals. The accident that occurred in 1984 killed and injured thousands, as well as leaving many families without a primary income earner. Thousands of jobs were lost, and the costs to the Indian Government in providing food, medical treatment and hospitals exceeded US$40 million. Union Carbide was held responsible for the disaster and made to pay US$470 million in damages in a landmark decision, which formally recognised that organisations have a significant responsibility to the society in which they operate. Robertson and Fadil (1998) suggest that the deficient safety programme that led to the Bhopal disaster can be viewed as a product of a faulty ethical compliance programme. They suggest that the communication of standards and procedures from the headquarters to the Indian plant was ineffective, codes and standards were not enforced and Union Carbide's senior management was not held responsible for the management programmes at Bhopal.

There is increasing pressure on organisations to demonstrate transparency and accountability for more than the traditional measure of financial performance. There has been an increasing public expectation that organisations will take responsibility for their non-financial impacts, including their environmental performance and impacts upon the community. This has led to the notions of social responsibility and Triple Bottom Line (TBL) reporting. A new suite of Australian Standards has just been published concerning business governance. AS 8003 (Standards Australia 2003) deals with the issue of corporate social responsibility (CSR). CSR is defined as: 'A mechanism for entities to voluntarily integrate social and environmental concerns into their operations and their interaction with their stakeholders, which are over and above the entity's legal responsibilities' (p. 4).

The standard provides for the establishment of a CSR policy, the establishment of management responsibilities for CSR, the identification of CSR issues relevant to the organisation, the establishment of operating procedures for CSR, stakeholder engagement, communication, education and training, and monitoring and reviewing CSR performance. Health and safety issues explicitly identified in the standard include:

- personal health
- food and nutrition
- potable water
- dormitories
- factory ventilation
- emergency evacuation
- fire safety
- use of chemicals and
- ergonomics.

In addition, employee-related issues that could impact upon OHS include unreasonable working hours, freedom of association and discrimination. The standard also lists supplier ethical issues and employment standards as issues of CSR.

TBL reporting requires organisations to report their performance in accordance with a range of financial, environmental and social indicators. OHS performance is an important component of these social indicators. At the moment, TBL reporting is voluntary but, if the trend continues, clients, shareholders and job-seekers will become more socially aware and begin to demand the reporting of non-financial performance. In the UK, HSC Strategy Statement contains an action point focusing on the public reporting of health and safety by large companies. The HSC have issued guidelines recommending that companies report on their health and safety principles, performance and targets. A recent study of the annual reports of leading British companies revealed that reporting OHS issues had increased considerably between 1995 and 2001 (Peebles *et al.* 2002). However, the information included in reports was sometimes poor and was often limited to a broad statement of policy. Few companies reported their OHS targets. There were only two construction companies in the sample, neither of which included any health and safety information in their annual reports. While the reporting of OHS information is currently voluntary, principles of corporate social responsibility support transparency and reporting of non-financial performance. Failure to be transparent, or poor performance in these areas, is likely to harm a company's reputation, and may also impact upon financial performance in the future. For example, the notion of ethical investment is becoming more significant and ethically aware graduates are less likely to be willing to work for organisations whose performance in the social and environmental domains is poor.

The issues of social responsibility and TBL reporting are grounded in stakeholder theory, which holds that managers have a responsibility to all those who have a stake in or claim on the organisation. This is counter to the shareholder value theory, which holds that managers have a special relationship with the shareholders of the organisation. This view was perhaps most famously stated by Milton Friedman who wrote in the *New York Times Magazine* that business had no responsibility beyond making profits for its shareholders. Employees are a key stakeholder group whose interests, stakeholder theory holds, should be taken into consideration in decision-making. This moral view of employees as stakeholders may necessitate a trading off of the economic benefit to shareholders against the interests of other stakeholders, including employees. Thus, in situations in which OHS risks are intolerable, stakeholder theory would require that profits be sacrificed in the interests of workers' well-being.

Rowan (2000) suggests that there is something morally significant about persons and it is this that makes it morally wrong to treat them in certain ways. The way in which organisations treat their employees is an issue that raises ethical concerns. In human resource management, Greenwood (2002) suggests management is a euphemism for 'use'. The use of others poses ethical problems. Kantian ethical theory holds that it is morally wrong to treat people entirely as a means to an end. However, Greenwood (2002) suggests that saying that a company should not use employees as a means to an end is untenable because the reason that people are employed is to contribute to the achievement of company goals. She goes on to suggest that it may be better to say that employees should not be used *exclusively* as a means to an end, nor used as a means to an end *under specific circumstances*. Workers exposed to unsafe work practices are being used as a means to achieving an end, in the case of construction of a building or other structure, in unacceptable circumstances. In terms of what managerial behaviour passes the ethics test, Greenwood (2002) recommends the development of some ethical principles for managing employees. Rowan (2000) suggests three principles, which are based upon an individual's right to pursue his or her own interests. These are the right to freedom, the right to well-being and the right to equality. If these rights are accepted as being universal human rights, it follows that workers have a moral right to safety in the workplace. This right is held against employers, who have a concomitant duty to provide a safe work environment.

Society has recognised this right and passed legislation to protect workers' health and safety in and arising from work. However, not all moral rights are legal rights. Rowan (2000) cites the example of African slaves in the

United States. While there was no law against slavery for many years, the treatment of people as slaves could not be said to be morally right. Thus, ethical conduct is more than remaining within the law (compare the definition of social responsibility provided above).

The next question is how can we ensure that managers will behave responsibly and manage employees in an ethical way. Ensuring that all organisational members understand a unified set of ethical guidelines is important because in modern large decentralised organisations, people at many different levels of the hierarchy and in many different locations are making decisions with ethical content on a daily basis. Organisational codes of ethics are one means by which a unified set of ethical guidelines can be provided to guide behaviour. Schwartz (2001) examined the mechanisms by which codes of ethics influenced behaviour and reports that they operate in a number of ways. These are:

- as a rule book, guiding conduct or action;
- as a signpost, pointing out sources of further guidance;
- as a mirror, validating employees' own behavioural intentions;
- as a magnifying glass, magnifying the importance of ethical issues;
- as a shield, increasing employees ability to resist unethical requests;
- as a smoke detector, providing a warning device to employees who tend towards unethical behaviour;
- as a fire alarm, signalling unethical conduct; and
- as a club, providing the threat of potential discipline for breaching the code.

Including the treatment of people with respect and OHS issues in codes of conduct or codes of ethics may therefore be a valuable mechanism for ensuring that managers, at all levels, make decisions that do not violate employees' right to well-being.

Conclusions

In concluding this book, we cannot overemphasise the need for the construction industry to learn effectively from its experiences and mistakes. The key questions do not arise from what we do not know, so much as how we can make better use of what we do know, to bring about sustainable improvements in OHS in the industry. We know much about how, why, when and to whom injuries and illnesses occur and we know that these occurrences are too frequent. We know that cultural change must be an integral part of becoming a learning organisation within a learning industry and we know of impediments to achieving the cultural change required to achieve continuing improvements in OHS in construction. We are beginning to recognise that existing work practices, such as long hours and inadequate attention to workers' psychological well-being and work-life balance, militate against better OHS performance in the industry. We understand that, in order to be effective in improving OHS performance, we need to carefully design and evaluate interventions, so as not to waste time or resources in fruitless endeavours. We have an appreciation of the ethical and moral, and increasingly legal, imperatives acting upon managers and organisations to protect, and even actively promote, the health and safety of their employees, contractors and other people who may be affected by the organisation's activities.

The construction industry must now make a concerted effort to transform the way that it operates. It needs to break the cycle of repeating the same mistakes. This will require technological measures, for example in implementing incident information systems, such as the HKHA example described in Chapter 9, to enable patterns in the underlying causes of incidents to be diagnosed and treated in a timely manner. But it will also require a significant cultural transformation to overcome the difficulty that industry participants currently have in discussing OHS issues openly and making realistic attributions about their causes. This cultural transformation might be supported by the use of tools, such as benchmarking OHS cultures and management systems. This benchmarking should not be restricted to choosing partners within the construction industry but would be more effective if benchmarking partners were chosen from other industries, for example petrochemicals, telecommunications and mining, in which more established OHS management systems are established and stronger safety cultures exist.

In particular, change must filter down to the level of construction projects, at which the difference between espoused OHS values and 'lived' safety becomes immediately apparent. The importance of the role of first-line supervisors in communicating project priorities, and the importance of OHS, must be recognised, and the responsibility of first-line supervisors for OHS should be clearly communicated. However, this should not result in supervisors adopting a rule-based corrective approach. Rather, it is better that supervisors adopt a participative management approach in which workers feel free to openly voice OHS concerns and have a direct input into team-level OHS decision-making.

At an industry level, the impact of work hours needs careful examination. The research linking fatigue to incidents coupled with the awareness of the mental health implications of work-life imbalance indicates a need to examine whether current job scheduling practices are ethical, and indeed, legal. All parties to a construction project must play a part in this analysis, including clients who must recognise that the cost and time constraints they impose can have implications for OHS. Clearly concerted effort is required and no one party should be held to be solely responsible for OHS. However, in many countries, there is still a need to examine the role played by construction clients, designers and suppliers and manufacturers of plant and materials in ensuring construction workers' OHS.

There is also a need for rigorous evaluation techniques for OHS interventions. Few interventions are evaluated adequately and, without such evaluation, there is no way to know whether they are having a positive, beneficial effect. Universities and research organisations can play an important role in working with industry to evaluate OHS initiatives, implemented either at an industry, organisation or project level.

We occupy a highly complex world in which technology plays a critical part. We need buildings, roads and infrastructure in order to enjoy the lifestyle we expect and we rely on the construction process to provide these things. However, in the course of providing facilities we continue to expose construction workers to unacceptable levels of risk of injury and illness. In doing so, we deny many workers each year their basic right to well-being. This cannot be justified on any grounds, and organisations involved in the construction process must come to understand that OHS is more than a compliance issue but is also an important moral issue.

Discussion and review questions

1 Describe the key characteristics of a learning organisation. What features of construction organisations need to change in order to facilitate organisational learning in OHS?
2 Explain the difference between safety culture and climate. Is it fair to assume that organisations have a single unified safety culture or climate? Why, or why not?
3 Why is it important that OHS interventions are rigorously evaluated? In organisational settings what types of evaluation designs can be applied? Why are some of these evaluation designs difficult to apply?
4 Describe the relationship between business ethics and OHS.

References

ABS (Australian Bureau of Statistics) (1997a) *Building Activity*, publication 8752, Australia.

ABS (Australian Bureau of Statistics) (1997b) *Engineering Construction Activity*, publication 8762, Australia.

ABS (Australian Bureau of Statistics) (1998) *Business Register Data*, Australian Bureau of Statistics Melbourne.

ABS (Australian Bureau of Statistics) (2003) *Employer Training Expenditure and Practices*, publication 6362.0.

AEGIS (1999) *Mapping the Building and Construction Product System in Australia*, AEGIS, University of Western Sydney, Sydney.

Ajzen, I. (1988) *Attitudes, Personality and Behavior*, Open University Press, Milton Keynes.

Akerstedt, T. (1995) Work hours, sleepiness and accidents: Introduction and summary, *Journal of Sleep Research*, 4 (Supplement 4), 1–3.

Akerstedt, T., Fredlund, P., Gillberg, M. and Jansson, B. (2002) A prospective study of fatal occupational accidents: Relationship to sleeping difficulties and occupational factors, *Journal of Sleep Research*, 11, 69–71.

Alarcon, L. (ed.) (1997) *Proceedings of International Conference on Lean Production*, Balkema, Rotterdam, ISBN 90 5410 648 4.

Algera, J. A. (1990) Feedback system in organisations, in C. L. Cooper and I. T. Robertson (eds), *International Review of Industrial and Organisational Psychology*, 5, John Wiley & Sons Ltd, London, pp. 169–193.

Amick, B. C. III, Habeck, R. V., Hunt, A., Fossel, A. H., Chapin, A., Keller, R. B. and Katz, J. N. (2000) Measuring the impact of organizational behaviors on work disability prevention and management, *Journal of Occupational Rehabilitation*, 10, 21–37.

Anderson, J. (1998a) Construction safety: Changes needed now to the CDM Regs, *The Safety and Health Practitioner*, May, 26–28.

Anderson, J. (1998b) Growing a safety culture, in M. Barnard (ed.), *Health and Safety for Engineers*, Thomas Telford, London.

Andersson, R. and Lagerloff, E. (1983) Accident data in the new Swedish information system on occupational injuries, *Ergonomics*, 26, 33–42.

Andreoni, D. (1986) *The Cost of Occupational Accidents and Diseases*, Occupational Safety and Health Series, No. 54, ILO, Geneva.

APESMA (2000) APESMA working hours and employment security survey report, PowerPoint presentation (personal communication).

Appels, A. and Schouten, E. (1991) Burnout as a risk factor for coronary heart disease, *Behavioral Medicine*, 17, 53–59.

Argyris, C. (1999) *On Organizational Learning* (2nd edition), Blackwell Business, Oxford.

Armstrong, T. J., Haig, A. J., Franzblau, A., Keyserling, W. M., Levine, S. P., Martin, B. A., Ulin, S. S. and Werner, R. A. (2000) Medical management and rehabilitation in the workplace: Emerging issues, *Journal of Occupational Rehabilitation*, 10, 1–6.

Arup, C. (1993) A critical review of workers' compensation, in M. Quinlan (ed.), *Work and Health: The Origins, Management and Regulation of Occupational Illness*, Macmillan, Melbourne.

AS/NZS 4360 (1997) *Risk Management*, Australian/New Zealand Standards, Homebush, NSW.

AS/NZS 3931 (1998) *Risk Analysis of Technological Systems – Application Guide*, Australian/New Zealand Standards, Homebush, NSW.

Ashton, M. C. (1998) Personality and job performance: The importance of narrow traits, *Journal of Organisational Behavior*, 19, 289–303.

Askenazy, P. (2001) Innovative work practices and occupational injuries and illnesses in the United States, *Economic and Industrial Democracy*, 22, 485–516.

Atkin, B. L., Rowlinson, S. M. and Matthews, J. (unpublished 2000) *Benchmarking Construction Safety*, Research Report.

Atkinson, J. W. (1964) *An Introduction to Motivation*, Van Nostrand, Princeton NJ.

Atkinson, W. (1999) Wake up! Fighting fatigue in the workplace, in *Risk Management*, Risk Management Society, New York, pp. 10–22.

Austen, A. D. and Neale, R. H. (eds) (1984) *Managing Construction Projects – A Guide to Processes and Procedures*, International Labour Office (ILO), Geneva.

Australian Financial Review, 29 January 2003, Work manslaughter laws dead but not buried, John Fairfax Publications Pty Ltd, p. 8.

Australian Financial Review, 17 February, 2003, Lunch injury widens scope for claims, John Fairfax Publications Pty Ltd, p. 6.

Australian Financial Review, 31 March, 2003, Building employers won't escape penalties: Abbott, John Fairfax Publications Pty Ltd, p. 3.

Australian Law Reform Commission (1994) Report No. 68, *Compliance with the Trade Practices Act 1994*, AGPS, Canberra.

Ayres, I. and Braithwaite, J. (1992) *Responsive Regulation, Transcending the Deregulation Debate*, Oxford University Press, Oxford.

Bacharach, S. B., Bamberger, P. and Conley, S. (1991) Work-home conflict among nurses and engineers: Mediating the impact of role stress on burnout and satisfaction at work, *Journal of Organizational Behavior*, 12, 39–53.

Bamber, L. (1996) Risk management techniques and practices, in J. Ridley (ed.), *Safety at Work* (4th edition), Butterworth-Heinneman, Oxford.

Bandura, A. (1986) *Social Foundations of Thought and Action: A Social Cognitive Theory*, Prentice-Hall, Englewood Cliffs, NJ.

Barabasi, A. (2002) *Linked: The New Science of Networks*, Perseus.

Barlow, D. H. and Hersen, M. (1984) *Single Case Experimental Designs: Strategies for Studying Behavior Change*, Pergamon Press, New York.

Baron, M. M. and Pate-Cornell, E. B. (1999) Designing risk-management strategies for critical engineering systems, *IEEE Transactions*, 46, 87–100.

Bartel, A. P. and Thomas, L. G. (1985) Direct and indirect effects regulation: A new look at OSHA's impact, *Journal of Law and Economics*, 28, 1–25.

Bartrip, P. (1985) The rise and decline of workmen's compensation, in P. Weindling (ed.), *The Social History of Occupational Health*, Croom Helm, London.

Baxendale, T. and Jones, O. (2000) Construction design and management safety regulations in practice: Progress on implementation, *International Journal of Project Management*, 18, 33–40.

Bedeian, R. C., Burke, B. G. and Moffett, R. C. (1988) Outcomes of work-family conflict among male and female professionals, *Journal of Management*, 14, 475–491.

Behrens, V. J. and Brackbill, R. M. (1993) Worker awareness of exposure: Industries and occupations with low awareness, *American Journal of Industrial Medicine*, 23, 695–701.

Belbin, R. M. (1981) *Management Teams: Why They Succeed or Fail*, Heinemann, London.

Belbin, R. M. (1993) *Team Roles at Work*, Butterworth-Heinemann, Oxford.

Berger, Y. (2001) Hot water, wartime solidarity and updraughts of disaster, *ANU Reporter*, http://www.anu.edu.au/mac/reporter/volume/31/16/longford.html.

Bernard, B. P. (1997) *Musculoskeletal Disorders (MSDs) and Workplace Factors*, second printing, US Department of Health and Human Services.

Bhattacharya, A., Mueller, M. and Putz-Anderson, V. (1985) Traumatogenic factors affecting the knees of carpet installers, *Applied Ergonomics*, 16, 243–250.

Biggins, D., Phillips, M. and O'Sullivan, P. (1991) Benefits of worker participation in health and safety, *Labour and Industry*, 4, 138–159.

Bird, F. E. and Loftus, R. G. (1976) *Loss Control Management*, Institute Press, Loganville, USA.

Blockley, D. (ed.) (1992) *Engineering Safety*, McGraw-Hill, Maiden head.

Bluff, L. (2003) *Regulating Safe Design and Planning of Construction Works*, Working Paper 19, National Research Centre for OHS Regulation, Australian National University, Canberra.

Bobick, T. G., Stanevich, R. L., Pizatella, T. J., Keane, P. R. and Smith, D. L. (1994) Preventing falls through skylights and roof openings, *Professional Safety*, 39, 33–37.

Bohle, P. and Quinlan, M. (2000) *Managing Occupational Health and Safety: A Multi-disciplinary Approach* (2nd edition), Macmillan Publishers Australia, Melbourne.

Bottomly and Associates (2003) National Occupational Health and Safety Commission Issues Paper 5, *An Assessment of Current Performance Indicators as a Measurement Tool and the Development of An Adaptive Model for Benchmarking Best Practice for Physical Handling in the Work Environment*, NOHSC, Canberra.

Bradley, G. L. (1989) The forgotten role of environmental control: Some thoughts on the psychology of safety, *Journal of Occupational Health and Safety: Australia and New Zealand*, 5, 501–508.

Brody, B., Letourneau, Y. and Poirier, A. (1990) An indirect cost theory of work accident prevention, *Journal of Occupational Accidents*, 13, 255–270.

Brook, M. (1993) *Estimating and Tendering for Construction Work*, Butterworth-Heinemann, Oxford.

Brooks, A. (1988) Rethinking occupational health and safety legislation, *Journal of Industrial Relations*, 30, 347–362.

Brooks, A. (1993) *Occupational Health and Safety Law in Australia* (4th edition), CCH Australia Ltd, Sydney.

Brown, P. E. (1996) Total integration of the safety professional into the project management team, in L. M. Aves Dias and R. J. Coble (eds), *Implementation of Safety and Health on Construction Sites*, A. A. Balkema, Rotterdam.

Brown, R. (1992) Administrative and criminal penalties in the enforcement of occupational health and safety legislation, *Osgoodehall Law Journal*, 30, 691–735.

Brown, R. L. and Holmes, H. (1986) The use of a factor analytic procedure for assessing the validity of an employee safety climate model, *Accident Analysis and Prevention*, 18, 455–470.

Bruce, C. and Aitkins, F. (1993) Efficiency effects of premium-setting regimes under workers' compensation: Canada and the United States, *Journal of Labor Economics*, Supplement, S38–S69.

Bruck, C. S., Allen, T. D. and Spector, P. E. (2002) The relation between work-family conflict and job satisfaction: A finer-grained analysis, *Journal of Vocational Behavior*, 60, 336–353.

Bruyere, S. and Shrey, D. (1991) Disability management in industry: A joint labor-management process, *Rehabilitation Counseling Bulletin*, 34, 227–242.

Budworth, N. (1997) The development and evaluation of a safety climate measure as a diagnostic tool in safety management, *Journal of the Institution of Occupational Safety and Health*, 1, 1, 19–29.

Burgess, C. (2000) The wilful traffic offender profile and its implications for education and training, PhD Research Summary, School of Psychology, University of Exeter.

Burgess-Limerick, R. (2003) National Occupational Health and Safety Commission Issues Paper 2, *Issues Associated with Force and Weight Limits and Associated Threshold Limit Values in the Physical Work Environment*, NOHSC, Canberra.

Burns, T. and Stalker, G. M. (1961) *The Management of Innovation*, Tavistock Publications, London.

Burtsyn, I., Kromhout, H. and Boffetta, P. (2000) Literature review of levels and determinants of exposure to potential carcinogens and other agents in the road construction industry, *American Industrial Hygiene Association Journal*, 61, 715–726.

Busby, J. S. (1999) The effectiveness of collective retrospection as a mechanism of organizational learning, *Journal of Applied Behavioral Science*, 35, 109–129.

Butler, R., Johnson, W. and Baldwin, M. (1995) Managing work disability: Why first return to work is not a measure of success, *Industrial and Labor Review*, 48, 452–467.

Campbell, D. T. and Stanley, J. C. (1963) Experimental and quasi-experimental designs for research, in D. T. Campbell and J. C. Stanley, Rand McNally, Chicago.

Campbell, J. P. and Pritchard, R. D. (1976) Motivation theory in industrial and organisational psychology, in M. D. Dunnette (ed.), *Handbook for Industrial and Organisational Psychology*, Rand McNally, USA, pp. 63–130.

Campbell, R. (1999) *Crisis Control: Preventing and Managing Corporate Crises*, Prentice-Hall, Australia.

Cannon, M. D. and Edmonson, A. C. (2001) Confronting failure: Antecedents and consequences of shared beliefs about failure in organisational work groups, *Journal of Organisational Behavior*, 22, 161–177.

Carpenter, J., Williams, P. and Charlton Smith, N. (2001) *Identification and Management of Risk in Undergraduate Construction Courses*, Health and Safety Executive, HMSO, Norwich.

Carson, W. G. (1979) The conventionalisation of early factory crime, *International Journal of the Sociology of Law*, 7, 37–60.

Cheadle, A., Franklin, G., Wolfhagen, C., Savarino, J., Liu, P. Y. and Weaver, M. (1994) Factors influencing the duration of work-related disability: A population-based study of Washington state workers' compensation, *American Journal of Public Health*, 84, 190–196.

Cheetham, G. and Chivers, G. (2001) How professionals learn in practice: An investigation of informal learning amongst people working in professions, *Journal of European Industrial Training*, 25, 5, 248–292.

Cheng, E. W. L., Li, H., Love, P. E. D. and Irani, Z. (2001) Network communication in the construction industry, *Corporate Communications: An International Journal*, 6, 61–70.

Cherns, A. B. and Bryant, D. T. (1984) Studying the client's role in construction management, *Construction Management and Economics*, 2, 177–184.

Cherrington, D. J. (1991) Need theories of motivation, in R. M. Steers and L. W. Porter (eds), *Motivation and Work Behavior*, McGraw-Hill, New York, pp. 31–44.

Chokkar, J. S. and Wallin, J. A. (1984) Improving safety through applied behaviour analysis, *Journal of Safety Research*, 15, 141–151.

Clarke, S. (1999) Perceptions of organisational safety: Implications for the development of safety culture, *Journal of Organisational Behavior*, 20, 185–198.

Clayton, A. (2002) *The Prevention of Occupational Injuries and Illness: The Role of Economic Incentives*, Working Paper 5, National Research Centre for Occupational Health and Safety Regulation, Australian National University, Canberra.

Clayton, A., Johnstone, R. and Sceats, S. (2002) The legal concept of work-related injury and disease in Australian OHS and workers' compensation systems, *Australian Journal of Labour Law*, 15, 105–153.

Coase, R. H. (1937) The nature of the firm, *Economica*, 4, 386–405.

Coglianese, C., Nash, J. and Olmstead, T. (2002) *Performance-based Regulation: Prospects and Limitations in Health, Safety and Environmental Protection*, John F. Kennedy School of Government Research Working Paper No. RWP02–050, Harvard University, Boston.

Cohen, A. (1977) Factors in successful occupational safety programs, *Journal of Safety Research*, 9, 168–178.

Cohen, A. V. (1996) Quantitative risk assessment and decisions about risk: An essential input into the decision process, in C. Hood and D. K. C. Jones (eds), *Accident and Design: Contemporary Debates in Risk Management*, UCL Press, London, pp. 87–98.

Colella, A. (2001) Coworkers distributive fairness judgements of the workplace accommodation of employees with disabilities, *Academy of Management Review*, 26, 100–116.

Commission of the European Communities (1993) *Safety and Health in the Construction Sector*, Office for Official Publications of the European Communities, Luxembourg.

Confederation of British Industry (1990) *Developing a Safety Culture*, CBI, London.

Cook, T. M. and Zimmerman C. L. (1992) A symptom and job factor survey of unionised construction workers, in S. Kumar (ed.), *Advances in Industrial Ergonomics and Safety* IV, Taylor & Francis, Philadelphia, PA.

Cooney Report (1984) Committee of Inquiry into the Victorian Workers' Compensation System 1983–1984, Report, Government Printer, Melbourne.

Cooper, M. (1998) Current issues in health and safety training in the UK, *Journal of European Industrial Training*, 22, 9, 354–361.

Cooper, M. and Cotton, D. (2000) Safety training: A special case? *Journal of European Industrial Training*, 24, 9, 481–490.

Cooper M. D. (1992) *An Examination of Participative and Assigned Goal-setting in Relation to the Improvement of Safety on Construction Sites*, Unpublished doctoral thesis, School of Management, University of Manchester Institute of Science & Technology (UMIST), UK.

Cooper, M. D. and Phillips, R. A. (1994) Validation of a safety climate measure, presented at the British Psychological Society Occupational Psychology Conference, January 1994, Birmingham.

Cooper, M. D., Phillips, R. A., Sutherland, V. J. and Makin, P. J. (1994) Reducing accidents with goal-setting and feedback: A field study, *Journal of Occupational and Organisational Psychology*, 67, 219–240.

Costa, P. T. Jr and McCrae, R. R. (1985) Hypochondriasis, neuroticism and aging: When are somatic complaints unfounded? *American Psychologist*, 40, 19–28.

Cox, S. and Cox, T. (1991) The structure of employee attitudes to safety: A European example, *Work and Stress*, 5, 93–106.

Cox, S. and Tait, R. (1993) *Safety Reliability & Risk Management: An Integrated Approach* (2nd edition), Butterworth-Heinneman, Oxford.

Coyle, I. R., Sleeman, S. D. and Adams, N. (1995) Safety climate, *Journal of Safety Research*, 26, 247–254.

Crabtree, M. (1994) Corporate culpability for industrial manslaughter: Finding the 'soul' of the Australian corporation, *Australian Business Law Review*, 22, 376–381.

Cranes (Suspended Personnel) Regulations (1993) State of Victoria, Melbourne.

Cranes Regulations (1989) State of Victoria, Melbourne.

Creighton, B. and Rozen, P. (1997) *Occupational Health and Safety Law in Victoria* (2nd edition), The Federation Press, Sydney.

Creighton, B. and Stewart, A. (1994) *Labour Law: An Introduction*, The Federation Press, Sydney.

Crook, J., Moldofsky, H. and Shannon, H. (1998) Determinants of disability after a work related musculoskeletal injury, *Journal of Rheumatology*, 25, 1570–1577.

Crossland, B., Bennett, P. A., Ellis, A. F., Farmer, F. R., Gittus, J., Godfrey, P. S., Hambley, E. C., Kletz, T. A. and Lees, F. P. (1993) Estimating engineering risk, in *Risk, Analysis, Perception and Management*, Royal Society, London.

Dahlman, C. J. (1979) The problem of externality, *Journal of Law and Economics*, 22, 141–162.

Dale, M. (1994) Learning organizations, in C. Mabey and P. Iles (eds), *Managing Learning*, Routledge, London, pp. 22–33.

Daniell, W. E., Fulton-Kehoe, D., Cohen, M., Swan, S. S. and Franklin, G. M. (2002) Increased reporting of occupational hearing loss: Workers' compensation in

Washington State 1984–1998, *American Journal of Industrial Medicine*, 45, 502–510.

Dasinger, L. K., Krause, N., Deegan, L. J., Brand, J. B. and Rudolph, L. (2000) Physical workplace factors and return to work after compensated lowback injury: A disability phase-specific analysis, *Journal of Occupational and Environmental Medicine*, 42, 323–333.

Davies, N. V. and Teasedale, P. (1994) *The Costs to the British Economy of Work Accidents and Work-Related Ill-Health*, Health and Safety Executive, HMSO, London.

Davies, V. J. and Tomasin, K. (1990) *Construction Safety Handbook*, Thomas Telford, London.

Davison, J. (2003) *The Development of a Knowledge Based System to Deliver Health and Safety Information to Designers in the Construction Industry*, Research Report 173, HSE Books, HMSO, Norwich.

Dawson, D., Willman, P., Clinton, A. and Bamford, M. (1988) *Safety at Work: The Limits of Self Regulation*, Cambridge University Press, Cambridge.

Dedobbeleer, N. and Beland, F. (1991) A safety climate measure for construction sites, *Journal of Safety Research*, 22, 97–103.

DeJoy, D. M. (1989) The optimism bias and traffic accident risk perception, *Accident Analysis and Prevention*, 21, 333–340.

DeJoy, D. M. (1994) Managing safety in the workplace: An attribution theory analysis and model, *Journal of Safety Research*, 22, 97–103.

Dekker, S. W. A. (2002) Reconstructing human contributions to accidents: The new view on error and performance, *Journal of Safety Research*, 33, 371–385.

Dembe, A. E. (2001) The social consequences of occupational injuries and illnesses, *American Journal of Industrial Medicine*, 40, 403–417.

Department of Employment, Training and Industrial Relations (2000) Health and safety in the building and construction industry: Workplace health and safety taskforce final report, DETIR, Brisbane.

Derr, J., Forst, L., Chen, H. Y. and Conroy, L. (2001) Fatal falls in the US construction industry 1990–1999, *Journal of Occupational and Environmental Medicine*, 4, 853–860.

Diaz, R. I. and Cabrera, D. D. (1997) Safety climate and attitude as evaluation measures of organizational safety, *Accident Analysis and Prevention*, 29, 643–650.

Dimov, M., Bhattacharya, A., Lemasters, G., Atterbury, M., Greathouse, L. and Ollola-Glenn, N. (2000) Exertion and body discomfort perceived symptoms associated with carpentry tasks: An on-site evaluation, *American Industrial Hygiene Association Journal*, 61, 685–691.

Dinges, D. F. (1995) An overview of sleepiness and accidents, *Journal of Sleep Research*, 4 (Supplement 2), 4–14.

Dong, S. (2002) Work-scheduling, overtime and work-related injuries in construction, presented at the 12th Annual Construction Safety Conference, Chicago, Illinois.

Donovan, P., Hannigan, K. and Crowe, D. (2001) The learning transfer system approach to estimating the benefits of training: Empirical evidence, *Journal of European Industrial Training*, 25, 221–228.

Doree, A. (1991) How professional clients obtain design, Chapter 11.

Dougherty, E. M. and Fragola, J. R. (1988) *Human Reliability Analysis*, John Wiley, New York.

Duff, A. R., Robertson, I. T., Cooper, M. D. and Phillips, R. A. (1993) *Improving Safety on Construction Sites by Changing Personnel Behaviour*, HSE Contract Research Report No. 51/1993, HMSO, London.

Duff, A. R., Robertson, I. T., Phillips, R. A. and Cooper, M. D. (1994) Improving safety by the modification of behaviour, *Construction Management and Economics*, 12, 67–78.

Dwyer, T. (1991) *Life and Death at Work: Industrial Accidents As a Case of Socially Produced Error*, Plenum, New York.

Eakins, J. (1992) Leaving it up to the workers: Sociological perspectives on the management of health and safety in small workplaces, *International Journal of Health Services*, 22, 689–704.

Eastham, R. A. (1990) The decision to tender within current contractual arrangements.

EC (1993) Europe for health and safety at work, *Social Europe 3*, DG V, European Commission, Office for Official Publications of the European Communities, Luxemourg.

Egan, J. (1998) *Rethinking Construction: The Report of the Construction Task Force*, Department of the Environment, Transport and the Regions, London.

Ellickson, R. C. (1987) A critique of economic and sociological theories of social control, *Journal of Legal Studies*, 16, 67–99.

Entec (2000) *Construction Health and Safety for the New Millennium*, Health and Safety Executive Contract Research Report 313/2000, HMSO, Norwich.

European Construction Institute (1996) *Total Project Management of Construction Safety, Health and Environment*, Thomas Telford, London.

Feehely, J. and Huntington, M. (2002) Royal commission into the building and construction industry: The story so far, *Australian Property Journal*, August, 200–201.

Fernandez, J. A., Daltuva, J. A. and Robins, T. G. (2000) Industrial emergency response training: An assessment of long term impact of a union-based program, *American Journal of Industrial Medicine*, 38, 598–605.

Ferris, R., Young, C. and Mayhew, C. (no date) *An Evaluation of the Legal Requirements for the Completion of Workplace Health and Safety Plans in Small Business Building*, Workplace Health and Safety, Department of Training and Industrial Relations (Queensland), Brisbane.

Festinger, L. (1950) Informal social communication, *Psychological Review*, 57, 271–282.

Field, S. and Jorg, N. (1991) Corporate liability and manslaughter: Should we be going Dutch? *Criminal Law Review*, pp. 156–171.

Fink, S. (1986) *Crisis Management: Planning for the Inevitable*, American Management Association, New York.

Fiol, C. M. and Lyles, M. A. (1985) Organizational learning, *Academy of Management Review*, 10, 803–813.

Fishbein, M. and Ajzen, I. (1975) *Belief, Attitude, Intention and Behavior: An Introduction to Theory and Research*, Addison-Wesley, Reading, Mass.

Fisse, B. (1994) Individual and corporate criminal responsibility and sanctions against corporations, in R. Johnstone (ed.), *Occupational Health and Safety Prosecutions in Australia*, Centre for Employment and Relations Law, The University of Melbourne, Melbourne.

Fisse, B. and Braithwaite, J. (1993) *Corporations, Crime and Accountability*, Cambridge University Press, Cambridge.

Flin, R., Mearns, K., O'Connor, P. and Bryden, R. (2000) Measuring safety climate: Identifying the common features, *Safety Science*, 34, 177–192.

Ford, D. N., Voyer, J. J. and Gould Wilkinson, J. M. (2000) Building learning organizations in engineering cultures: Case study, *ASCE Journal of Management in Engineering*, July/August, 72–83.

Foreman, P. and Murphy, G. (1996) Work values and expectancies in occupational rehabilitation: The role of cognitive variables in the return-to-work process, *Journal of Rehabilitation*, July/August/September, 44–48.

Franche, R.-L. and Krause, N. (2002) Readiness for return to work following injury or illness: Conceptualizing the interpersonal impact of health care, workplace and insurance factors, *Journal of Occupational Rehabilitation*, 12, 233–255.

Frick, K. and Walters, D. (1998) Worker representation on health and safety in small enterprises: Lessons from a Swedish approach, *International Labour Review*, 137, 367–389.

Frick, K. and Wren, J. (2000) Reviewing occupational health and safety management: Multiple roots, diverse perspectives and ambiguous outcomes, in K. Frick, P. L. Jensen, M. Quinlan and T. Wilthagen (eds), *Systematic Occupational Health and Safety Management*, Pergamon, Amsterdam.

Friedman, M. and Booth-Kewley, S. (1987) *The 'Disease-Prone Personality'*, Knopf, New York.

Frone, M. R., Russell, M. and Cooper, M. L. (1992) Antecedents and outcomes of work-family conflict: Testing a model of the work-family interface, *Journal of Applied Psychology*, 77, 65–78.

Furnham, A. (1990) The type A behaviour pattern and the perception of self, *Personality and Individual Differences*, 11, 841–851.

Furnham, A. (1992) *Personality at Work*, Routledge, London.

Gadd, S., Keeley, D. and Balmforth, H. (2003) *Good Practice and Pitfalls in Risk Assessment*, Research Report 151, Health and Safety Executive, HMSO, Norwich.

Gaines, B. R. (1990) Knowledge acquisition systems, in A. Hojjat (ed.), *Knoweledge Engineering Fundamentals* (volume 1), McGraw-Hill, pp. 52–70.

Gambatese, J. and Hinze, J. (1999) Addressing construction worker safety in the design phase: Designing for construction worker safety, *Automation in Construction*, 8, 643–649.

Ganora, A. and Wright, G. (1987) Occupational rehabilitation: Costs and benefits, *Journal of Occupational Health and Safety – Australia and New Zealand*, 3, 331–337.

Garvin, D. (1998) Building a learning organisation, *Harvard Business Review on Knowledge Management*, Harvard Business School Press, Boston, MA.

Geier, J. and Schnuch, A. A. (1995) A comparison of contact allergies among construction and non construction workers attending contact dermatitis clinics in Germany: Results of the information network of departments of dermatology from November 1989 to July 1993, *American Journal of Contact Dermatitis*, 6, 86–94.

Geller, E. S. (1991) If only more would actively care, *Journal of Applied Behavior Analysis*, 24, 607–612.

Geller, E. S. (2001) *The Psychology of Safety Handbook*, Lewis Publishers, Boca Raton.

Geller, E. S., Roberts, D. S. and Gilmore, M. R. (1996) Predicting propensity to actively care for occupational safety, *Journal of Safety Research*, 27, 1–8.

Gillen, M., Baltz, D., Gassel, M., Kirsch, L. and Vaccaro, D. (2002) Perceived safety climate, job demands, and coworker support among union and nonunion injured construction workers, *Journal of Safety Research*, 33, 33–51.

Gillen, M., Faucett, J. A., Beaumont, J. J. and McLoughlin, E. (1997) Injury severity associated with nonfatal construction falls, *American Journal of Industrial Medicine*, 32, 647–655.

Gilliland, D. I. and Manning, K. C. (2002) When do firms conform to regulatory control? The effect of control processes on compliance and opportunism, *Journal of Public Policy & Marketing*, 21, 319–331.

Glendon, A. I. and McKenna, E. F. (1995) *Human Safety and Risk Management*, Chapman & Hall, London.

Glezer, H. and Wolcott, I. (2000) Conflicting commitments: Working mothers and fathers in Australia, in L. L. Hass, P. Hwang and G. Russell (eds), *Organisational Change and Gender Equity: International Perspectives on Fathers and Mothers at the Workplace*, Sage Publications, California.

Gmelch, W. H. and Gates, G. (1998) The impact of personal, professional and organizational characteristics on administrator burnout, *Journal of Educational Administration*, 36, 2, 146–159.

Goldstein, I. L. (1993) *Training in Organizations* (3rd edition), Brookes/Cole, Pacific Grove, CA.

Gordon, J. E. (1949) The epidemiology of accidents, *American Journal of Public Health*, 39, 504–515.

Grandjean, E. (2000) *Fitting the Task to the Man* (5th edition), Taylor & Francis, London.

Gray, C. and Flanagan, R. (1989) *The Changing Role of Specialist and Trade Contractors*, Chartered Institute of Building, Ascot, UK.

Gray, W. B. and Scholz, J. (1991) *Do OSHA Inspections Reduce Injuries? A Panel Analysis*, National Bureau of Economic Research Working Paper Series, Cambridge, Mass.

Gray, W. B. and Scholz, J. T. (1993) Does regulatory enforcement work? A panel analysis of OHSA enforcement, *Law and Society Review*, 27, 177–213.

Green, C. H. and Brown, R. A. (1978) Counting lives, *Journal of Occupational Accidents*, 2, 55–70.

Green, C. H., Tunstall, S. M. and Fordham, M. (1991) The risks from flooding: Which risk and whose perception? *Disasters*, 15, 227–236.

Green, S. D. (2002) The human resource management implications of lean construction: Critical perspectives and conceptual chasms, *Journal of Construction Research*, 3, 147–165.

Greene, C. N. (1989) Cohesion and productivity in work groups, *Small Group Behavior*, 20, 70–86.

Greenspan, C. A., Moure-Eraso, R. and Wegman, D. H. (1995) Occupational hygiene characterisation of a highway construction project: A pilot study, *Applied Occupational and Environmental Hygiene*, 10, 50–58.

Greenwood, M. and Woods, H. M. (1919) *The Incidence of Industrial Accidents upon Individuals with Special Reference to Multiple Accidents*, Industrial Fatigue Research Board, Report 4, HMSO, London.

Greenwood, M. R. (2002) Ethics and HRM: A review and conceptual analysis, *Journal of Business Ethics*, 36, 261–278.

Griffin, M. and Neal, A. (2000) Perceptions of safety at work: A framework for linking safety climate to safety performance, knowledge and motivation, *Journal of Occupational Health Psychology*, 5, 347–358.

Grimshaw, J. (1999) *Employment and Health: Psychosocial Stress in the Workplace*, The British Library, London.

Grzywacz, J. G. and Marks, N. F. (2000) Family, work, work-family spillover and problem drinking during midlife, *Journal of Marriage and the Family*, 62, 336–348.

Guadalupe, M. (2002) The hidden cost of fixed term contracts: The impact on work accidents, *Labour Economics*, 292, 1–19.

Guberan, E. and Usel, M. (1998) Permanent work incapacity, mortality and survival without incapacity among occupations and social classes: A cohort study of aging men in Geneva, *International Journal of Epidemiology*, 27, 1026–1032.

Guldenmund, F. W. (2000) The nature of safety culture: A review of theory and research, *Safety Science*, 215–257.

Gun, R. T. (1992) Regulation or self-regulation: Is Robens-style legislation a formula for success? *Journal of Occupational Health and Safety – Australia and New Zealand*, 8, 383–388.

Gunningham, N. (1996) From compliance to best practice in OHS: The roles of specification, performance and systems-based standards, *Australian Journal of Labour Law*, 9, 221–246.

Gunningham, N. (1998) Towards innovative occupational health and safety regulation, *Journal of Industrial Relations*, 40, 203–231.

Gunningham, N., Johnstone, R. and Burritt, P. (2000) *Safe Design Project: Review of Occupational Health and Safety Legal Requirements for Designers, Manufacturers, Suppliers, Importers and Other Relevant Obligation Bearers*, National Occupational Health and Safety Commission, Sydney.

Gunningham, N., Sinclair, D. and Burritt, P. (1998) *Impact of On-the-Spot Fines on Prevention Outcomes for OHS in Australian Workplaces*, National Occupational Health and Safety Commission, Sydney.

Hadikusumo, B. H. W. and Rowlinson, S. M. (2002) Integration of virtually real construction model and design-for-safety-process database, *Automation in Construction*, Elsevier Science B. V., 11, 501–509.

Hale, A. R. and Hovden, J. (1998) Management and culture: The third age of safety. A review of approaches to organisational aspects of safety, health and environment, in A. M. Feyer and A. Williamson (eds), *Occupational Injury-risk Prevention and Intervention*, Taylor & Francis, London.

Hampson, K. (1997) *Technology Strategy and Competitive Performance*, PhD Thesis, Stanford University.

Hansen, C. (1989) A causal model of the relationship among accidents, biodata, personality and cognitive factors, *Journal of Applied Psychology*, 74, 81–90.

HASTAM (1999) *CHASE for Windows*, HASTAM, UK.

Hawkins, F. H. (1993) *Human Factors in Flight*, 2nd edition, Harry W. Orlady (ed.), Ashgate Publishing Co. Ltd, Aldershot, England.

Hayes, B. E., Perander, J., Smecko, T. and Trask, J. (1998) Measuring perceptions of workplace safety: Development and validation of the work safety scale, *Journal of Safety Research*, 29, 145–161.

Health and Safety at Work Act (1974) HMSO, London.

Health and Safety Commission (UK) (1994), *Review of Health and Safety Regulation – Main Report*, Health and Safety Executive, HMSO, London.

Health and Safety Executive (HSE) (1994) Selection and training of offshore installation managers for crisis management, *Health and Safety Executive*, HSE Books.

Health and Safety Executive (HSE) (1997) *The Costs of Accidents at Work*, HMSO, London.

Hecker, S. and Gambatese, J. (2003) Safety in design: A proactive approach to construction worker safety and health, *Applied Occupational and Environmental Hygiene*, 18, 339–342.

Heinrich, H. W. (1959) *Industrial Accident Prevention* (4th edition), McGraw-Hill, New York.

Helander, M. G. (1991) Safety hazards and motivation for safe work in the construction industry, *International Journal of Industrial Ergonomics*, 8, 205–23.

Henderson, J., Whittington, C. and Wright, K. (2001) *Accident Investigation – The Drivers, Methods and Outcomes*, Contract Research Report 344/2001, Health and Safety Executive, HMSO, Norwich.

Herrero, S. G., Saldana, M. A. M., Manzanedo del Campo, M. A. and Ritzel, D. O. (2002) From the traditional concept of safety management to safety integrated with quality, *Journal of Safety Research*, 33, 1–20.

Herzberg, F. (1974) *Work and the Nature of Man*, Crosby Lockwood Staples, London.

Hibberd, P. (1991) Key factors in procurement, Chapter 8.

Hill, F. (1991) Urban development corporation in England and Wales, Chapter 5.

Hinze, J. (1976) *The Effect of Middle Management on Safety in Construction*, Technical Report no 209, Department of Civil Engineering, Stanford University.

Hinze, J. (1981) Human aspects of construction safety, *Journal of the Construction Division, Proceedings of the American Society of Civil Engineers*, 107, No. Co1, 61–72.

Hinze, J. and Gambatese, J. A. (1994) Design decisions that impact construction worker safety, in Proceedings of the Fifth Annual Rinker International Conference on Construction Safety and Loss Control, University of Florida, Gainsville.

HKLEGCO (Hong Kong Legislative Council) (2003) http://www.legco.gov.hk/yr02-03/english/sc/sc_bldg/reports/rpt_1.htm.

Hofmann, D. A. and Morgeson, F. P. (1999) Safety-related behaviour as a social exchange: The role of perceived organisational support and leader-member exchange, *Journal of Applied Psychology*, 84, 286–296.

Hofmann, D. A. and Stetzer, A. (1996) A cross-level investigation of factors influencing unsafe behaviours and accidents, *Personnel Psychology*, 49, 307–339.

Hogg-Johnson, S., Frank, J. W. and Rael, E. (1994) *Prognostic Risk Factor Models for Low Back Pain: Why They Have Failed and a New Hypothesis*, Working Paper 19, Ontario Workers' Compensation Institute, Toronto.

Holmes, N. and Gifford, S. M. (1996) Social meanings of risk in OHS: Consequences for risk control, *Journal of Occupational Health and Safety Australia and New Zealand*, 12, 443–450.

Holmes, N., Lingard, H., Yesilyurt, Z. and DeMunk, F. (1999) An exploratory study of meanings of risk control for long term and acute effect occupational health and safety risks in small business construction firms, *Journal of Safety Research*, 30, 4, 251–261.

Holstrom, E., Moritz, U. and Engholm, G. (1995) Musculoskeletal disorders in construction workers, *Occupational Medicine*, 10, 295–312.

Holton, E. F. III (1996) The flawed four-level evaluation model, *Human Resource Development Quarterly*, 7, 5–25.

Hopkins, A. (1994a) Making prosecutions more effective, in R. Johnstone (ed.), *Occupational Health & Safety Prosecutions in Australia: Overview & Issues*, pp. 22–38.

Hopkins, A. (1994b) Patterns of prosecution, in R. Johnstone (ed.), *Occupational Health & Safety Prosecutions in Australia: Overview & Issues*, pp. 3–12.

Hopkins, A. (1995) *Making Safety Work*, Allen & Unwin, Sydney.

Horlick-Jones, T. (1996) Is safety a by-product of quality management, in C. Hood and D. K. C. Jones (eds), *Accident and Design: Contemporary Debates in Risk Management*, UCL Press, London, pp. 144–154.

Houghton, S. M., Imon, M., Aquino, K. and Goldberg, C. B. (2000) No safety in numbers: Persistence of biases and their effects on team risk perception and team decision-making, *Group and Organisation Management*, 21, 325–353.

HSC (Health and Safety Commission (Advisory Committee on the Safety of Nuclear Installations – Human Factors Study Group)) (1993). Third Report: Organising for Safety, HSE Books, Sudbury.

HSE (Health and Safety Executive) (1991) *Successful Health and Safety Management* (HS(G) 65), HSE Books, London.

HSE (Health and Safety Executive) (1993) *The Costs of Accidents at Work*, HMSO, London.

HSE (Health and Safety Executive) (1995) *Designing for Health and Safety in Construction*, HMSO, London.

HSE (Health and Safety Executive) (1996) *Manual Handling – Guidance on Regulations* (Manual Handling Operations Regulations 1992) (L23), HSE Books, London.

HSE (Health and Safety Executive) (1999) *Reducing Error and Influencing Behaviour*, HSG48, HMSO, London.

HSE (Health and Safety Executive) (2000) *Successful Health and Safety Management*, HMSO Books, London.

HSE (Health and Safety Executive) (2003a) *Work-related Deaths: A Protocol for Liaison*, www.hse.gov.au, Ist April 2003.

HSE (Health and Safety Executive) (2003b) *Causal Factors in Construction Accidents*, Research Report 156, HMSO Books, Norwich.

Hunt, A. and Habeck, R. (1993) *The Michigan Disability Prevention Study: Research Highlights*, W. E. Upjohn Institute for Employment Research, Kalamazoo.

Hynes, M. and VanMarke, E. (1976) Reliability of embankment performance prediction, *Proceedings of the ASCE Engineering Mechanics Division Speciality Conference*, University of Waterloo Press, Waterloo, Canada.

Industrial Fatigue Research Board (1922) *Report 19: Two contributions to the study of accident causation*, HMSO, London.

Industry Commission (1995) *Work, Health and Safety: Inquiry into Occupational Health and Safety (volume II)*, Commonwealth of Australia, Canberra.

Infante-Rivard, C. and Lortie, M. (1996) Prognostic factors for return to work after a first compensated episode of back pain, *Journal of Occupational and Environmental Medicine*, 53, 488–494.

Institute of Employment Rights (1999) *Regulating Health and Safety at Work: The Way Forward*, IER, London.

Ireland, V. (1984) Virtually meaningless distinctions between nominally different procurement systems, *CIB W65 Proceedings of the 4th International Symposium on Organisation and Management of Construction*, University of Waterloo, Canada, 1, 203–212.

Ireland, V. (1984) Virtually meaningless distinctions between nominally different procurement forms, *Proceedings of 4th International Symposium on Organisation and Management of Construction*, Waterloo, Ontario, Canada, Vol. 1, pp. 203–212.

Iversen, H. and Rundmo, T. (2002) Personality, risky driving and accident involvement among Norwegian drivers, *Personality and Individual Differences*, 33, 1251–1263.

James, P., Cunningham, I. and Dibben, P. (2003) *Job Retention and Vocational Rehabilitation: The Development and Evaluation of a Conceptual Framework*, Research Report 106, Health and safety Executive, HMSO, Norwich.

Janicak, C. A. (1998) Fall-related deaths in the construction industry, *Journal of Safety Research*, 29, 35–42.

Janis, I. L. (1972) *Victims of Groupthink*, Houghton Mifflin, Boston.

Jensen, P. L., Alstrop, L. and Thoft, E. (2001) Workplace assessment: A tool for occupational health and safety management in small firms? *Applied Ergonomics*, 32, 433–440.

Jensen, R. S. (1982) Pilot judgement: Training and evaluation, *Human Factors*, 34, 61–73.

Job, R. F. S. (1990) The application of learning theory to driving confidence: The effect of age and the impact of random breath testing, *Accident Analysis and Prevention*, 22, 97–107.

Johnstone, R. (1993) The legal regulation of pre-employment health screening, in M. Quinlan (ed.), *Work and Health*, Macmillan Australia, Melbourne, pp. 191–238.

Johnstone, R. (1999a) Paradigm crossed? The statutory occupational health and safety obligations of the business undertaking, *Australian Journal of Labour Law*, 12, 73–112.

Johnstone, R. (1999b) Improving worker safety: Reflections on the legal regulation of OHS in the 20th century, *Journal of Occupational Health and Safety – Australia and New Zealand*, 15, 521–526.

Johnstone, R. (2000a) *Evaluation of Queensland Construction Safety 2000 Initiative*, National Occupational Health and Safety Commission, Sydney.

Johnstone, R. (2000b) Occupational health and safety prosecutions in Victoria: An historical study, *Australian Journal of Labour Law*, 13, 113–142.

Johnstone, R. (2002) *Safety, Courts and Crime*, Working Paper 6, National Research Centre for OHS Regulation, The Australian National University, Canberra.

Kant, I. (1981) *Grounding for the Metaphysics of Morals*, translated by James W. Ellington, Hackett, Indianapolis.

Kaplan, R. S. and Norton, D. P. (1996) *The Balanced Scorecard: Translating Strategy into Action*, Harvard Business School Press, Cambridge, Mass.

Karakowsky, L. and Elangovan, A. R. (2001) Risky decision making in mixed-gender teams: Whose risk tolerance matters? *Small Group Research*, 32, 94–111.

Karhu, *et al.* (1977) Correcting working postures in industry: A practical method for analysis, *Applied Ergonomics*, 8, 4, pp. 199–201.

Karhu, *et al.* (1981) Observing working postures in industry: Examples of OWAS application, *Applied Ergonomics*, 12, 1, pp. 13–17.

Kasperson, R. E., Renn, O., Slovic, P., Brown, H. S., Emel, J., Goble, R., Kasperson, J. X. and Ratick, S. (1988) The social amplification of risk: A conceptual framework, *Risk Analysis*, 8, 177–187.

Kennedy, R. J. (1997) *The Development of a HAZOP-based Methodology to Identify Safety Management Vulnerabilities and Their Associated Safety Cultural Factors*, unpublished PhD Thesis, University of Birmingham.

Kenny, D. (1995) Common themes, different perspectives: A systematic analysis of employer-employee experiences of occupational rehabilitation, *Rehabilitation Counseling Bulletin*, 39, 54–77.

Kenny, D. (1999) Employer's perspectives on the provision of suitable duties in occupational rehabilitation, *Journal of Occupational Rehabilitation*, 4, 267–276.

Kenny, D. T. (1994) Determinants of time lost from workplace injury: The impact of the injury, the injured, the industry, the intervention and the insurer, *International Journal of Rehabilitation Research*, 17, 333–342.

Khurana, A., *Understanding and Using SWOT Analysis* [Accessed online 5 August 2003] http://businessmajors.about.com/library/weekly/aa123002a.htm.

Kines, P. (2001) Occupational injury risk assessment using severity odds ratios: Male falls from heights in the Danish construction industry 1993–1999, *Human and Ecological Risk Assessment*, 7, 1929–1943.

Kines, P. (2002) Construction workers' falls through roofs: Fatal versus serious injuries, *Journal of Safety Research*, 33, 195–208.

King, R. W. and Hudson, R. (1985) *Construction Hazard and Safety Handbook*, Butterworths, London.

Kisner, S. M. and Fosbroke, D. E. (1994) Injury hazards in the construction industry, *Journal of Occupational Medicine*, 36, 137–143.

Kirwan, B. (1994) *A Guide to Practical Human Reliability Assessment*, Burgess Science Press, Basingstoke.

Kivi, P. and Mattila, M. (1991) Analysis and improvement of work postures in the building industry: Application of the computerised OWAS method, *Applied Ergonomics*, 22, 43–48.

Kjellberg, A. (1990) Subjective behavioural and psychophysiological effects to noise, *Scandinavian Journal of Work, Environment and Health*, 16, 29–38.

Kletz, T. A. (1985) *An Engineer's View of Human Error*, Institution of Chemical Engineers, Warwickshire, England.

Kletz, T. A. (1993) Organisations have no memory when it comes to safety: A thoughtful look at why plants don't learn from the past, *Hydrocarbon Processing*, 6, 88–95.

Koberg, C. S., Boss, R. W., Senjem, J. C. and Goodman, E. A. (1999) Antecedents and outcomes of empowerment, *Group & Organisation Management*, 24, 71–91.

Koeske, G. F. and Kirk, S. A. (1995) Direct and buffering effects of internal locus of control among mental health professionals, *Journal of Social Service Research*, 20, 1–28.

Kohn, M. and Schooler, C. (1983) *Work and Personality: An Inquiry into the Impact of Social Stratification*, Ablex, Norwood NJ.

Komaki, J. and Jensen, M. (1986) Within-group designs: An alternative to traditional control-group designs, in M. F. Cataldo and T. J. Coates (eds), *Health and Industry: A Behavioral Medicine Perspective*, John Wiley & Sons, New York.

Komaki, J., Barwick, K. D. and Scott, L. R. (1978) A behavioural approach to occupational safety: Pinpointing and reinforcing safe performance in a food manufacturing plant, *Journal of Applied Psychology*, 63, 4, 434–445.

Kopelman, R. E., Brief, A. P. and Guzzo, R. A. (1990) The role of climate and culture in productivity, in B. Schneider (ed.), *Organisational Climate and Culture*, Jossey-Bass, San Francisco.
Kossek, E. E. and Ozeki, C. (1998) Work-family conflict, policies and the job-life satisfaction relationship: A review and directions for organizational behavior-human resources research, *Journal of Applied Psychology*, 83, 139–149.
Kozlowski, S. W. J. and Doherty, M. L. (1989) Integration of climate and leadership: Examination of a neglected issue, *Journal of Applied Psychology*, 74, 546–553.
Krause, N., Dasinger, L. K. and Neuhauser, F. (1998) Modified work and return to work: A review of the literature, *Journal of Occupational Rehabilitation*, 8, 113–139.
Krietner, R. and Luthans, F. (1991) A social learning approach to behavioral management: Radical behaviorists 'mellowing out', in R. M. Steers and L. W. Porter (eds), *Motivation and Work Behavior*, McGraw-Hill, New York.
Kroemer, K. H. E. and Grandjean, E. (2000) *Fitting the Task to the Human: A Textbook of Occupational Ergonomics* (5th edition), Taylor & Francis, London.
Kululanga, G. K., McCaffer, R., Price, A. D. F. and Edum-Fotwe, F. (1999) Learning mechanisms employed by construction contractors, *Journal of Construction Management and Engineering*, July/August, 215–223.
Kumaraswamy, M. M. (1994) Growth strategies for less developed construction industries, *10th Annual Conference of the Association of Researchers in Construction Management*, Loughborough University, pp. 154–163.
Kumaraswamy, M. M. and Dissanayaka, S. M. (1997) Synergising construction research with industry development, *First International Conference on Construction Industry Development*, National University of Singapore, Singapore, December, 1, pp. 182–189.
Kurtz, J. R., Robins, T. G. and Schork, M. A. (1997) An evaluation of peer and professional trainers in a union-based occupational health and safety training program, *Journal of Occupational and Environmental Medicine*, 39, 661–671.
La Trobe/Melbourne Occupational Health and Safety Project (1989) *Victorian Occupational Health and Safety – An Assessment of Law in Transition*, Department of Legal Studies, LaTrobe University, Melbourne.
Lam, R. H. C. (2003) *An Investigation into the Implementation of Safety Management Systems by Hong Kong Construction Contractors*, unpublished PhD Thesis, The University of Hong Kong.
Landy, F. J. (1989) *Psychology of Work Behavior*, Wadsworth Inc., Belmont CA.
Lansley, P., Sadler, P. and Webb, T. (1974) Organisation structure, management style and company performance, *Omega*, 2, 4, 467–485.
Larsson, T. and Field, B. (2002) The distribution of occupational risks in the Victorian construction industry, *Safety Science*, 40, 439–456.
Latham, M. (1994) *Constructing the Team*, HMSO, London.
Latour, B. (1987) *Science in Action*, Open University Press, Milton Keynes.
Laufer, A., Tucker, R. L., Shapira, A. and Shenhar, A. J. (1994) The multiplicity concept in construction planning, *Construction Management and Economics*, 11, 53–65.
Laukkanen, T. (1999) Construction work and education: Occupational health and safety reviewed, *Construction Management and Economics*, 17, 53–62.
Lavers, A. (1991) The implementation of the EC construction products directive in the UK, Chapter 7.

Law Reform Commission of Victoria (1991) *Homicide*, Report No. 40, pp. 7–114 and Appendix 3.

Leather, P. J. (1987) Safety and accidents in the construction industry: A work design perspective, *Work and Stress*, 1, 167–174.

Leavitt, H. (1972) Some effects of certain communication patterns on a group performance, *Journal of Abnormal and Social Psychology*, 46, 38–50.

Lee, R. T. and Ashforth, B. E. (1996) A meta-analytic examination of the correlates of the three dimensions of job burnout, *Journal of Applied Psychology,* 81, 2, 123–133.

Lee, T. (1997) *Proceedings of a Conference on 'Safety Culture in the Energy Industries'*, Cookham: Energy Logistics International Ltd, Chapter: How can we monitor the safety culture and improve it where necessary?

Lehto, M. R. (1998) The influence of chemical warning label content and format on information retrieval speed and accuracy, *Journal of Safety Research*, 29, 43–56.

Lemos-Giraldez, S. and Fidalgo-Aliste, A. M. (1997) Personality dispositions and health-related habits and attitudes: A cross sectional study, *European Journal of Personality*, 11, 197–209.

Lenard, D. and Mohsini, R. (1997) Recommendations from the organisational workshop, in C. H. Davidson (ed.), *Proceedings of CIB W92 Montreal Conference, Procurement: The Way Forward*, International Council for Building Research, Studies and Documentation.

Lenard, D. and Mohsini, R. (1998) Recommendations from the organisational workshop, ISBN 0–9682215–1–3.

Lichtenstein, S., Slovic, P., Fischhoff, B., Layman, M. and Combs, B. (1978) Judged frequency of lethal events, *Journal of Experimental Psychology, Human Learning and Memory*, 4, 551–578.

Likert, R. (1967) *The Human Organization: Its Management and Value*, McGraw-Hill, New York.

Lilley, B. (1993) Can you be charged with murder by accident? *Australian Safety News*, November, 24–30.

Lilley, R., Feyer, A.-M., Kirk, P. and Gander, P. (2002) A survey of forest workers in New Zealand. Do hours of work, rest and recovery play a role in accidents and injury? *Journal of Safety Research*, 33, 53–71.

Lindell, M. K. (1994) Motivational and organisational factors affecting implementation of worker safety training, *Occupational Medicine: State of the Art Reviews*, Vol. 9, No. 2.

Lingard, H. (1993) The epidemiology of accidents: The Hong Kong Housing Authority Approach, *Proceedings of the Ninth Annual Conference of the Association of Researchers in Construction Management*, Oxford, UK.

Lingard, H. (2002) The effect of first aid training on Australian construction workers' occupational health and safety knowledge and motivation to avoid work-related injury or illness, *Construction Management and Economics*, 20, 263–273.

Lingard, H. (2003) The impact of individual and job characteristics on 'burnout' among civil engineers in Australia and implications for employee turnover, *Construction Management and Economics*, 21, 69–80.

Lingard, H. and Francis, V. (2002) *Work-life Issues in the Australian Construction Industry: The Results of a Pilot Study* (Report prepared for and submitted to the Construction Industry Institute of Australia).

Lingard, H. and Holmes, N. (2001) Meanings of occupational health and safety risk control in small business construction firms: Barriers to implementing technological controls, *Construction Management and Economics*, 19, 216–227.

Lingard, H. and Rowlinson, S. (1991) OH&S in Hong Kong's Construction Industry, *The Hong Kong Engineer*, 11–23.

Lingard, H. and Rowlinson, S. (1994) Construction site safety in Hong Kong, *Construction Management and Economics*, 12, 501–510.

Lingard, H. and Rowlinson, S. (1997) Behavior-based safety management in Hong Kong's construction industry, *Journal of Safety Research*, 28, 243–256.

Lingard, H. and Rowlinson, S. (1998) Behaviour-based safety management: The results of a field study, *Construction Management and Economics*, 16, 481–488.

Lingard, H. and Sublet, A. (2002) The impact of job and organisational demands on marital and relationship quality among Australian civil engineers, *Construction Management and Economics*, 20, 507–521.

Lingard, H. and Yesilyurt, Z. (2003) The effect of attitudes on the occupational safety actions of Australian construction workers: The results of a field study, *Journal of Construction Research*, 4, 59–69.

Lingard, H., Hughes, W. and Chinyio, E. (1998) The impact of contractor selection method on transaction costs: A review, *Journal of Construction Procurement*, 4, 89–102.

Lippin, T., Eckman, A., Calkin, K. R. and McQuiston, T. H. (2000) Empowerment-based health and safety training: Evidence of workplace change from four industrial sectors, *American Journal of Industrial Medicine*, 38, 697–706.

Lloyd-Schut, W. S. M. (1991) Recent developments in EEC construction law, Chapter 6.

Locke, E. A. and Latham, G. P. (1990) *A Theory of Goal-Setting and Task Performance*, Prentice-Hall, Englewood Cliffs, New Jersey.

Loosemore, M. (1994) Problem behaviour, *Construction Management and Economics*, 12, 511–520.

Loosemore, M. and Teo, M. M. M. (2002) The crisis management practices of Australian construction companies, *The Australian Journal of Construction Economics and Building*, 2, 2, 15–26.

Loosemore, M., Dainty, A. and Lingard, H. (2003) *Human Resource Management in Construction Projects: Strategic and Operational Approaches*, Spon Press, London.

Lott, A. J. and Lott, B. E. (1965) Group cohesiveness as interpersonal attraction: A review of relationships with antecedent and consequent variables, *Psychological Bulletin*, 64, 259–309.

Lowery, J. T., Borgerdin, J. A., Boguang, Z., Glazner, J. E., Bondy, J. and Kreiss, K. (1998) Risk factors for injury among construction workers at Denver International Airport, *American Journal of Industrial Medicine*, 34, 113–120.

Luntz, H. (1981) The role of compensation in health and safety at work, *Journal of Industrial Relations*, 23, 382–396.

Ludborzs, B. (1995) Surveying and assessing 'safety culture' within the framework of safety audits, in J. J. Mewis, H. J. Pasman, E. E. De Rademaecker (eds), *Loss Prevention and Safety Promotion in the Process Industries*, 1, 83–92.

Manzella, J. C. (1997) Achieving safety performance through total quality management, *Professional Safety*, 42, 5, 26–28.

Marmaras, N., Poulakakis, G., Papakostopoulos, V. (1999) Ergonomic design in ancient Greece, *Applied Ergonomics,* 30, 361–368.

Martens, N. (1997) The Construction (Design and Management) Regulations 1994: Considering the competence of the planning supervisor, *Journal of the Institution of Occupational Safety and Health*, 1, 41–49.

Martens, N. (1998) Competence and skills of the planning supervisor: The UK Construction regulations, *Professional Safety*, September, 30–34.

Martins, R. and Taylor, R. (1996) Cultural identities and procurement, Late Paper.

Maslach, C., Jackson, S. E. and Leiter, M. P. (1996) *Maslach Burnout Inventory Manual* (3rd edition), Consulting Psychologists Press, Palo Alto, CA.

Maslach, C., Schaufeli, W. B. and Leiter, M. P. (2001) Job burnout, *Annual Review of Psychology*, 52, 397–422.

Matthews, J. (1993) *Health and Safety at Work*, Pluto Press, Sydney.

Mathurin, C. (1991) A European construction agency, Chapter 1.

Mattila, M. (1985) Job load and hazard analysis: A method for the analysis of workplace conditions for occupational health care, *British Journal of Industrial Medicine*, 42, 656–666.

Mattila, M. (1989) Improvement in the occupational health programme in a Finnish construction company by means of systematic workplace investigations of job load and hazard analysis, *American Journal of Industrial Medicine*, 15, 61–72.

Mattila, M. and Hyodynmaa, M. (1988) Promoting job safety in building: An experiment on the behaviour analysis approach, *Journal of Occupational Accidents*, 9, 255–267.

Mauno, S. and Kinnunen, U. (1999) The effects of job stressors on marital satisfaction in Finnish dual earner couples, *Journal of Organizational Behavior*, 20, 879–895.

Mayhew, C. (1995) *An evaluation of the impact of Robens style legislation on the OHS decision-making of Australian and United Kingdom builders with less than five employees [Worksafe Australia Research Grant Report]*, National Occupational Health and Safety Commission, Canberra.

Mayhew, C. (1995) An evaluation of the impact of Robens style legislation on the OHS decision-making of Australian and United Kingdom builders with less than five employees, Griffith University, Brisbane.

Mayhew, C. (1997) Barriers to implementation of known occupational health and safety solutions in small business, Australian Government Publishing Service, Canberra.

Mayhew, C., Quinlan, M. and Bennett, L. (1996) *The Effects of Subcontracting Outsourcing on Occupational Health and Safety*, Industrial Relations Research Monograph, University of New South Wales.

McAfee, R. B. and Winn, A. R. (1989) The use of incentives/feedback to enhance work place safety: A critique of the literature, *Journal of Safety Research*, 20, 7–19.

McColgan, A. (1994) The law commission consultation document on involuntary manslaughter: Heralding corporate liability, *Criminal Law Review*, 547–557.

McDermott, P. and Jaggar, D. (1991) Towards establishing the criteria for a comparative analysis of procurement methods in different countries: Social systems in construction, Chapter 9 in Anon (1991) *Procurement Systems Symposium*, Las Palmas, Gran Canaria, Spain, CIB publication NI. 145.

McDermott, P. and Jaggar, D. (1991) Towards establishing the criteria for a comparative analysis of procurement methods in different countries: Social systems in construction, Chapter 9.

McKenna, S. P. and Hale, A. R. (1981) The effect of emergency first aid training on the incidence of accidents in factories, *Journal of Occupational Accidents*, 3, 101–114.

McKenna, S. P. and Hale, A. R. (1982) Changing behaviour towards danger: The effect of first aid training, *Journal of Occupational Accidents*, 4, 47–59.

McLellan, R. K., Pransky, G. and Shaw, W. S. (2001) Disability management training for supervisors: A pilot intervention program, *Journal of Occupational Rehabilitation*, 11, 33–41.

McVittie, D., Banikin, H. and Brocklebank, W. (1997) The effect of firm size on injury frequency in construction, *Safety Science*, 27, 19–23.

McWilliams, G., Rechnitzer, G., Deveson, N., Fox, B., Clayton, A., Larsson, T. and Cruickshank, L. (2001) *Reducing Serious Injury Risk in the Construction Industry*, Policy Research Report No. 9, Monash University Accident Research Centre, Melbourne.

Merlino, L., Rosencrance, J., Anton, D. and Cook, T. (2003) Symptoms of musculoskeletal disorders among apprentice construction workers, *Applied Occupational and Environmental Hygiene*, 18, 1–8.

Merrill, M. (1994) Trust in training: The Oil, Chemical and Atomic Workers International Union worker-to-worker training program, *Occupational Medicine: State of the Art Reviews*, 9, 341–354.

Miller, D. P. and Swain, A. D. (1987) Human error and human reliability, in G. Salvendy (ed.), *Handbook of Human Factors*, Wiley, New York.

Miller, G. and Agnew, N. (1973), First aid training and accidents, *Occupational Psychology*, 47, 209–218.

Miller, I. and Cox, S. (1997) Benchmarking for loss control, *The Institution of Occupational Safety and Health Journal*, 11, 39–47.

Miner, J. B. (1992) *Industrial-Organisational Psychology*, McGraw-Hill Inc, New York.

Mitroff, I. and Pearson, C. (1993) *Crisis Management: A Diagnostic Guide for Improving Your Organisation's Crisis Preparedness*, Jossey-Bass Publishers, San Francisco.

Mohamed (2000) Empirical investigation of construction safety management activities and performance in Australia, *Safety Science*, 33(3), 129–142.

Mohamed (2002) Safety climate in construction site environments, *ASCE Journal of Construction Engineering and Management*, October.

Mohamed, S. (2003) Scorecard approach to benchmarking organizational safety culture in construction, *Journal of Construction Engineering and Management*, 129, 1, 80–88.

Mohamed, S. (ed.) (1997) *Construction Process Re-engineering, International Conference*, Gold Coast, Australia, 14–15 July.

Mohsini, R. and Davidson, C. H. (1989) Building procurement – key to improved performance, 83, in D. Cheetham, D. Carter, T. Lewis and D. M. Jaggar (eds), *Contractual Procedures for Building, Proceedings of the International Workshop*, 6–7 April, The University of Liverpool.

Monk, V. (1998) Postural assessment of building industry tasks using the Ovako Working Posture Analysing System, *Journal of Occupational Health and Safety Australia and New Zealand*, 14, 149–155.

Moodely, K. and Preece, C. N. (1996) Implementing community policies in the construction industry, in D. A. Langford and A. Retik (eds), *The Organisation and Management of Construction*, E&FN Spon, London.

Moorehead, A., Steele, M., Alexander, M., Stephen, K. and Duffin, L. (1997) *Changes at Work: The 1995 Australian Workplace Industrial Relations Survey*, Longman, Australia.

Morgan, G. (1985) *Images of Organisation*, Sage, Beverly Hills, USA.

Morris, G., Cook, C., Creyke, R., Geddes, R. and Holloway, I. (1996) *Laying Down the Law: The Foundations of Legal Reasoning, Research and Writing in Australia*, Butterworths, Sydney.

Munro, W. D. (1996) The implementation of the Construction (Design and Management) Regulations 1994 on UK construction sites, in L. M. Aves Dias and R. J. Coble (eds), *Implementation of Safety and Health on Construction Sites*, A. A. Balkema, Rotterdam.

Myers, J. R. and Trent, R. B. (1988) Hand tool injuries at work: A surveillance perspective, *Journal of Safety Research*, 19, 165–176.

Naoum, S. G. and Coles, D. (1991) Procurement method and project performance, Chapter 10.

National Occupational Health and Safety Commission (1999) *OHS Performance Measurement in the Construction Industry: The Development of Positive Performance Indicators*, NOHSC, Canberra.

National Research Council (1989) *Improving Risk Communication*, National Academy Press, Washington.

Naylor, J. C., Pritchard, R. D. and Ilgen, D. R. (1980) *A Theory of Behavior in Organizations*, Academic Press, New York.

Neal, D. (1996) Corporate manslaughter, *Victorian Law Institute Journal*, 30, 10, 39–41.

Neitzel, R., Seixas, N. S., Camp, J. and Yost, M. (1999) An assessment of occupational noise exposures in four construction trades, *American Industrial Hygiene Association Journal*, 60, 807–817.

Nichol, J. (2001) Have Australia's Major Hazard facilities learnt from the Longford disaster? An evaluation of the impact of the 1998 Esso Longford explosion on Major Hazard Facilities in 2001, Institution of Engineers, Australia.

Nichols, T. (1997) *The Sociology of Industrial Injury*, Mansell Publishing, London.

Nichols, T. and Armstrong, P. (1973) *Safety or Profit: Industrial Accidents and the Conventional Wisdom*, Falling Wall Press, Bristol.

Niskanen, T. (1994) Assessing the safety environment bin work organisation of road maintenance jobs, *Accident Analysis and Prevention*, 26, 27–39.

NOHSC: 1001 (1990) National Standard for Manual Handling, Australian Government Publishing Service, Canberra.

NOHSC: 2005 (1990) National Code of Practice for Manual Handling, Australian Government Publishing Service, Canberra.

NOHSC: 2013 (1994) National Code of Practice for the Prevention of Occupational Overuse Syndrome, Australian Government of Occupational Overuse Syndrome, Canberra.

Nurminen, M. (1997) Reanalysis of the occurrence of back pain among construction workers: Modeling for the interdependent effects of heavy physical work, earlier back accidents and aging, *Occupational and Environmental Medicine*, 54, 11, 807–811.

O'Brien, R. M., Dickinson, A. M. and Rosnow, M. P. (1982) *Industrial Behaviour Modification*, Pergamon, New York.

O'Dea, A. and Flin, R. (2003) *The Role of Managerial Leadership in Determining Workplace Safety Outcomes*, Research Report 044, Health and Safety Executive, HMSO, London.

O'Toole, M. (2002) The relationship between employees' perceptions of safety and organisational culture, *Journal of Safety Research*, 33, 231–243.

O'Toole, M. F. (1999) Successful safety committees: Participation not legislation, *Journal of Safety Research*, 30, 39–65.

Olsen, D. K. and Gerberich, S. G. (1986) Traumatic amputations in the workplace, *Journal of Occupational Medicine*, 28, 480–485.

Oluwoye, J. and MacLennan, H. (1994) Preplanning safety in project buildability, in *Proceedings of the Fifth Annual Rinker International Conference on Construction Safety and Loss Control*, University of Florida, Gainsville.

Ones, D. S. and Viswesvaran, C. (1996) Bandwidth-fidelity dilemma in personality measurement for personnel selection, *Journal of Organisational Behavior*, 17, 609–626.

Otway, H. J. and Wynne, B. (1989) Risk communication: Paradigm and paradox, *Risk Analysis*, 9, 141–145.

Parkes, K. R. (1991) Locus of control as moderator: An explanation for additive versus interactive findings in the demand-discretion model of work stress? *British Journal of Psychology*, 82, 291–312.

Payne, R. (1988) Individual differences in the study of occupational stress, in C. L. Cooper and R. Payne (eds), *Causes, Coping and Consequences of Stress at Work*, Wiley, Chichester.

Pedler, M., Boydell, T. and Burgoyne, J. (1988) *Learning Company Project Report*, Training Agency, Sheffield.

Peebles, L., Kupper, A. and Heasman, T. (2002) *A Study of the Provision of Health and Safety Information in the Annual Reports of the Top UK Companies*, HSE Contract Research Report 446/2002, HMSO, London.

Peters, T. J. and Waterman, R. H. (1982) *In Search of Excellence*, Harper & Row, New York.

Petersen, D. (1989) *Safe Behavior Reinforcement*, Aloray, Goshen NY.

Phillips, C. C., Wallace, B. C., Hamilton, C. B., Prsley, R. T., Petty, G. C. and Bayne, C. K. (1999) The efficacy of material safety data sheets and worker acceptability, *Journal of Safety Research*, 30, 113–122.

Pidgeon, N. (1996) Technocracy, democracy, secrecy and error, in C. Hood and D. K. C. Jones (eds), *Accident and Design: Contemporary Debates in Risk Management*, UCL Press, London, pp. 164–171.

Pidgeon, N. F. (1991) Safety culture and risk management in organisations, *Journal of Cross Cultural Psychology*, 22, 129–140.

Pidgeon, N., Hood, C., Jones, D., Turner, B. and Gibson, R. (1993) Risk perception, from Risk: Analysis, perception, management, *The Royal Society*, London, pp. 89–134.

Polk, K., Haines, F. and Perrone, S. (1993) Homicide, negligence and work death: The need for legal change, in M. Quinlan (ed.), *Work and Health*, Macmillan, Melbourne, Australia.

Porter, M. (1985) *Competitive Advantage – Creating and Sustaining Superior Performance*, The Free Press, New York.

Preece, C. N., Moodley, K. and Cavina, C. (1999) The role of the planning supervisor under new health and safety legislation in the United Kingdom, in A. Singh, J. Hinze and R. J. Coble (eds), *Implementation of Safety and Health on Construction Sites*, A. A. Balkema, Rotterdam.

Price, V. (1982) *Type A Behaviour Pattern: A Model for Research and Practice*, Academic Press, London.

Prussia, G. E., Browm, K. A. and Willis, P. G. (2003) Mental models of safety: Do managers and employees see eye to eye? *Journal of Safety Research*, 34, 143–156.

Purse, K. (2002) Workers compensation-based employment security for injured workers: A review of legislation and enforcement, *Journal of Occupational Health and Safety – Australia and New Zealand*, 18, 61–66.

Pybus, R. (1996) *Safety Management: Strategy and Practice*, Butterworth-Heinemann, Oxford.

Quinlan, M. (1994) Trends in occupational health and safety prosecutions and penalties, a comment, in R. Johnstone (ed.), *Occupational Health & Safety Prosecutions in Australia: Overview & Issues*, pp. 13–20.

Quinlan, M. and Bohle, P. (1991) *Managing Occupational Health and Safety in Australia: A multi-disciplinary approach*, Macmillan Education Australia, Melbourne.

Rasmussen, J., Pejtersen, A. M. and Goodstein, L. P. (1994) *Cognitive Systems Engineering*, John Wiley, New York.

Reading Construction Forum (RCF) (1998) *The Seven Pillars of Partnering: A Guide to Second Generation Partnering*, Center for Strategic Studies in Construction Reading, England.

Reason, J. (1990) *Human Error*, Cambridge University Press, New York.

Reason, J. (1997) *Managing the Risks of Organisational Accidents*, Ashgate, Aldershot.

Redinger, C. F. and Levine, S. P. (1998) Analysis of third party certification approaches using an occupational health and safety conformity-assessment model, *American Industrial Hygiene Association Journal*, 59, 802–812.

Reichers, A. E. and Schneider, B. (1990) Climate and culture: An evolution of constructs, in B. Schneider (ed.), *Organisational Climate and Culture*, Jossey-Bass, San Francisco, pp. 5–39.

Reid, J. L. (2000) *Crisis Management: Planning and Media Relations for the Design and Construction Industry,* John Wiley & Sons, New York.

Reilly, M. J., Rosenman, K. D. and Kalinowski, D. J. (1998) Occupational noise-induced hearing loss surveillance in Michigan, *Journal of Occupational and Environmental Medicine*, 40, 667–674.

Rhoton, W. W. (1980) A procedure to improve compliance with coal mine safety regulations, *Journal of Organisational Behaviour Management*, 4, 243–249.

Ridley, A. and Dunford, L. (1997) Corporate killing – legislating for unlawful death? *Industrial Law Journal*, 26, 2, 99–113.

Ridley, J. and Channing, J. (1999) *Risk Management*, Butterworth-Heinemann, Oxford.

Rigby, N. (2003) *Designer Initiative 17th March 2003: Final Report*, Scotland and Northern England Unit, Construction Division, Health and Safety Executive.

Ringen, K., Englund, A., Welch, L., Weeks, J. L. and Seegal, J. L. (1995) Why construction is different, *Occupational Medicine: State of the Art Reviews*, 10, 2, 255–259.

Robens Committee (Committee on Safety and Health at Work) (1972), *Health and Safety at Work: Report of the Committee 1970–72*, HMSO, London.

Robertson, C. and Fadil, P. A. (1998) Developing corporate codes of ethics in multinational firms: Bhopal revisited, *Journal of Managerial Issues*, 10, 545–568.

Robertson, D. and Fox, J. (2000) *Industrial Use of Safety-related Expert Systems*, Contract Research Report 296/2000, HSE Books, HMSO, Norwich.

Robertson, I. T., Smith, M. and Cooper, M. D. (1992) *Motivation*, IPM, London.

Robertson, I. T., Duff, A. R., Marsh, T. W., Phillips, R. A., Weyman, A. K. and Cooper, M. D. (1995) Improving safety on construction site by changing personnel behaviour – phase two, *Health and Safety Executive*, HMSO, London.

Robson, L. S., Shannon, H. S., Goldenhar, L. M. and Hale, A. R. (2001) *Guide to Evaluating the Effectiveness of Strategies for Preventing Injuries: How to Show Whether a Safety Intervention Really Works*, National Institute for Occupational Safety and Health, Cincinnati.

Rosa, R. R. (1995) Extended workshifts and excessive fatigue, *Journal of Sleep Research*, 4 (Supplement 2), 51–56.

Rosa, R. R. and Colligan, M. J. (1988) Long workdays versus restdays: Assessing fatigue and alertness with a portable performance battery, *Human Factors*, 30, 305–317.

Rosen, R. H. and Freeman, S. (1992) Occupational contact dermatitis in New South Wales, *Australian Journal of Dermatology*, 33, 1–10.

Rosenbloom, R. S. and Burgelman, R. A. (ed.) (1989) *Research on Technological Innovation, Management and Policy: Tech, Competition, and Organization Theory, Vol. 4*, JAI Press Inc., Greenwich, CT, USA.

Rowan, J. J. (2000) The moral foundation of employee rights, *Journal of Business Ethics*, 25, 355–361.

Rowan, J. R. (2000) The moral foundation of employee rights, *Journal of Business Ethics*, 24, 355–361.

Rowlinson, S. (2003) Virtually real construction components and processes for design-for-safety-process (DFSP), in M. Phair (ed.), *Developments and Applications: The Visual Construction Enterprise*, Prentice-Hall, pp. 211–222.

Rowlinson, S. (ed.) (2003) *Health and Safety on Construction Sites: A Guide to Best Practice*, Taylor & Francis Books Ltd, United Kingdom.

Rowlinson, D. and McDermott, P. (eds) (1999) *Procurement Systems: A Guide to Good Practice*, E&FN Spon, London, United Kingdom, July, pp. 308+xxvi.

Rowlinson, S. (1987) Design build, *Chartered Institute of Building (UK) Occasional Paper, Design Build – Its Development and Present Status*, May, p. 36.

Rowlinson, S., Mohamed, S. and Lam, S. W. (2003) Hong Kong construction Foremen's Safety Responsibilities: A case study of management oversight, *Engineering, Construction and Architectural Management*, Blackwell Science: Oxford United Kingdom, 10, 1.

Rowlinson and Matthews, J. (1999) Partnering: Incorporating safety management, *Engineering, Construction and Architectural Management*, Blackwell Science: Oxford, United Kingdom, 6, 4, 347–357.

Rowlinson, S., Matthews, J., Phua, F. T. T., McDermott, P. and Chapman, T. (1999) Emerging issues in procurement systems, *Harmony and Profit in Construction Procurement*, CIB Publication 246, Asian Institute of Technology, Thailand, 13.

Ruser, J. (1991) Workers' compensation and occupational injuries and illnesses, *Journal of Labor Economics*, 9, 325–350.

Russell, J. S., Hancher, D. E. and Skibniewski, M. J. (1992) Contractor prequalification data for construction owners, *Construction Management and Economics*, 10, 117–129.

Saarela, K. L., Saari, J. and Aaltonen, M. (1989) The effects of an information safety campaign in the shipbuilding industry, *Journal of Occupational Accidents*, 10, 255–266.

Safely building NSW report (2001) *Priority Issues for Construction Reform*, New South Wales WorkCover Authority.

Sarvela, P. D. and McDermott, R. J. (1993) *Health Education Evaluation and Measurement*, Brown and Benchmark, Madison Wisconsin.

Scaffolding Regulations (1992) State of Victoria, Melbourne.

Schaufeli, W. and Enzmann, D. (1998) *The Burnout Companion to Study and Practice: A Critical Analysis*, Taylor & Francis, London, Philadelphia, PA.

Schein, E. (1992) *Organisational Culture and Leadership*, Jossey-Bass, San Francisco.

Schneider B. (1975) Organisational climates: An essay, *Personnel Psychology*, 28, 447–479.

Schneider, B. (1983) Interactional psychology and organisational behavior, *Research in Organisational Behavior*, 5, 1–31.

Schneider, S., Griffin, M. and Chowdhury, R. (1998) Ergonomic exposures of construction workers: An analysis of the US Department of Labor Employment and Training Administration database on job demands, *Applied Occupation Environment and Hygiene*, 13, 238–241.

Schneider, S., Punnett, L. and Cook, T. M. (1995) Ergonomics: Applying what we know, *Occupational Medicine*, 10, 385–394.

Schwartz, M. (2001) The nature of the relationship between corporate codes of ethics and behavior, *Journal of Business Ethics*, 32, 247–262.

Seashore, S. E. (1967) Group cohesiveness in the industrial work group, in W. A. Faunce (ed.), *Readings in Industrial Sociology*, Appleton-Century Crofts, New York.

Seixas, N. S., Ren, K., Neitzel, R., Camp, J. and Yost, M. (2001) Noise exposure among construction electricians, *American Industrial Hygiene Association Journal*, 62, 615–621.

Senge, P. M. (1994) The leaders' new work: Building learning organizations, in C. Mabey and P. Iles (eds), *Managing Learning*, Routledge, London, pp. 5–21.

Shannon, H. S., Robson, L. S. and Sale, J. E. M. (2001) Creating safer and healthier workplaces: Role of organisational factors and job characteristics, *American Journal of Industrial Medicine*, 40, 319–334.

Shaw, L. and Sichel, H. (1970) *Accident Proneness*, Oxford, Pergamon.

Shaw, L. and Sichel, H. S. (1971) *Accident Proneness: Research in the Occurrence, Causation, and Prevention of Road Accidents*, Pergamon Press, oxford.

Shaw, W. S., Robertson, M. M., Pransky, G. and McLellan, R. K. (2003) Employee perspectives on the role of supervisors to prevent workplace disability after injuries, *Journal of Occupational Rehabilitation*, 13, 129–142.

Sheehy, N. P. and Chapman, A. J. (1987) Industrial accidents, in C. L. Cooper and I. T. Roberston (eds), *International Review of Industrial and Organizational Psychology*, John Wiley & Sons, Chichester.

Sheehy, N. P. and Chapman, A. J. (1987) Industrial accidents, in C. L. Cooper and I. T. Roberston (eds), *International Review of Industrial and Organizational Psychology*, John Wiley & Sons, Chichester.

Sheeran, P. and Silverman, M. (2003) Evaluation of three interventions to promote workplace health and safety: Evidence for the utility of implementation intentions, *Social Science & Medicine*, 2153–2163.

Shimmin, S., Leather, P. J. and Wood, J. (1981) *Attitudes and Behaviour About Safety on Construction Work*, Report to the Building Research Establishment by the Department of Behaviour in Organisations, University of Lancaster.

Simard, M. and Marchand, A. (1994) The behaviour of first-line supervisors in accident prevention and effectiveness on occupational safety, *Safety Science*, 17, 169–185.

Simard, M. and Marchand, A. (1995) A multi-level analysis of organisational factors related to the taking of safety initiatives by work groups, *Safety Science*, 21, 113–129.

Simard, M. and Marchand, A. (1997) Workgroups' propensity to comply with safety rules: The influence of micro-macro organisational factors, *Ergonomics*, 40, 172–188.

Simonds, T. H. and Shafari-Sahrai, Y. (1977) Factors apparently affecting injury frequency in eleven matched pairs of companies, *Journal of Safety Research*, 9, 120–127.

Skinner, B. F. (1969) *Contingencies of Reinforcement: A Theoretical Analysis*, Appleton Century-Crofts, New York.

Slappendel, C., Laird, I., Kawachi, I., Marshall, S. and Cryer, C. (1993) Factors affecting work-related injury among forestry workers: A review, *Journal of Safety Research*, 24, 19–32.

Slovic, P., Fischhoff, B. and Lichtenstein, S. (1978) Accident probabilities and seatbelt usage: A psychological perspective, *Accident Analysis and Prevention*, 17, 10–19.

Slovic, P., Fischhoff, B. and Lichtenstein, S. (1980) Facts and fears: Understanding perceived risk, in R. C. Schwing and W. A. Albers (eds), *Societal Risk: How Safe is Safe Enough?* Plenum Press, New York, pp. 181–213.

Smith, D. (2000) On a wing and a prayer? Exploring the human components of technological failure, *Systems Research and Behavioral Science*, 17, 543–559.

Smith, D., Hunt, G. and Green, C. (1998) *Managing Safety the BS 8800 Way*, British Standards Institution, London.

Smith, M., Cohen, H. H., Cohen, A. and Cleveland, R. J. (1978) Characteristics of successful safety programmes, *Journal of Safety Research*, 10, 87–88.

Smith, N. J. (1991) Some aspects of civil engineering procurement in the UK within the single European market, Chapter 3.

Smith-Crowe, K., Burke, M. J. and Landis, R. S. (2003) Organizational climate as a moderator of safety knowledge-safety performance relationships, *Journal of Organizational Behavior*, 24, 861–876.

Solomon, R. C. (1992) *Ethics and Excellence*, Oxford University Press, Oxford.

Sparks, K., Cooper, C., Fried, Y. and Shirom, A. (1997) The effects of hours of work on health: A meta-analytic review, *Journal of Occupational and Organizational Psychology*, 70, 391–408.

Sporrong, H., Sandsjo, L., Kadefors, R., and Herberts, P. (1999) Assessment of workload and arm position during different work sequences: A study with portable devices on construction workers, *Applied Ergonomics*, 30, 495–503.

Spreitzer, G. M. (1995) Psychological empowerment in the workplace: Dimensions, measurement and validation, *Academy of Management Journal*, 38, 1442–1465.

Spreitzer, G. M. (1996) Social structural characteristics of psychological empowerment, *Academy of Management Journal*, 39, 483–504.

Standards Australia (1998) AS/NZS 3931:1998 *Risk Analysis of Technological systems: Application Guide*, Standards Australia, Sydney.

Standards Australia (1999) AS/NZS 4360:1999 *Risk Management*, Standards Australia, Sydney.

Standards Australia (2001) AS/NZS 4804:2001 *Occupational Health and Safety Management Systems – General Guidelines on Principles, Systems and Supporting Techniques*, Standards Australia, Sydney.

Standards Australia (2003) *AS 8003:2003 Corporate Social Responsibility*, Standards Australia, Sydney.

Starr, C. (1969) Social benefit versus technological risk, *Science*, 165, 1232–1238.

Steers, R. M. (1981) *Introduction to Organisational Behavior*, Scott Foresman & Company, Dallas.

Steers, R. M. and Porter, L. W. (1991) *Motivation and Work Behavior*, McGraw-Hill Inc, New York.

Stetzer, A. and Hofmann, D. A. (1996) Risk compensation: Implications for safety interventions, *Organisational Behavior and Human Decision Processes*, 66, 73–88.

Strunin, L. and Boden, L. I. (2000) Paths of reentry: Employment experiences of injured workers, *American Journal of Industrial Medicine*, 38, 373–384.

St-Yves, A., Freeston, M. H., Godbout, F. and Poulin, L. (1989) Externality and burnout among dentists, *Psychological Reports*, 65, 755–758.

Suchman, E. A. (1961) A conceptual analysis of the accident phenomenon, in *Behavioral Approaches to Accident Research*, Association for the Aid of Crippled Children, New York.

Surry, J. (1979) *Industrial Accident Research: A Human Engineering Appraisal*, Ontario Labor Safety Council, Toronto.

Suruda, A., Fosbroke, D. and Braddee, R. (1995) Fatal work-related falls from roofs, *Journal of Safety Research*, 26, 1–8.

Sveiby K. E. (1997) The intangible assets monitor, *Journal of Human Resource Costing & Accounting*, Vol. 2, Number 1, Spring 1997.

Sydney Morning Herald, 16 November 2002, *Building Industry Split over 36-hour Week Deal*, John Fairfax Publications Pty Ltd, p. 11.

Sydney Morning Herald, 28 March 2003, *Change in Attitude Vital to Lift Safety Rates*, John Fairfax Publications Pty Ltd, p. 10.

Sydney Morning Herald, 29 March 2003, *Work and Stress: Judge Finds a Deathly Link*, John Fairfax Publications Pty Ltd, p. 27.

Tait, N. R. S. and Cox, S. (1998) *Safety Reliability and Risk Management*, Butterworth-Heinemann, London.

Tang, H. (2001) Construct for excellence: Report of the Construction Industry Review Committee, Hong Kong.

Tate, D. G. (1992) Factors influencing injured workers' return to work, *Journal of Applied Rehabilitation Counseling*, 23, 17–20.

Tate, R. B., Yassi, A. and Cooper, J. (1999) Predictors of time loss after back injury in nurses, *Spine*, 24, 1930–1936.

Terborg, J. R. (1981) Interactional psychology and research on human behavior in organisations, *Academy of Management Review*, 6, 569–576.

The Age (Melbourne), 16 November 2002, When work is a killer, The Age Company Ltd, p. 40.

The Age (Melbourne), 5 December 2002, So now, what about the workers, The Age Company Ltd, p. 15.

The Age (Melbourne), 21 February 2003, Record payout for work fall, The Age Company Ltd, p. 4.

The Age, 15 November 2002, Coroner blames Esso for Longford disaster, Melbourne, Australia.

The Australian, Tuesday 29 June 1999, *Gas Blast Esso's Fault*, Sydney, Australia.

The Australian, 4 December 2002, Death law backdown angers grieving mother, Nationwide News Pty Ltd, p. 6.

The Consultancy Company Ltd (1997) *Evaluation of the Construction (Design and Management) Regulations 1994*, HSE Books, Sudbury.

The Guardian, Guilty Esso, 4 July 2001.

The Guardian, The Longford disaster: Esso to face mass court action, 15 November 2000.

The Health and Safety Practitioner (2000) HSE report lists convicted firms, 15 November, p. 9.

The National Code of Practice for the Prevention of Occupational Overuse Syndrome (1994) National Occupational Health and Safety Council: 2013, Australian Government Publishing Service, Canberra.

The National Standard for Manual Handling (1990) National Occupational Health and Safety Council: 1001, Australian Government Publishing Service, Canberra.

The Weekend Australian, 22 March 2003, WorkCover to be overhauled, Nationwide News Pty Ltd, p. 22.

Training Agency (1989) *Training in Britain, a Study of Funding, Activity and Attitudes*, HMSO, London.

Trethewy, R., Cross, J., Marosszeky, M. and Gavin, I. (2000) Safety measurement: A 'positive' approach towards best practice, *Journal of Occupational Health and Safety Australia and New Zealand*, 16, 237–245.

Triandis, H. C. (1971) *Attitude and Attitude Change*, Wiley, New York.

Turner, B. A. (1978) *Man-made Disasters*, Wykeham Press, London.

Turner, B. A. (1991) The development of a safety culture, *Chemistry and Industry*, 4, 241–243.

Turner, H. A., Fosh, P. and Ng, S. H. (1991) *Between Two Societies: Hong Kong Labour in Transition*, Centre of Asian Studies, University of Hong Kong, Hong Kong.

Varonen, U. and Mattila, M. (2000) The safety climate and its relationship to safety practices, safety of the work environment and occupational accidents in eight wood-processing companies, *Accident Analysis and Prevention*, 32, 761–769.

Vassie, L. H. and Lucas, W. R. (2001) An assessment of health and safety management within working groups in the UK manufacturing sector, *Journal of Safety Research*, 32, 479–490.

Veld, J. and Peeters, W. A. (1989) Keeping large projects under control: The importance of contract type selection, *Project Management*, Butterworth & Co. (Publishers) Ltd, August, 7, 3, UK.

Veljanovski, C. (1981) The economic theory of tort liability: Toward a corrective justice approach, in P. Burrows and C. Veljanovski (eds), *The Economic Approach to Law*, Butterworths, London.

Veltri, A. (1990) An accident cost impact model: The direct cost component, *Journal of Safety Research*, 21, 67–73.

Victorian WorkCover Authority (2002) *The Return to Work Guide for Victorian Employers*, Melbourne.

Viner, D. (1991) *Accident Analysis and Risk Control*, Derek Viner Pty Ltd, Melbourne.

Vlek, C. A. J. and Stallen, P. J. (1981) Judging risks and benefits in the small and in the large, *Organizational Behavior and Human Performance*, 28, 235–271.

Vojtecky, M. A. and Schmitz, M. F. (1986) Program evaluation and health and safety training, *Journal of Safety Research*, 17, 57–63.

Vollrath, M., Knoch, D. and Cassano, L. (1999) Personality, risky health behaviour and perceived susceptibility to health risk, *European Journal of Personality*, 13, 39–50.

Vredenburgh, A. G. (2002) Organisational safety: Which management practices are most effective in reducing employee injury rates? *Journal of Safety Research*, 33, 259–276.

Vroom, V. H. (1964) *Work and Motivation*, John Wiley & Sons, New York.

Wahba, M. A. and Bridwell, L. G. (1976) Maslow reconsidered: A review of research on the Need Hierarchy Theory, *Organisational Behavior and Human Performance*, 15, 212–240.

Walker, A. (1996) *Project Management in Construction,* 3rd edition, Oxford, Blackwell Science, UK.

Walker, A. and Newcombe, R. (1998) The positive use of power to facilitate the completion of a major construction project: A case study, *Construction Management and Economics*, forthcoming.

Walker, D. and Hampson, K. (2003) Enterprise networks, partnering and alliancing, in D. Walker and K. Hampson (eds), *Procurement Strategies: A Relationship-based Approach*, Blackwell Science Ltd Press, UK.

Walker, D. H. T. (1994a) *An Investigation into Factors that Determine Building Construction Time Performance*, Department of Building and Construction Economics, RMIT University, Melbourne, Australia.

Walters, D. (2001) *Prescription to Process: Convergence and Divergence in Health and Safety Regulation in Europe*, Seminar Paper 1, National Research Centre for OHS Regulation, Australian National University, Canberra.

Walters, D. R. (1998) Health and safety strategies in a changing Europe, *International Journal of Health Services*, 28, 305–331.

Waring, A. E. and Glendon, A. I. (1998) *Managing Risk*, International Thomson Business.

Warner, F. (1993) Introduction, *Risk*, a report prepared by the Royal Society Risk Study Group, The Royal Society, London.

Warren-Langford, P., Biggins, D. and Phillips, M. (1993) Union participation in occupational health and safety in Western Australia, *Journal of Industrial Relations*, 35, 585–606.

Watson, D. and Pennebaker, J. W. (1989) Health complaints, stress and distress: Exploring the central role of negative affectivity, *Psychological Review*, 96, 234–254.

Weidner, B. L., Grotsch, A. R., Delnevo, C. D., Newman, J. B. and McDonald, M. (1998) Worker health and safety training: Assessing impact among responders, *American Journal of Industrial Medicine*, 33, 241–246.

Weil, D. (1999) Are mandated health and safety committees substitutes for or supplements to labor unions? *Industrial and Labor Relations Review*, 52, 339–360.

Weinstein, N. D. (1980) Unrealistic optimism about future life events, *Journal of Personality and Social Psychology*, 39, 806–820.

Welch, L. S., Hunting, K. L. and Nessel-Stephens, L. (1999) Chronic symptoms in construction workers treated for musculoskeletal injuries, *American Journal of Industrial Medicine*, 36, 532–540.

Wells, C. (1989) Manslaughter and corporate crime, *New Law Journal*, July 7, 931–934.

Wilde, G. J. S. (1982) The theory of risk homeostatis: Implications for safety and health. *Risk Analysis*, 2, 209–258.

Wilkinson, A. (1998) Empowerment: Theory and practice, *Personnel Review*, 27, 40–56.

Williams, M. A. (1998) Designing for safety, in M. Barnard (ed.), *Health and Safety for Engineers*, Thomas Telford, London.

Williams, P. (1991) The new engineering contract, Chapter 4.

Williamson, A. M. and Feyer, A. M. (1995) Causes of accidents and time of day, *Work and Stress*, 9, 158–164.

Williamson, O. (1975) *Markets and Hierarchies: Analysis of Anti-trust Implications: A Study in the Economics of Internal Organisation*, Free Press, New York.

Williamson, O. E. (1975) *Markets and Hierarchies: Analysis and Antitrust Implications: A Study in the Economics of Internal Organisation*, The Free Press, London.

Willis, E. (1989) The industrial relations of occupational health and safety: A labour process approach, *Labour and Industry*, 2, 317–333.

Wilson, H. A. (1989) Organizational behaviour and safety management in the construction industry, *Construction Management and Economics*, 7, 303–319.

Wogalter, M. S., Sojourner, R. J. and Brelsford, J. W. (1997) Comprehension and retention of safety pictorials, *Ergonomics*, 40, 531–542.

Wogalter, M. S., Young, S. L., Brelsford, J. W. and Barlow, T. (1999) The relative contributions of injury severity and likelihood information on hazard-risk judgements and warning compliance, *Journal of Safety Research*, 30, 151–162.

Wood, R. E., Mento, A. J. and Locke, E. A. (1987) Task complexity as a moderator of goal effects: A meta-analysis, *Journal of Applied Psychology*, 72, 416–425.

WorkCover Authority (1999) Workers Compensation Statistical Bulletin 1999/2000, publication number 520.2. Published by WorkCover Publications, NSW, Australia.

WorkCover New South Wales (2001) *CHAIR: Safety in Design Tool*, Sydney.

WorkCover NSW (1999) *Safety in Design: Guidelines for Major Buildings and Civil Projects*, WorkCover, Sydney.

WorkCover NSW (2001) *CHAIR Safety in Design Tool*, WorkCover NSW, Sydney.

WorkSafe Victoria (1998) *Applying the 1991 NIOSH Lifting Equation*, Guidance note, GN3/98, Issued August 1998, http://www.workcover.vic.gov.au/dir090/vwa/alerts.nsf.

WorkSafe Victoria (2001a) *Cement Bag Handling*, Guidance note, GN13/2001, Issued June 2001, http://www.workcover.vic.gov.au/dir090/vwa/alerts.nsf.

WorkSafe Victoria (2001b) *Job Rotation Doesn't Eliminate Manual Handling Risk*, Alert A08/002, Issued July 2002, http://www.workcover.vic.gov.au/dir090/vwa/alerts.nsf.

WorkSafe Victoria (2002) *Back Belts are Not Effective in Reducing Back Injuries*, Guidance note, GN06/2002, Issued July 2002, http://www.workcover.vic.gov.au/dir090/vwa/alerts.nsf.

Yamnill, S. and McLean, G. N. (2001) Theories supporting transfer of training, *Human Resource Development Quarterly*, 12, 195–208.

Yates, C., Lewchuk, W. and Stewart, P. (2001) Empowerment as a Trojan horse: New systems of work organisation in the North American Automobile Industry, *Economic and Industrial Democracy*, 22, 517–541.

Young, S. (1996) Construction OH&S: A vision for the future, *Journal of Management in Engineering*, July/August, 33–36.

Young, S. L. and Wogalter, M. S. (1990) Comprehension and memory of instruction manual warnings: Conspicuous print and pictorial icons, *Human Factors*, 32, 637–649.

Zohar, D. (1980) Safety climate in industrial organisations: Theoretical and applied implications, *Journal of Applied Psychology*, 65, 96–102.

Zohar, D. (2000) A group-level model of safety climate: Testing the effect of group climate on micro-accidents in manufacturing jobs, *Journal of Applied Psychology*, 85, 587–596.

Zohar, D. (2002a) The effect of leadership dimensions, safety climate and assigned priorities on minor injuries in work groups, *Journal of Organisational Behavior*, 23, 75–92.

Zohar, D. (2002b), Modifying supervisory practices to improve sub-unit safety: A leadership-based intervention model, *Journal of Applied Psychology*, 87, 156–163.

Zohar, D. and Fussfield, N. (1981) Modifying earplug wearing behavior by behavior modification techniques: An empirical evaluation, *Journal of Organisational Behaviour Management*, 3, 41–52.

Zuckerman, M. (1979) *Sensation-seeking: Beyond the Optimal Level of Arousal*, Hillsdale NJ, Erlbaum.

Zwerling, C., Miller, E. R., Lynch, C. F. and Torner, J. (1996) Injuries among construction workers in rural Iowa: Emergency Department Surveillance, *Journal of Occupational & Environmental Medicine*, 38, 698–704.

Proceedings of CIB W92 – procurement systems

Cheetham, D., Carter, D., Lewis, T. and Jaggar, D. M. (eds) (1989) *Contractual Procedures for Building, Proceedings of the International Workshop*, 6–7 April, The University of Liverpool.

Bosanac, B. (ed.) (1990) *International Symposium on Procurement Systems*, Gradevinski Institut, Zagreb, CIB Publication 132.

Anon (1991) *Procurement Systems Symposium*, Las Palmas, Gran Canaria, Spain, CIB Publication NI. 145.

Rowlinson, S. M. (ed.) (1994) *East Meets West, Proceedings of CIB W92 Procurement Systems Symposium*, University of Hong Kong, CIB Publication 175.

Taylor, R. (ed.) (1996) *North Meets South, Proceedings of CIB W92 Procurement Systems Symposium*, University of Natal, Durban.

Davidson, C. H. and Abdel-Meguid, T. A. (eds) (1997) *Procurement – A Key to Innovation, Proceedings of CIB W92 Symposium*, Universite de Montreal, Canada, CIB Publication 203.

Davidson, C. H. (ed.) (1998) *Procurement – The Way Forward, Proceedings of CIB W92 Symposium*, Universite de Montreal, Canada, CIB Publication 203.

Note: The references given below are from the Proceedings of CIB W92 – Procurement Systems

Abdel-Meguid, T. and Davidson, C. (1996) Managed claims procurement strategy (MCPS): A preventive approach, pp. 11–20.

Agrilla, J. A. (1999) Construction safety management formula for success, *Proceedings of the 2nd International Conference of the International Council for Research and Innovation in Building and Construction (CIB) Working Commission W99*, Hawaii, pp. 33–36.

Akintola, S. A. (1994) Design and build procurement method in the UK construction industry, pp. 1–10.

Akintoye, A. and Taylor, C. (1997) Risk prioritisation of private sector finance of public sector projects, pp. 1–10.

Allin, S. (1990) Registration of contractors, p. 8.

Alsagoff, A. and McDermott, P. (1994) Relational contracting: A prognosis for the UK construction industry, pp. 11–19.

Askenazy, P. (2001) Innovative workplace practices and occupational injuries and illnesses in the United States, *Economic and Industrial Democracy*, 22, 485–516.

Aziz, A. and Ofori, G. (1996) Developing world-beating contractors through procurement policies: The case of Malaysia, pp. 1–10.

Baxendale, T. *et al.* (1996) Simultaneous engineering and its implications for procurement, pp. 21–30.

Bowen, P. A., Hindle, Robert D. and Pearl, Robert G. (1997) The effectiveness of building procurement systems in the attainment of client objectives, pp. 39–49.

Brochner, J. (1989) Building procurement – Key to improved performance, p. 83.

Carter, D. J. (1990) The use of the concept of activity profiles in contractual systems, pp. 30–39.

Chan, A. and Tam, C. M. (1994) Design and build through motivation, pp. 27–33.

Chau, K. W. and Walker, A. (1994) Institutional costs and the nature of subcontracting in the construction industry, pp. 371–378.

Cheung, S. O. (1997) Planning for dispute resolution in construction contracts, pp. 71–80.

Conlin, J. T., Langford, D. A. and Kennedy, P. (1996) The relationship between construction procurement strategies and construction disputes, pp. 66–82.

Construction Industry Institute (CII) (1991) *In Search of Partnering Excellence*, Special Publication, 17–1, Austin, Texas.

Cottrell, G. P. (1989) Sub-contracting, p. 117.

Craig, R. (1996) International public procurement systems, pp. 83–92.

Davenport, D. (1994) Assessing the efficiency of international procurement systems in order to improve client satisfaction with construction investment – The French experience, pp. 43–51.

Davidson, C. H. and Tarek, A. M. (eds) (1997) *Procurement – A Key to Innovation*, Montreal, IF Research Corporation, 850 pp.

Dimkic, Z. (1990) Some aspects of Yugoslav contractual arrangements and comparison with international practice in the construction industry, pp. 52–58.

Dulaimi, M. F. and Dalziel, R. C. (1994) The effects of the procurement method on the level of management synergy in construction projects, pp. 53–59.

Edwards, P. and Bowen, P. (1996) Building procurement in the 'new' South Africa: The communication imperative, pp. 120–129.

Egbu, C., Torrance, V. and Young, B. (1996) The procurement of project works in the ship refurbishment and construction industries, pp. 130–140.

Elliot, C. and Palmer, A. (1997) A measuring sustainability using traditional procurement systems, pp. 175–185.

Fellows, R. (1989) Development of British building contracts, pp. 14–19.
Fenn, P. (1989) Settling disputes in construction projects, p. 79.
Gidado, K. (1996) Political and economic development in Nigeria, what procurement system is suitable?, pp. 160–168.
Goldenhar, L. M. (2003) The 'Goldilocks model' of overtime in construction: Not too much, not too little but just right, *Journal of Safety Research*, 215–226.
Gow, H. A. and Fenn, F. P. (1989) The UK experience: Scotland, p. 61.
Green, S. (1994) Sociological paradigms and building procurement, pp. 89–97.
Green, S. (1997) Rhetoric and reality: A social constructivist research agenda for business re-engineering in construction, pp. 203–212.
Guldenmund, F. W. (2000) The nature on safety culture, *Safety Science*, 34, 215–257.
Gunning, J. G. and McDermott, M. A. (1997) Development in design and build contract practice in Northern Ireland, pp. 213–222.
Haddon, W. (1980) Advances in the epidemiology of injuries as a basis for public policy, *Public Health Reports*, 95, 411–421.
Hadikusumo, B. H. W. and Rowlinson (2002) Integration of virtually real construction model and design-for-safety-process database, *Automation in Construction*, Elsevier Science B. V., 11, 501–509.
Hamilton, N. (1990) A review of UK project procurement methods, pp. 100–109.
Harrington, J. M. (2001) Health effects of shiftwork and extended hours of work, *Occupational and Environmental Medicine*, 58, 68–72.
Hashim, M. (1997) Clients criteria on the choice of procurement systems – A Malaysian experience, pp. 273–284.
Heath, B. and Berry, M. (1996) An examination of the issues arising from the use of standard and non-standard procurement methods and the implications of same for project success, pp. 200–212.
Hibberd, P., Basden, A., Brandon, P. S., Brown, A. J., Kirkham, J. A. J. and Ttetlow, S. (1994) Intelligent authoring of contracts, pp. 115–124.
Hodgson, G. (1994) Cross border trading through contractor design: The United Kingdom/European community highway procurement model, pp. 125–131.
Hofmann, D. A. and Stetzer, A. (1998) The role of safety climate and communication in accident interpretation: Implications for learning from negative events, *Academy of Management Journal*, 41, 644–657.
Hughes, W. (1990) Designing flexible procurement systems, pp. 110–119.
Hutton, W. (1994) *The State We're in, Jonathan Cape*, London, p. 25.
Jennings, I. and Kenley, R. (1996) The social factor of project organisation, pp. 239–250.
Jensen, R. S. (1982) Pilot judgement: Training and evaluation, *Human Factors*, 34, 61–73.
Klimov, V. A., Didkovski, V. M. and Rekitar, Y. A. (1990) Contracts and tenders in construction in the USSR, pp. 121–134.
Kosaba, S. C., Maddi, S. R. and Zola, M. A. (1983) Type A and hardiness, *Journal of Behavioral Medicine*, 6, 41–51.
Lahdenpera, P. (1994) Increasing use of the product conception and its incorporation into various procurement systems, pp. 149–158.
Lahdenpera, P. (1996) Re-engineering the construction process-formulation of a model-based research approach, pp. 275–286.

Lam, P. and Chan, A. P. C. (1994) Construction management as a procurement method: A new direction for Asian contractors, pp. 159–168.

Laner, S. and Sell, R. G. (1960) An experiment on the effects of specially designed safety posters, *Occupational Psychology*, 34, 3, 153–169.

Leung, H. F., Chau, K. W. and Ho, D. C. W. (1996) Features in the Hong Kong airport core projects general conditions of contract – A comparison with the Hong Kong Government general conditions of contract for civil engineering works, pp. 287–300.

Lindsay, N. (2003) Union steps up worker safety push, *Australian Financial Review*, 10 February 2003, John Fairfax Holdings Ltd, Sydney, p. 7.

Lingard, H. and Lin, J. (2003) Managing motherhood in the Australian construction industry, *Australian Journal of Construction Economics and Building*, 3, 15–24.

Liu, A. M. M. (1994) From act to outcome – A cognitive model of construction procurement, pp. 169–178.

Liu, A. M. M. and Fellows, R. (1996) Towards an appreciation of cultural factors in the procurement of construction projects, pp. 301–310.

Ludborzs, B. (1995) Surveying and assessing 'Safety Culture' within the framework of safety audits, *Loss Prevention and Safety Promotion in the Process Industry*, 1, 83–92.

Matthews, J. (1996) *A Project Partnering Approach to the Main Contractor–Subcontractor Relationship*, Unpublished PhD, Loughborough University, p. 200.

Matthews, J., Tyler, A. and Thorpe, A. (1996) Pre-construction project partnering: Developing the process, *Engineering Construction and Architectural Management*, 3, 1 & 2, 117–131.

McCarthy, C. and Wang, W. (1994) A flexible knowledge based system for the new engineering contract, pp. 195–202.

McDermott, P. (1996) Role conflict within the United Kingdom contracting system, pp. 353–367.

McDermott, P., Melaine, Y. and Sheath, D. (1994) Construction procurement systems: What choice for the Third World?, pp. 203–211.

McDermott, P., Rowlinson, S. M. and Jaggar, D. (1997) Foreword, pp. xv–xxii.

Mohsini, R. and Davidson, C. H. (1989) *Building Procurement – Key to Improved Performance*, p. 83.

Moss, A. (1994) The HKCEC: An unusual and highly successful procurement example, highly successful procurement example, pp. 213–220.

Moss, A. and Rowlinson, S. M. (1996) Project management lessons identified from the development of the Hong Kong Convention and Exhibition Centre and the extension project, pp. 408–418.

Mukalula, P. (1996) The effects of the structural adjustment programme (SAP) on maintenance procurement contracts (Zambia's case), pp. 419–429.

Mustapha, F. H., Naoum, S. G. and Aygun, T. (1994) Public sector procurement methods used in the construction industry in Turkey, pp. 229–234.

Naoum, S. G. (1990) Management contracting – Review and analysis, p. 135.

Naoum, S. G. and Mustapha, F. H. (1994) Influences of the client, designer and procurement methods on project performance, pp. 221–228.

Ndekugri, I. (1989) Sub-contracting in the UK construction industry, p. 139.

Newcombe, R. (1994) Procurement paths – A power paradigm, pp. 243–250.
Ng, W. F. (1994) Procurement methods for rural housing projects in the poverty-stricken areas of Guangxi in the People's Republic of China, pp. 251–258.
Ofori, G. and Pin, T. (1996) Linking project procurement to construction industry development: The case of Singapore, pp. 473–482.
Ogunlana, S. (1997) Build operate transfer procurement traps: Examples from transportation projects in Thailand, pp. 585–594.
Ogunlana, S. and Malmgren, C. (1996) Experience with professional construction management in Bangkok, Thailand, pp. 483–491.
Pasquire, C. (1994) Early incorporation of specialist design capability, pp. 259–267.
Pasquire, C. (1997) The implications of environmental issues on construction procurement, pp. 603–615.
Peckitt, S. J., Glendon, A. I. and Booth, R. T. (2002) A comparative study of safety culture in the construction industry of Britain and the Caribbean, in M. Lewis (ed.), *Proceedings of the International Symposium of the Working Commission CIB W92, Procurement Systems and Technology Transfer*, The Engineering Institute, University of West Indies, Trinidad & Tobago, pp. 195–220.
Pollock, R. W. and Rees, K. (1989) The potential for expert systems in the interpretation of building contracts, pp. 73–82.
Potts, K. and Toomey, D. (1994) East and West compared: A critical review of two alternative payment systems, pp. 269–276.
Potts, K. and Weston (1996) Risk analysis estimation and management on major construction works, pp. 522–531.
Ramsay-Dawber, P. J. (1996) Business performance measures of the UK construction companies: An aid to pre-selection, pp. 532–539.
Robinson, N. M. (1990) Towards a multi-purpose modular form of construction contract, pp. 144–153.
Ross, S. E. and Reynolds, J. R. (1996) The effect of power, knowledge, and trust on income disclosure surveys, *Social Science Quarterly*, 77, 899–911.
Saito, T. (1994) The comparative study of procurement system in the UK and Japan, pp. 389–401.
Sharif, A. and Morledge, R. (1994) A functional approach to modelling procurement systems internationally and the identification of necessary support frameworks, pp. 295–305.
Singh, S. (1990) A rational procedure for the selection of appropriate procurement systems, pp. 175–182.
Slappendel, C., Laird, I., Kawachi, I., Marshall, S. and Cryer, C. (1993) Factors affecting work-related injury among forestry workers: A review, *Journal of Safety Research*, 24, 19–32.
Smith, A. and Wilkins, B. (1994) Procurement of major publicly funded health care projects, pp. 307–314.
Smith, N. J. (1989) Changes in contractual procedure prior to 1992, p. 153.
Stetzer, A. and Hofmann, D. A. (1996) Risk compensation: Implications for safety interventions, *Organizational Behavior and Human Decision Processes*, 66, 73–88.
Streff, F. M. and Geller, E. S. (1998) An experimental test of risk compensation: Between-subject versus within-subject analyses, *Accident Analysis and Prevention*, 20, 277–287.

Swanston, R. (1989) United Kingdom procurement procedures, pp. 23–36.

Tam, C. M., Li, W. Y. and Chan, A. (1994) BOT applications in the power industry of Southeast Asia: A case study in China, pp. 315–322.

Taylor, R. and Norval, G. (1994) Developing appropriate procurement systems for developing communities, pp. 323–334.

Tennant, C. (1996) Experimental stress and cardiac function, *Journal of Psychosomatic Research*, 40, 569–583.

Thomson, A. L. B. (1989) The construction industry in West Germany, pp. 38–51.

Torkornoo, G. A. E. (1991) Procuring construction in Ghana, p. 19.

Vukovic, S. and Marinic, I. (1994) Organizational technological and economic development of construction companies under the conditions of privatization in Vojvodina, pp. 335–342.

Wahlstrom, O. (1989) Simplified tender documents gibing an unambiguous representation of the finished building, p. 107.

Walker, D. (1996) Characteristics of a good client's representative, pp. 614–624.

Walker, D. H. T. (1994b) Procurement systems and construction time performance, pp. 343–351.

Wickstrom, G., Hanninen, K., Lehtinen, M. and Riihimaki, H. (1978) Previous back syndromes and present back syndromes in concrete reinforcement workers, *Scandinavian Journal of Work, Environment and Health*, 4 (Supplement 1), pp. 20–29.

Yates, C., Lewchuk, W. and Stewart, P. (2001) Empowerment as a Trojan horse: New systems of work organization in the North American Automobile Industry, *Economic and Industrial Democracy*, 22, 517–541.

Zohar, D. (2002) Modifying supervisory practices to improve sub-unit safety: A leadership-based intervention model, *Journal of Applied Psychology*, 87, 156–163.

Zohar, D. and Fussfield, N. (1981) Modifying earplug wearing behavior by behavior modification techniques: An empirical evaluation, *J Org Behav Mgmt*, 3, 41–52.

Index

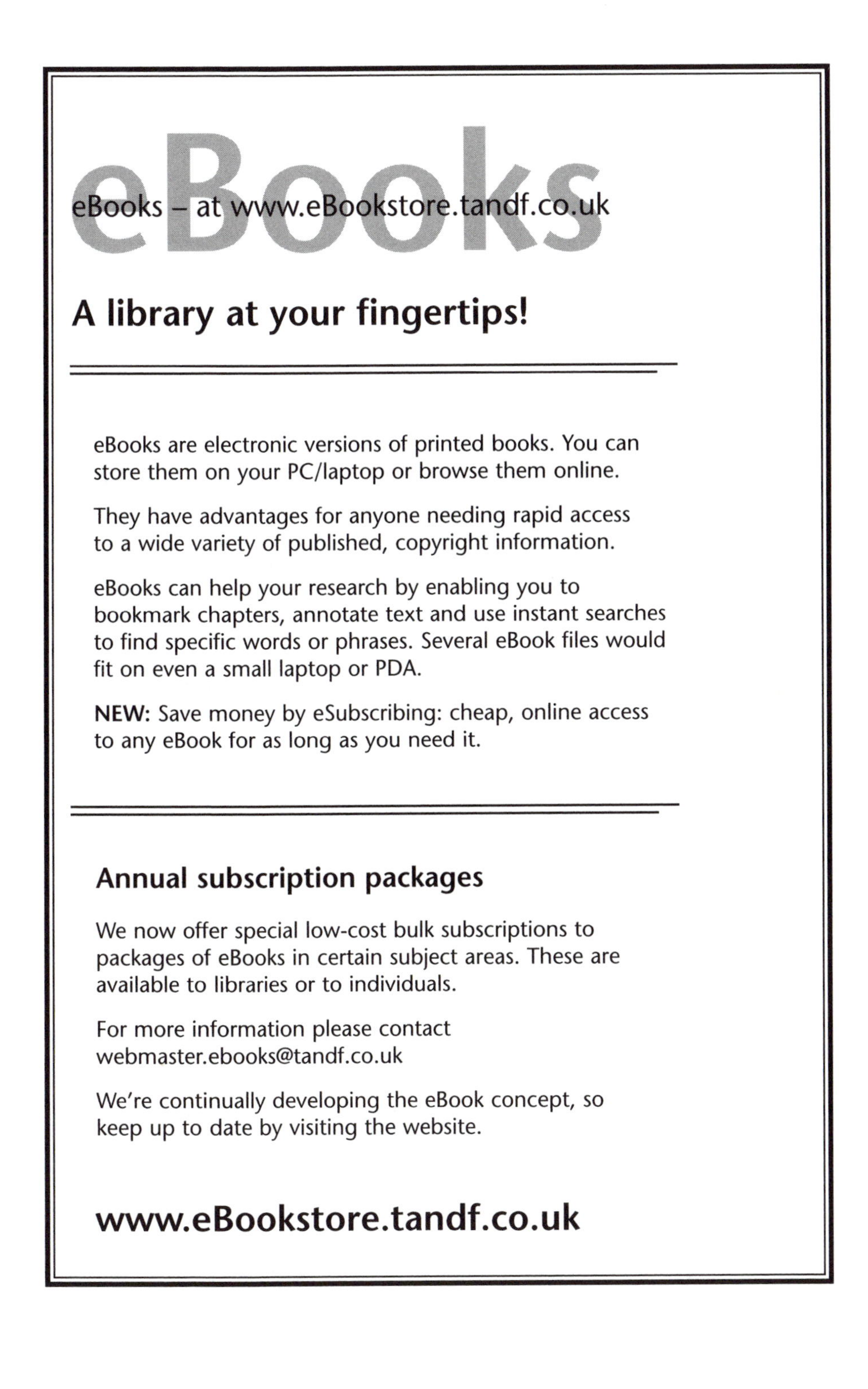
eBooks
eBooks – at www.eBookstore.tandf.co.uk
A library at your fingertips!
eBooks are electronic versions of printed books. You can store them on your PC/laptop or browse them online.
They have advantages for anyone needing rapid access to a wide variety of published, copyright information.
eBooks can help your research by enabling you to bookmark chapters, annotate text and use instant searches to find specific words or phrases. Several eBook files would fit on even a small laptop or PDA.
NEW: Save money by eSubscribing: cheap, online access to any eBook for as long as you need it.
Annual subscription packages
We now offer special low-cost bulk subscriptions to packages of eBooks in certain subject areas. These are available to libraries or to individuals.
For more information please contact webmaster.ebooks@tandf.co.uk
We're continually developing the eBook concept, so keep up to date by visiting the website.
www.eBookstore.tandf.co.uk